SAGE Outcomes: **Measure Results, Track Success**

FOR STUDENTS, understanding the objectives for each chapter and the goals for the course is essential for getting the grade you deserve!

FOR INSTRUCTORS, being able to track your students' progress allows you to more easily pinpoint areas of improvement and report on success.

This title was crafted around specific chapter objectives and course outcomes, vetted by experts, and adapted from renowned syllabi. Tracking student progress can be challenging. Promoting and achieving success should never be. We are here for you.

COURSE **OUTCOMES** FOR INTRODUCTION TO CRIMINAL JUSTICE:

ARTICULATE the foundations of criminal justice, including definitions, theories, typologies, measurement issues, and the law.

EXAMINE the development of policing organizations and strategies and the challenges faced by the police.

EXPLAIN the structure and processes of the judiciary at the local, state, and federal levels.

EXPLAIN the goals, methods, and effectiveness of various forms of corrections.

ANALYZE important criminal justice issues and their impact on society.

Want to see how these outcomes tie in with this book's chapter-level objectives? Visit us at edge.sagepub.com/peakbrief for complete outcome-to-objective mapping.

Sara Miller McCune founded SAGE Publishing in 1965 to support the dissemination of usable knowledge and educate a global community. SAGE publishes more than 1000 journals and over 600 new books each year, spanning a wide range of subject areas. Our growing selection of library products includes archives, data, case studies and video. SAGE remains majority owned by our founder and after her lifetime will become owned by a charitable trust that secures the company's continued independence.

Los Angeles | London | New Delhi | Singapore | Washington DC | Melbourne

A Brief Introduction to Criminal Justice

A district judge once said, "When we pull back the layers of government services, the most fundamental and indispensable virtues are public safety and social order." This book is dedicated to those industrious students of today seeking to ensure that those virtues are upheld later in their chosen profession.

—K. J. P.

To Joel D. Lieberman, a great friend and colleague who has given me opportunities to live an extraordinary life.

—T. D. H.

A Brief Introduction to Criminal Justice

Practice and Process

Kenneth J. Peak
University of Nevada, Reno

Tamara D. Madensen-Herold
University of Nevada, Las Vegas

Los Angeles | London | New Delhi
Singapore | Washington DC | Melbourne

FOR INFORMATION:

SAGE Publications, Inc.
2455 Teller Road
Thousand Oaks, California 91320
E-mail: order@sagepub.com

SAGE Publications Ltd.
1 Oliver's Yard
55 City Road
London EC1Y 1SP
United Kingdom

SAGE Publications India Pvt. Ltd.
B 1/I 1 Mohan Cooperative Industrial Area
Mathura Road, New Delhi 110 044
India

SAGE Publications Asia-Pacific Pte. Ltd.
18 Cross Street #10–10/11/12
China Square Central
Singapore 048423

Acquisitions Editor: Jessica Miller
Content Development Editors: Adeline Grout and Laura Kearns
Editorial Assistant: Sarah Manheim
Marketing Manager: Jillian Ragusa
Production Editor: Veronica Stapleton Hooper
Copy Editor: Amy Marks
Typesetter: C&M Digitals (P) Ltd.
Proofreader: Dennis W. Webb
Indexer: Sheila Bodell
Cover Designer: Scott Van Atta

Printed in Canada

Library of Congress Cataloging-in-Publication Data

Names: Peak, Kenneth J., 1947- author. | Madensen, Tamara, author.

Title: A brief introduction to criminal justice : practice and process / Kenneth John Peak, University of Nevada, Reno, Tamara D. Madensen-Herold, University of Nevada, Las Vegas.

Description: Thousand Oaks : SAGE Publishing, 2020. | Includes bibliographical references and index.

Identifiers: LCCN 2019030469 | ISBN 9781544373294 (paperback) | ISBN 9781544373270 (epub) | ISBN 9781544373263 (epub) | ISBN 9781544373256 (pdf)

Subjects: LCSH: Criminal justice, Administration of—United States. | Criminal procedure—United States.

Classification: LCC HV9950 .P43 2020 | DDC 364.973—dc23
LC record available at https://lccn.loc.gov/2019030469

This book is printed on acid-free paper.

19 20 21 22 23 10 9 8 7 6 5 4 3 2 1

/// BRIEF CONTENTS

/// DETAILED CONTENTS

[illegible]

[illegible] has many years of combined experience in [illegible] and academic environments, including several positions in criminal justice administration, policing, corrections, and research and training for police agencies, major sports leagues, and private industries. Therefore its chapters contain a unique [illegible] that college and university [illegible]

[illegible] is likely to be [illegible] for students [illegible] and practitioners [illegible]

[illegible]

[illegible] why it [illegible] of crime [illegible]

/// PREFACE

A Unique Approach

Famed educator John Dewey advocated the "learning by doing" approach to education, or problem-based learning. This book is written, from start to finish, with that philosophy in mind. Its approach also comports with the popular learning method espoused by Benjamin Bloom, known as Bloom's taxonomy, in which he called for "higher-order thinking skills"—critical and creative thinking that involves analysis, synthesis, and evaluation.

This book also benefits from the authors having many years of combined practitioners' and academic experience, including several positions in criminal justice administration, policing, corrections, and research and training for police agencies, major sports leagues, and private industries. Therefore, its chapters contain a palpable, real-world flavor that college and university criminal justice students typically find missing from their textbooks.

It is hoped that readers will put to use the several features of the book that are intended to help them accomplish the overall goal of learning by doing. In addition to chapter opening questions (which allow students to assess their knowledge of the chapter materials), learning objectives, and a chapter summary, each chapter contains a number of boxed features such as case studies and "You Be the . . ." (Judge, Prosecutor, Defense Attorney, and so on, as the specific situation may require) exhibits, as well as a list of key terms, review questions, and "Learn by Doing" activities at the end of each chapter. Also provided are "Practitioner's Perspectives" in which people in the field describe their criminal justice occupation. Taken together, these supplemental materials should greatly enhance readers' critical analysis, problem-solving, and communication capabilities, allowing them to experience the kinds of decisions that must be made in the field.

In today's competitive job market, students who possess these kinds of knowledge, skills, and abilities will have better opportunities to obtain employment as a criminal justice practitioner and to succeed in the field. Although the book certainly delves into some theoretical, political, and sociological subject matter, it attempts to remain true to its *practical,* applied focus throughout and to the extent possible.

Distinctive Chapter Contents

This book also contains chapters devoted to topics not typically found in introductory criminal justice textbooks. For example, Chapter 3 is devoted to criminal justice ethics. Ethics is always a timely topic for our society and especially for today's criminal justice students and practitioners. Chapter 12 also describes four unique and contemporary issues on the U.S. criminal justice policy-making agenda: drug abuse, sex trafficking, terrorism, and immigration. Finally, several chapters also discuss the technologies employed in the system.

In sum, this book introduces the student to the primary individuals, theorists, practitioners, processes, concepts, technologies, and terminologies as they work within or are applied to our criminal justice system. Furthermore, the concepts and terms learned in this introductory textbook will serve as the basis for more complex criminal justice studies of police, courts, and corrections in later course work.

Chapter Organization

To facilitate the book's goals, we first need to place the study of criminal justice within the big picture, which is accomplished in the three chapters composing Part I. **Chapter 1** briefly examines why it is important to study criminal justice, foundations and politics of criminal justice, an overview of the criminal justice process and the offender's flow through the system, and how discretion and ethics apply to the field. **Chapter 2** considers the sources and nature of law (including substantive and procedural, and criminal and civil), the elements of criminal acts, felonies and misdemeanors, offense definitions and categories, and prevailing methods in use for trying to measure how many crimes are committed. **Chapter 3** concerns ethics and includes definitions and problems, with emphases on the kinds of ethical problems that confront the police, the courtroom work group, and corrections staff. Included are an ethical decision-making process; legislative enactments; and judicial decisions involving ethics at the federal, state, and local levels.

Part II consists of three chapters that address federal and state law enforcement and local policing in the United States. **Chapter 4** discusses the organization and operation of law enforcement agencies at the federal, state, and local (city and county) levels. Included are discussions of their English and colonial roots, the three eras of U.S. policing, and brief considerations of INTERPOL and the field of private security. **Chapter 5** focuses on the kinds of work that police do. After looking at their recruitment and training, considered next are the roles, styles, and tasks of the police as well as patrol and investigative functions, generally. Also included are the dangers of the job, the use of police discretion, and the work of criminal investigators. We also broadly examine several policing issues that exist today, including calls for police reform regarding the use of force, greater diversity in police agencies, and body-worn cameras. **Chapter 6** examines the constitutional rights of the accused (as per U.S. Supreme Court decisions) as well as limitations placed on the police under the Fourth, Fifth, and Sixth Amendments; the focus is on arrest, search and seizure, the right to remain silent, and the right to counsel.

Part III consists of three chapters that generally examine the courts. **Chapter 7** examines court structure and functions at the federal, state, and trial court levels; included are discussions of pretrial preparations, the actual trial process, the jury system, and some court technologies. **Chapter 8** looks at the courtroom work group, including the judges, prosecutors, and defense attorneys, and legal defenses that are allowed under the law. Finally, **Chapter 9** discusses sentencing, punishment, and appeals. Included are the types and purposes of punishment, types of sentences convicted persons may receive, federal sentencing guidelines, victim impact statements, and capital punishment.

Part IV includes two chapters and examines many aspects of correctional organizations and operations. **Chapter 10** examines federal and state prisons and local jails in terms of their evolution and organization, inmate population trends (including mass incarceration) and classification, and some technologies. **Chapter 11** reviews community corrections and alternatives to incarceration: probation, parole, and several other diversionary approaches. Included are discussions of the origins of probation and parole, functions of probation and parole officers, and several intermediate sanctions (e.g., house arrest, electronic monitoring).

Finally, Part V contains a single chapter that considers problems and challenges that span the criminal justice system. Specifically, **Chapter 12** provides an in-depth view of four particularly challenging problems and policy issues confronting society and the criminal justice system today: drug abuse, sex trafficking, terrorism, and immigration.

A Note on Juvenile Justice

This first edition of *A Brief Introduction to Criminal Justice: Process and Practice* does not contain a standalone chapter covering the philosophy and practices of the U.S. juvenile justice system. However, because our society has long been concerned with its minors (usually defined as persons between the ages of 10 and 18) who violate the law, the majority of whom have faced adverse and traumatic home situations, several chapters in this book will briefly discuss different facets of this system (in boxed "Juvenile Justice Journal" features). These discussions will emphasize the differences in how juvenile and adult offenders are treated by the criminal justice system as well as the different philosophical approaches to these populations.

RESOURCES FOR INSTRUCTORS AND STUDENTS

edge.sagepub.com/peakbrief

SAGE COURSEPACKS FOR INSTRUCTORS makes it easy to import our quality content into your school's learning management system (LMS)*. Intuitive and simple to use, it allows you to

Say NO to…

- required access codes
- learning a new system

Say YES to…

- using only the content you want and need
- high-quality assessment and multimedia exercises

*For use in: Blackboard, Canvas, Brightspace by Desire2Learn (D2L), and Moodle

Don't use an LMS platform? No problem, you can still access many of the online resources for your text via SAGE edge.

SAGE coursepacks includes:

- Our content delivered directly into your LMS
- An intuitive, simple format that makes it easy to integrate the material into your course with minimal effort
- Assignable SAGE Premium Video (available on the SAGE vantage platform, linked through SAGE coursepacks) that is tied to learning objectives, and curated and produced exclusively for this text to bring concepts to life and appeal to diverse learner, featuring:
    - Student on the Street videos that breathe life into concepts with stories drawn from common student misconceptions of criminal justice
    - Criminal Justice in Practice videos that offer students career advice, provide insights into a variety of career paths, and discuss challenges and misconceptions of each profession.
    - Corresponding multimedia assessment options that automatically feed to your gradebook
    - Assessment tools that foster review, practice, and critical thinking, and offer a more complete way to measure student engagement, including:
    - Diagnostic chapter quizzes that identify opportunities for improvement, track student progress, and ensure mastery of key learning objectives
    - Test banks built on Bloom's Taxonomy that provide a diverse range of test items with ExamView test generation
    - Activity and quiz options that allow you to choose only the assignments and tests you want
    - Instructions on how to use and integrate the comprehensive assessments and resources provided
- Chapter-specific discussion questions to help launch engaging classroom interaction while reinforcing important content
- Editable, chapter-specific PowerPoint® slides that offer flexibility when creating multimedia lectures so you don't have to start from scratch but you can customize to your exact needs

- Sample course syllabi with suggested models for structuring your course that give you options to customize your course in a way that is perfect for you
- Lecture notes that summarize key concepts on a chapter-by-chapter basis to help you with preparation for lectures and class discussions
- All tables and figures from the textbook

Student Study Site

SAGE EDGE FOR STUDENTS enhances learning, it's easy to use, and offers:

- mobile-friendly flashcards that strengthen understanding of key terms and concepts, and make it easy to maximize your study time, anywhere, anytime
- mobile-friendly practice quizzes that allow you to assess how much you've learned and where you need to focus your attention
- learning objectives that reinforce the most important material
- video resources that bring concepts to life, are tied to learning objectives, and make learning easier

/// ACKNOWLEDGMENTS

This first edition of *A Brief Introduction to Criminal Justice: Practice and Process* is the culmination of dedicated efforts on the part of several key individuals at SAGE Publications, Inc., and they should be recognized. First, this team effort was led by Jessica Miller, acquisitions editor; we greatly appreciate her ongoing commitment to this publishing effort and her astute experience and assistance toward that end. Also providing stellar performances in their roles were Laura Kearns and Adeline Grout, content development editors; Sarah Manheim, editorial assistant; Amy Marks, copy editor; and Veronica Stapleton Hooper, production editor.

—K. J. P. and T. D. H.

REVIEWERS

Thomas J. Alisankus, Rock Valley College
Bruce Bayley, Weber State University
Teresa R. Bean, Hood College
Sylvia Blake-Larson, Tarrant County College
James C. Brown, Utica College
John Brown, LA Valley College
Joel Cox, Liberty University
Jean Dawson, Franklin Pierce University
Lee DeBoer, Collin College
Tauya Forst, College of Dupage
Matilda Foster, University of North Georgia
LaRhonda Hamilton, Tyler Junior College
Matasha Harris, Bowie State University
James Hartley, Arkansas Northeastern College
Tracy Hoilman, Lees-McRae College
Patricia Hovis, York Technical College
Jason Jimerson, Franklin College
Theresa Leopold, Lake Superior College
Elizabeth Lewis, Roane State Community College
Gregory Lindsteadt, Missouri Western State University
Holly Miller, University of North Florida
Kerry Muehlenbeck, Mesa Community College
Jacqueline Mullany, Triton College
J. Brian Murphy, Valencia College
Jessica Noble, Lewis and Clark Community College
Cheryl North, Tarrant County College
Michelle Nuneville, George Mason University

Robert Overall, Nashville State Community College

Elizabeth Perkins, Morehead State University

Wayne Posner, East Los Angeles College

Lawrence Presley, Liberty University

Michael Reid, Los Angeles Harbor College

Greg Scott, Utica College

Tim Seguin, Cochise College

Jill Shelley, Northern Kentucky University

Celia Sporer, Queensborough Community College

Don Stemen, Loyola University Chicago

Karin Storm, Brandman University

Anne Stouth, NC State College

Steve Sullivan, Normandale

Jerry Trunillo, Santa Fe Community College

Travis Zimmerman, West Virginia Wesleyan College

/// ABOUT THE AUTHORS

Kenneth J. Peak is emeritus professor and former chair of the Department of Criminal Justice, University of Nevada, Reno, where he was named "Teacher of the Year" by the university's Honor Society. Following four years as a municipal police officer in Kansas, he subsequently held positions as a nine-county criminal justice planner for southeast Kansas; director of a four-state technical assistance institute for the Law Enforcement Assistance Administration (based at Washburn University in Topeka); director of university police at Pittsburg State University (Kansas); acting director of public safety, University of Nevada, Reno; and assistant professor of criminal justice at Wichita State University. He has authored or coauthored 37 additional textbooks (relating to general policing, community policing, criminal justice administration, police supervision and management, and women in law enforcement), two historical books (on Kansas temperance and bootlegging), and more than 60 journal articles and invited book chapters. He is past chair of the Police Section of the Academy of Criminal Justice Sciences and past president of the Western Association of Criminal Justice. He received two gubernatorial appointments to statewide criminal justice committees while residing in Kansas and holds a doctorate from the University of Kansas.

Tamara D. Madensen-Herold is an associate professor of criminal justice and graduate director at the University of Nevada, Las Vegas (UNLV). She holds a doctorate from the University of Cincinnati. Her research interests include crime opportunity structures, place management, and crowd violence. She is the recipient of UNLV's Spanos Distinguished Teaching Award, Faculty Excellence Award, and Greenspun College of Urban Affairs Teaching Award. Her publications propose, extend, or test crime science theoretical models. They also help to translate research findings into practice and policy. Her work has appeared in various outlets, including *Criminology* and *Justice Quarterly*. She has published numerous practitioner-focused research papers, including two Problem-Oriented Policing Guides funded by the COPS Office and research monographs for which she received Herman Goldstein Excellence in Problem-Oriented Policing Awards. Her book *Preventing Crowd Violence* (coedited with Johannes Knutsson) has been translated into two foreign languages. Dr. Madensen-Herold serves as director of UNLV's Crowd Management Research Council and conducts research and training for police agencies, major sports leagues, and private industries.

CRIMINAL JUSTICE AS A SYSTEM

The Basics

1

This part consists of three chapters. **Chapter 1** briefly examines why it is important to study criminal justice, the foundations and politics of criminal justice, the criminal justice process and the offender's flow through the system, and how discretion and ethics apply to the field.

Chapter 2 considers the sources and nature of law (including substantive and procedural, and criminal and civil), the elements of criminal acts, felonies and misdemeanors, offense definitions and categories, and prevailing methods in use for trying to measure how many crimes are committed.

Chapter 3 looks at definitions and types of ethics in general and then examines ethical dilemmas that confront the police, the courtroom work group, and correctional staff. Included are legislative enactments and judicial decisions involving ethics at the federal, state, and local levels.

©iStockphoto.com/janniswerner

CHAPTER 1

FUNDAMENTALS OF CRIMINAL JUSTICE

Essential Themes and Practices

The true administration of justice is the firmest pillar of good government.

—Inscription on the New York State Supreme Court, Foley Square, Manhattan, New York

When we pull back the layers of government services, the most fundamental and indispensable virtues are public safety and social order.

—Hon. David A. Hardy, Washoe County District Court, Reno, Nevada

LEARNING OBJECTIVES

As a result of reading this chapter, you will be able to

1. Explain the importance of studying and understanding our criminal justice system
2. Describe the foundations of our criminal justice system, including its legal and historical bases and the difference between the consensus and conflict theories of justice
3. Define the crime control and due process models of criminal justice
4. Describe the importance of discretion throughout the justice system
5. Describe the fundamentals of the criminal justice process—the offender's flow through the police, courts, and corrections components, and the functions of each component
6. Explain the wedding cake model of criminal justice
7. Discuss the importance of ethics and character in criminal justice

ASSESS YOUR AWARENESS

Test your knowledge of criminal justice fundamentals by responding to the following six true-false items; check your answers after reading this chapter.

1. Under the U.S. system of justice, people basically join together, form governments (thus surrendering their rights of self-protection), and receive governmental protection in return.
2. Very little, if any, political or discretionary behavior or authority exists in the field of criminal justice; its fixed laws and procedures prevent such influences.
3. All prosecutions for crimes begin with a grand jury indictment.
4. Police make the final decisions concerning the actual crimes with which a suspect will be charged.
5. *Parolee* is the term used to describe one who has been granted early release from prison.
6. The U.S. system of criminal justice is intended to function, and indeed does function in all respects, like a "well-oiled machine."

<< Answers can be found on page 293.

In 2015, Kalief Browder, a 22-year-old man from New York City, hanged himself in his parents' home. This suicide drew national attention because much had been publicized, just a year prior, about Mr. Browder and his time spent in detention at Rikers Island. He was arrested at the age of 16 after being accused of stealing a backpack. He was detained on Rikers Island for three years, despite never being tried or convicted for his alleged crime. While enduring repeated delays in the clogged Bronx court system, he was beaten by correctional officers and spent nearly two years in solitary confinement.[1] He was eventually released after the prosecutor determined that there wasn't enough evidence to try Mr. Browder for theft.

Many people claimed that Mr. Browder's suicide was the result of the mental and physical abuse he experienced in prison. His experience inspired New York City to end the use of solitary confinement for juveniles and influenced President Barack Obama's decision to ban juvenile solitary confinement in federal prisons. In 2019, New York City agreed to pay $3.3 million to settle a lawsuit brought on behalf of Mr. Browder's estate.[2]

Mr. Browder, like others with similar experiences of mistreatment, became a symbol of criminal justice system abuse in the United States. His experience, death, and subsequent calls for reform followed years of justice system policies and legislation that promoted a "tough-on-crime" approach to managing crime and criminals. This tough-on-crime approach resulted in backlogged courts and extreme prison overcrowding.

Kalief Browder

As you read this chapter, think about how the criminal law is influenced, and how new laws are made and ultimately changed, as society and its norms change. What types of laws help to keep residents safe while also protecting taxpayers and prisoners from the costs associated with an unnecessarily punitive justice system? What political pressures would lawmakers feel in crafting such laws, and what risks do we face when our justice system does not function as intended?

INTRODUCTION

The U.S. criminal justice system as we know it has existed for more than a century and a half. Interest in crime and our justice system has generated thousands of television series and movies that examine how and why people commit crimes, how the justice system responds, the roles of people who work within our system, various punishments for criminal behavior, and our protections under the Bill of Rights. Some are fictional (*The Wire, CSI, Law & Order,* and *NCIS*), while others are based on true stories or follow real-life events (*The People v. O. J. Simpson: American Crime Story, The First 48, Unsolved Mysteries,* and *Cops*). These programs have contributed to society's general knowledge about our justice system, but fictional stories and dramatic editing can produce misleading or inaccurate information.

The criminal justice system is a critical part of our free, democratic society, and for that reason alone, we need to study and understand it. It serves to define our culture and how we live. It also influences the way in which we interact with the rest of the world. The terrorist attacks that occurred on September 11, 2001, changed the way Americans felt about homeland safety and security. The events of 9/11 made people more aware that crime is an international problem. Crime regularly transcends national borders, and the manner in which our federal, state, and local criminal justice agencies must organize and plan in order to deal with crime has changed in many ways, as later chapters will show.

Another reason for carefully studying the criminal justice system is that, odds are, you and most Americans will be affected by crime during your lifetime. About 9 million serious criminal offenses are reported to the Federal Bureau of Investigation (FBI) each year, with about 1.3 million of them involving violence and 7.7 million involving damaged or stolen property.[3] Millions more offenses occur that are less serious in nature, and many crimes go unreported. Given the far-reaching and significant impact of criminal activity, Americans should understand how offenders are processed through the three major components of our system—police, courts, and corrections—and know their legal rights.

Finally, by studying the criminal justice system, you will understand how your tax dollars support criminal justice in federal, state, and local governments (which now spend about $284 billion annually and employ over 2.4 million persons).[4] A tremendous amount of resources are required to support our criminal justice system. But as the French novelist Alain-René Lesage stated several centuries ago, "Justice is such a fine thing that we cannot pay too dearly for it."[5] This chapter provides an overview of the foundations of the criminal justice system. You will learn about the legal and historical bases of the system, the crime control and due process models of crime, the stages of the criminal justice process, a four-tier model used to categorize and describe different types of criminal cases, and how discretion and ethics permeate the system.

FOUNDATIONS OF CRIMINAL JUSTICE: LEGAL AND HISTORICAL BASES

The foundation of our criminal justice system is the criminal law: laws that define criminal acts and how such acts will be punished. Indeed, enforcing these laws is what sets in motion the entire criminal justice process. But like most things in our dynamic society, the law is not static. Enactment of new criminal laws and changes to those laws are almost always triggered by social, political, and economic changes. New ways to commit crimes are discovered, new illegal drugs make their way to the marketplace, new weapons and technology (for criminals and police alike) come on the scene, and suddenly, lawmakers and law enforcement officials find themselves needing new tools to prevent and prosecute crimes. We turn first to how the law changes and the historical principles that still guide—and sometimes challenge—that process.

The Criminal Law: How It Changes and How It Changes the System

In the wake of news reports indicating abuses resulting from an overtaxed criminal justice system (such as the one outlined at the beginning of this chapter), many lawmakers and criminal justice officials have been exploring ways to divert offenders away from traditional system responses. Attempts to find alternatives to arrest, prosecution, and incarceration have included providing treatment to help offenders deal with underlying issues that cause criminal behavior (e.g., mental health or substance abuse issues), as well as decriminalizing nonviolent, low-level crimes (consider the discussion of cannabis law changes that follows later in this chapter). You will learn more about recent diversion laws and programs throughout this book.

The search for alternative approaches to crime and justice indicates a significant shift away from the many "tough-on-crime" laws that were enacted in the 1990s. Most politicians wanted to appear tough on crime, particularly following high-profile violent cases, like the kidnapping, sexual assault, and murder of 12-year-old Polly Klaas. Polly was abducted from her home in Petaluma, California, by Richard Allen Davis in 1993. California lawmakers responded just months after Davis confessed to his crimes by proposing the nation's first **three-strikes law**—a seemingly simple solution giving violent offenders only two chances to turn themselves around. If they did not, and they committed another crime, the third crime would be the final "strike" and the state could lock them up and throw away the key for 25 years to life.[6] California voters overwhelmingly approved the measure, and within two years more than 20 states and the federal government had done the same.[7]

Three-strikes law: a crime control strategy whereby an offender who commits three or more violent offenses will be sentenced to a lengthy term in prison, usually 25 years to life.

Supporters predicted the new law would curb crime and protect society by incapacitating the worst offenders for a long period of time, while opponents argued that offenders facing their third strike would demand trials (rather than plea bargain) and send prison populations skyrocketing.[8]

The law that was finally enacted in California was vastly different from what was originally intended—and with many negative and unanticipated repercussions.[9] According to the *New York Times,* the law was unfairly punitive and created a cruel and unfair criminal justice system that lost all sense of proportion, doling out life sentences disproportionately to black defendants. Under the statute, the third offense that could result in a life sentence could be any number of low-level felony convictions, like stealing a jack from the back of a tow truck, shoplifting a pair of work gloves from a department store, pilfering small change from a parked car, or passing a bad check.[10]

Other studies of the California law found that prisoners added to the prison system in one decade's time would cost taxpayers an additional $8.1 billion in prison and jail expenditures.[11] Furthermore, three-strikes inmates sentenced for nonviolent offenses would serve 143,439 more years behind bars than if they had been convicted prior to the law's passage.[12] A nationwide study of three-strikes laws conducted about a decade after many states had adopted the law found no credible evidence to suggest that the law reduced crime.[13] Nearly 19 years after adoption of the three-strikes law, in November 2012, Californians voted to soften the sentencing law, to impose a life sentence only when the third felony offense is serious or violent, as defined in state law. The law also authorizes the courts to resentence thousands of people who were sent away for low-level third offenses and who present no danger to the public,[14] and it provides redress to mentally ill inmates—who were

©iStockphoto.com/LightFieldStudios

Three-strikes laws, while differing in content somewhat from state to state, all have a simple premise: making violent offenders with three qualifying convictions serve lengthy prison sentences.

estimated to compose up to 40 percent of those inmates with life sentences under the three-strikes rule.[15]

As you will read throughout this book, the adoption of laws and the processing of cases through our justice system is heavily influenced by politics. Many lawmakers and other politicians want to do what is right for society, but their decisions can also be influenced by their desire to be reelected, made with limited or inaccurate information, or prompted by a "knee-jerk" response to a high-profile event, like Kalief Browder's detention experience or Polly Klaas's murder.

The Browder and Klaas cases provide excellent illustrations of the national impact of injustice and crime, the legislative process, and the democratic system of criminal justice that exists to deal with offenders. These cases also prompt us to consider questions about the interaction of government and the justice system: What is the source of legislative and law enforcement powers? How can governments presume to maintain a system of laws that effectively governs its people and, moreover, a legal and just system that exists to punish persons who willfully violate those laws? We now consider those questions.

The Consensus- Versus-Conflict Debate

The criminal justice system plays a central role in our democratic society. We enact criminal laws to maintain order and to punish those who violate the democratically decided rules. But is order maintained through consensus—agreement—or is it preserved through conflict,

Sir Godfrey Kneller

John Locke, an English philosopher and physician and one of the most influential thinkers of his day, developed two influential theories concerning government and natural law: social contract and tacit consent.

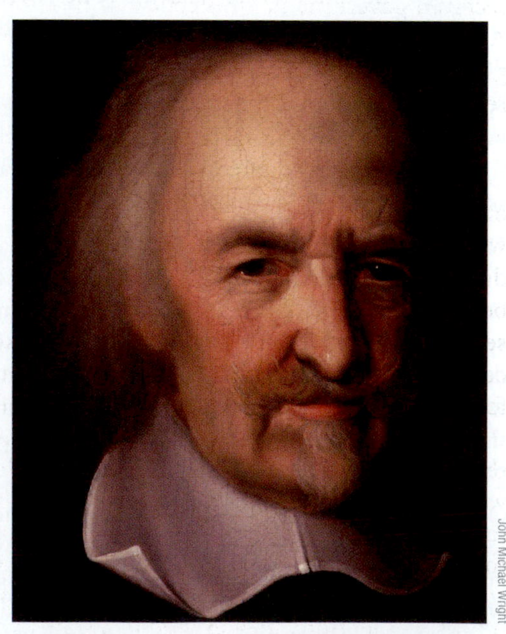

John Michael Wright

Another English philosopher and social contract theorist, Thomas Hobbes, believed in individual rights and representative government.

the exercise of power by certain groups over others? This debate is important because it forces us to look at how laws are created, to whom they are meant to apply, and the impact of our justice system in light of these competing perspectives.

Consensus Theory of Justice

Our society contains innumerable lawbreakers—many of whom are more violent than Richard Allen Davis. Most of them consent to police power in a cooperative manner, without challenging the legitimacy of the law if arrested and incarcerated. Nor do they challenge the system of government that enacts the laws or the justice agencies that carry them out. The stability of our government for more than 200 years is a testament to the existence of a fair degree of consensus as to its legitimacy.[16] Thomas Jefferson's statements in the Declaration of Independence are as true today as when he wrote them and are accepted as common sense:

> We hold these truths to be self-evident, that all men are created equal, that they are endowed by their Creator with certain inalienable Rights, that among these are Life, Liberty and the pursuit of Happiness—That to secure these rights, Governments are instituted among Men, deriving their just powers from the consent of the governed. That whenever any Form of Government becomes destructive of these ends, it is the Right of the People to alter or abolish it.

Consensus theory of justice: explains how a society creates laws as a result of common interests and values, which develop largely because people experience similar socialization.

The principles of the Declaration are almost a paraphrase of John Locke's *Second Treatise on Civil Government,* which justifies the acts of government on the basis of Locke's social contract theory. In the state of nature, people, according to Locke, were created by God to be free, equal, and independent, and to have inherent inalienable rights to life, liberty, and property. Each person had the right of self-protection against those who would infringe on these liberties. In Locke's view, although most people were good, some would be likely to prey on others, who in turn would constantly have to be on guard against those who might harm them. To avoid this brutish existence, people joined together, forming governments to which they surrendered their rights of self-protection. In return, they received governmental protection of their lives, property, and liberty. As with any contract, each side has benefits and considerations; people give up their rights to protect themselves and receive protection in return. Governments give protection and receive loyalty and obedience in return.[17]

Maurice Quentin de La Tour

Jean-Jacques Rousseau, a Genevan conflict theorist, argued that while the coexistence of human beings in equality and freedom is possible, it is unlikely that humanity can escape alienation, oppression, and lack of freedom: "Everywhere he is in chains."

Locke believed that the chief purpose of government was the protection of property. Properties would be joined together to form a commonwealth. Once the people unite into a commonwealth, they cannot withdraw from it, nor can their lands be removed from it. Property holders become members of that commonwealth only with their express consent to submit to its government. This is Locke's famous theory of tacit consent: "Every Man . . . doth hereby give his tacit consent, and is as far forth obliged to Obedience to the Laws of the Government."[18] Locke's theory essentially describes an association of landowners.[19] Another theorist connected with the social contract theory is Thomas Hobbes, who argued that all people were essentially irrational and selfish. He maintained that people had just enough rationality to recognize their situation and to come together to form governments for self-protection, agreeing "amongst themselves to submit to some Man, or Assembly of men, voluntarily, on confidence to be protected by him against all others."[20] Therefore, they existed in a state of consensus with their governments.

The **consensus theory of justice** assumes that most citizens in society share similar values and beliefs. It is based on the premise that even a diverse population of individuals hold the same morals or views concerning what should be labeled as "right" or "wrong" behavior.

These views are reflected in the law. Society comes together to protect itself from those who threaten the well-being of others by passing laws that prohibit harmful behavior and outlining the punishments for engaging in these acts.

Consensus theory acknowledges that perspectives and values can change over time. Some behaviors that were previously illegal are legal today. For example, adultery, same-sex marriage, drinking alcohol, and conducting business on Sundays were all labeled as criminal activity in the past.[21] Further, what is illegal today might be deemed legal tomorrow. Public attitudes toward marijuana use are shifting. In 2000, only 31 percent of Americans supported the legalization of marijuana, but recent polls show that two-thirds (66 percent) now support it.[22] This shift in public opinion has resulted in the legalization of marijuana use in numerous states, as depicted in Figure 1.1.

Conflict Theory of Justice

Jean-Jacques Rousseau, a conflict theorist, differed substantively from both Hobbes and Locke, arguing that "man is born free, but everywhere he is in chains."[23] Like Plato, Rousseau associated the loss of freedom and the creation of conflict in modern societies with

Figure 1.1 /// Cannabis Laws in the United States, as of 2019

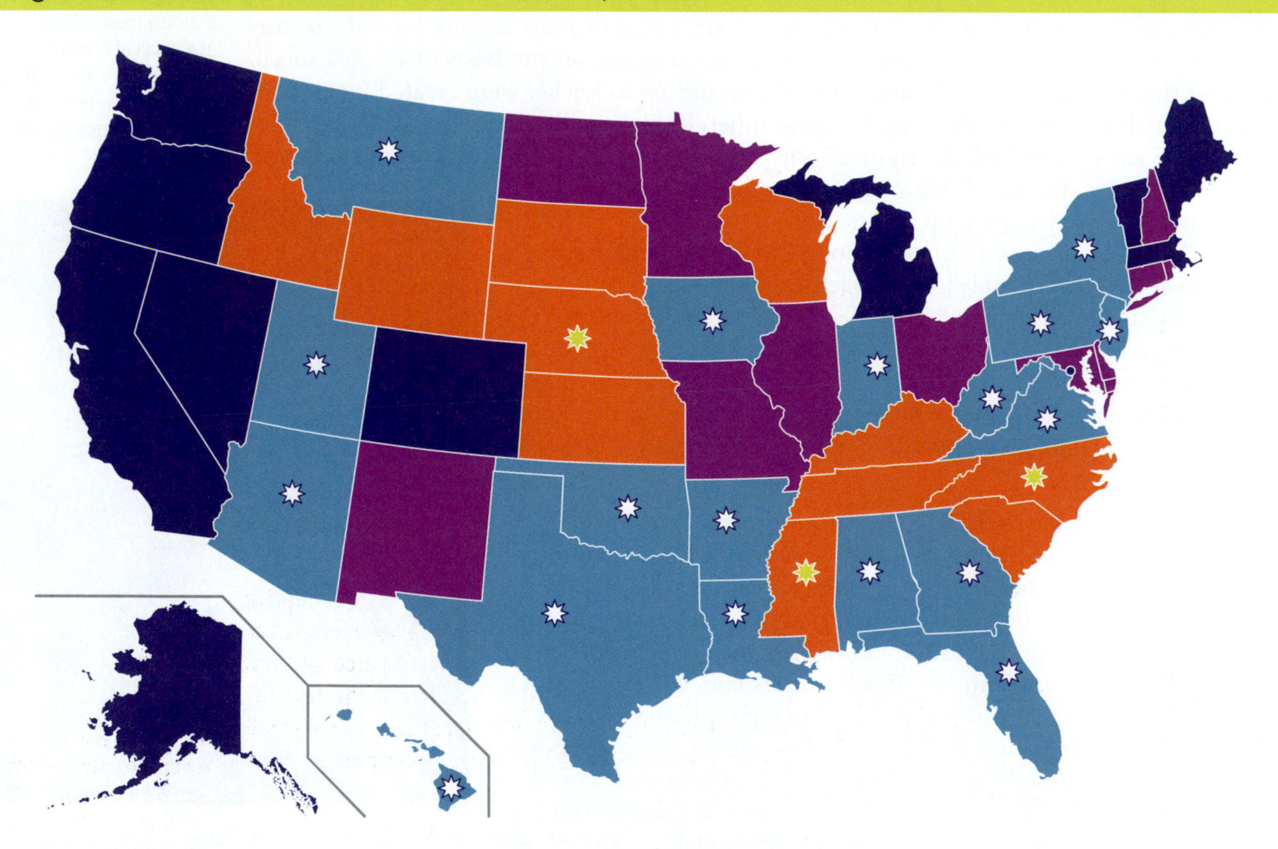

Cannabis Laws in the United States

- Jurisdiction with legalized cannabis.
- Jurisdiction with both medical and decriminalization laws.
- Jurisdiction with legal psychoactive medical cannabis.
- Jurisdiction with no decriminalized cannabis possession laws.
- Jurisdiction with reduced decriminalized cannabis possession laws.
- Jurisdiction with cannabis prohibition.

the development of private property and the unequal distribution of resources. Rousseau described conflict between the ruling group and the other groups in society, whereas Locke described consensus within the ruling group and the need to use force and other means to ensure the compliance of the other groups.[24]

The **conflict theory of justice** is based on the assumption that there will always be competing interests and viewpoints among members of society. People's beliefs can differ greatly and, as such, consensus is not possible. Women's right to abortion has been heavily debated since it was made legal across the country following the 1973 U.S. Supreme Court ruling in *Roe v. Wade*.[25] Yearly surveys reveal that opinions regarding abortion have remained relatively stable over time, and the country remains divided on this issue.[26]

Conflict theory of justice: explains how powerful groups create laws to protect their values and interests in diverse societies.

Since people do not share a common belief in what should or should not be legal, conflict theory maintains that laws reflect the interests of the most powerful people in society. Characteristics of certain groups make them more or less likely to be subjected to laws and criminal sanctions. These characteristics include age, class, race, and gender, as well as combinations of these attributes. The difference in laws governing crack cocaine and powder cocaine use is an example commonly used to support conflict theory. Prior to the Fair Sentencing Act of 2010, the penalty for possession of 5 grams of crack cocaine (commonly associated with poor, minority, and inner-city drug users) was a five-year mandatory prison sentence, but a person in possession of powder cocaine (commonly associated with wealthy, white drug users) would be subject to the same penalty only if the individual was carrying 500 grams of the drug.[27]

CRIME CONTROL AND DUE PROCESS: DO ENDS JUSTIFY MEANS?

Before learning about the specifics of the criminal justice process, it is important to consider how our system can serve the two seemingly competing goals of controlling crime and protecting the rights of persons accused of crime. How do we operate a system that is tough on crime and criminals while preserving the constitutional rights of those being accused? Crime control is an obvious and understandable goal for any society, but in a country founded on the ideals of freedom, liberty, and equality, we must be concerned about violating individual rights in our quest to achieve justice—we must ask whether the ends justify the means. For example, if the police illegally search a house and find clear evidence of several crimes, should the state be able to use that evidence to convict someone or should the evidence be excluded because the police violated the defendant's constitutional rights in the pursuit of crime control?

You will become familiar with these debates and conflicting objectives within the justice system as you learn more about the criminal justice process and the laws that govern police actions and ethics. By considering two different approaches to our criminal justice system, we can better understand how those accused of crimes move through the system and the different results—intended and otherwise—we might expect. In 1968, Herbert Packer described the two now-classic models of the criminal justice process in terms of two competing value systems: crime control and due process (see Table 1.1).[28]

Crime control model: a model by Packer that emphasizes law and order and argues that every effort must be made to suppress crime and to try, convict, and incarcerate offenders.

Due process model: a model by Packer that advocates defendants' presumption of innocence, protection of suspects' rights, and limitations placed on police powers to avoid convicting innocent persons.

The **crime control model** follows a highly traditional philosophy that Packer likened to an assembly line. The primary goal of this model is to deter criminal conduct and thus protect society. The accused is presumed guilty, police and prosecutors should have extensive freedom to exercise their own discretion (judgment) in the interest of crime control, legal loopholes should be eliminated, and offenders should be punished swiftly. This model views crime as a breakdown of individual responsibility; as such, only swift and certain punishment will deter and control crime.

In contrast, the **due process model**—likened to an obstacle course by some authors—focuses on fairness as its primary goal. In this model, criminal defendants should be presumed innocent, the courts' first priority is protecting the constitutional rights of the accused, and law

You Be the . . .

Legislator

Gun violence is a major concern in the United States. In addition to recent school shootings and mass shootings at public events and workplaces, many communities (particularly disadvantaged communities) suffer from high numbers of gun-related violent crimes. We do not have a complete database of all U.S. gun violence incidents, but some estimates suggest that firearms are involved in more than 22,000 suicides, just under 13,000 homicides, and almost 500 unintentional deaths in America annually. Further, the average number of injuries per year involving guns is almost 100,000.

Concerns over gun violence must be balanced with the rights afforded by the Second Amendment of our U.S. Constitution, which reads, "A well regulated Militia, being necessary to the security of a free State, the right of the people to keep and bear Arms, shall not be infringed." While there is considerable debate regarding the amendment's intended scope, lawmakers across the county are considering whether to enhance citizen rights to bear arms (e.g., by supporting laws that allow citizens to openly carry firearms) or to introduce legislation focused on enhancing gun control (e.g., to require background checks to purchase firearms).

1. Do laws that establish background check requirements, such as those proposed by the Bipartisan Background Checks Act of 2019, follow the consensus or conflict theory of justice?
2. What recent news stories might make lawmakers more likely to introduce gun rights or gun control legislation?
3. Does gun control legislation align more closely with the crime control model or the due process model?

Sources: Everytown for Gun Safety, "Gun Violence in America," https://everytownresearch.org/gun-violence-america/; H.R.8—Bipartisan Background Checks Act of 2019, 116th Congress (2019–2020), https://www.congress.gov/bill/116th-congress/house-bill/8.

enforcement officials—the police and prosecutors—must be held in check to preserve freedom and civil liberties for all Americans. As such, this model is designed to present "obstacles" for government actors at every stage, slowing down the process and affording an opportunity to uncover mistakes made in the pursuit of justice. This view also stresses that crime is not a result of individual moral failure but is driven by social influences (such as unemployment, racial discrimination, and other factors that disadvantage the poor); thus, courts that do not follow this philosophy are fundamentally unfair to these defendants. Furthermore, rehabilitation aimed at individual problems is embraced as a strategy to prevent future crime.

Table 1.1 /// Packer's Crime Control and Due Process Models

	Crime Control Model	Due Process Model
Views criminal justice system as an . . .	Assembly line	Obstacle course
Goal of criminal justice system	Controlling crime	Protecting rights of defendants
Values emphasized	Efficiency, speed, finality	Reliability
Process of adjudication	Informal screening by police and prosecutor	Formal, adversarial procedures
Focuses on . . .	Factual guilt	Legal guilt

Packer indicated that neither of these models would be found to completely dominate national or local crime policy.[29] To argue that one of these models is superior to the other requires an individual to make a value judgment. How much leeway should be given to the police? Should they be allowed to "bend" the laws in order to get criminals off the streets? Do the ends justify the means? Or, should criminal justice officials be required to follow rules and be held to higher standards than criminals? Is it important that our system be seen as fair and impartial?

The Kalief Browder Case: President Barack Obama signed an executive order to ban the solitary confinement of juveniles in federal prisons in 2016. Does this ban fall under the crime control model or the due process model, and what arguments would you make about why it falls under either category? What types of cases would make new laws—new laws focused on crime control or new laws focused on due process—more likely to be attractive to both the public and politicians?

DISCRETION: MAKING AND APPLYING THE LAW

After considering the two competing criminal justice models, you may wonder why the two models are even possible—in other words, how can criminal justice professionals follow different procedures in different situations when ours is a nation under the rule of law and due process? The answer is that players throughout the system exercise **discretion**, making decisions based on their own judgments in particular situations. As you consider the processes and cases presented throughout this book, you will see discretion at work in many different ways.

Discretion: authority to make decisions in enforcing the law based on one's observations and judgment ("spirit of the law") rather than the letter of the law.

For example, lawmakers understand that they cannot anticipate the range of circumstances surrounding each crime or create laws that reflect all local attitudes and priorities concerning crime control. Further, they cannot possibly enact enough laws to cover all potentially harmful behavior, with exceptions described for every possible scenario where strict application of the law would result in injustice. The report of the President's Commission on Law Enforcement and Administration of Justice (published in 1967) made a pertinent comment in this regard:

> Crime does not look the same on the street as it does in a legislative chamber. How much noise or profanity makes conduct "disorderly" within the meaning of the law? When must a quarrel be treated as a criminal assault: at the first threat, or at the first shove, or at the first blow, or after blood is drawn, or when a serious injury is inflicted? How suspicious must conduct be before there is "probable cause," the constitutional basis for an arrest? Every [officer], however sketchy or incomplete his education, is an interpreter of the law.[30]

Accordingly, enacting laws is just a first step, and persons charged with the day-to-day response to crime must exercise their own judgment within the limits set by those laws. Basically, they must decide whether to take action, which official response is appropriate, and how the community's attitude toward specific types of criminal acts should influence decisions. For example:

- Police officers exercise extensive discretion in deciding whether to stop, search, or arrest someone (discussed in Chapter 5).
- Prosecuting attorneys decide whether to bring criminal charges against an arrestee, thus making one of the most important judgment calls in the system (Chapter 8).
- Judges exercise discretion in setting or denying bail, and in imposing sentences (even with sentencing guidelines, discussed in Chapter 9).
- Corrections officials decide key issues of where to house convicted criminals, how to discipline them for rules violations on the inside, and whether to grant them early release on parole (Chapters 10 and 11).

The Kalief Browder Case: How does a law that bans the use of solitary confinement for juveniles affect a corrections official's discretion in managing inmate housing assignments? What might be the long-term impacts of solitary confinement bans on inmate mental health? Alternatively, how might limiting the discretion of corrections officials affect the overall safety of a correctional facility? When deciding when and under what conditions an inmate should be separated from other inmates, with whom should the discretion lie?

THE CRIMINAL JUSTICE PROCESS: AN OVERVIEW OF FLOW AND FUNCTIONS

Criminal justice flow and process: the movement of defendants and cases through the criminal justice process, beginning with the commission of a crime, and including stages that involve actions of criminal justice actors working within police, courts, and correctional agencies.

What follows is a brief description of the **criminal justice flow and process** in the United States. Figure 1.2 shows a flowchart of that system and summarizes the major stages of the process, including entry into the criminal justice system, prosecution and pretrial services, adjudication, sentencing and sanctions, and corrections. Note that *all* of the discussions in the following chapters of this book are based on the *people* and *processes* included in the depicted sequence of events.

The Offender's Pathway Through the Process

As we follow the path of the offender through the process, note that Figure 1.2 also depicts vertical pathways out of the criminal justice system. That is because many crimes fall out of the system for one of a variety of reasons: The crime is not discovered or reported to the police (the so-called dark figure of crime); no perpetrator is identified or apprehended; or, in some instances, a suspect is arrested, but the police later determine that no crime was committed, and the suspect is released from custody.

Law Enforcement: Investigation/Arrest

The flowchart in Figure 1.2 begins with "reported and observed crime." Police agencies learn about crime from the reports of victims or other citizens, from discovery by a police officer in the field, from informants, or from investigative and intelligence work. Once a law enforcement agency has established that a crime has been committed, the perpetrator must be identified and apprehended in order for the case to proceed through the system. Sometimes the offender is apprehended at the scene, but in other cases the police must conduct an investigation to find the perpetrator. Either way, the first formal step for most offenders in the criminal justice system is when the police take a suspect into custody for purposes of charging that person with a crime, known as an **arrest**.

Arrest: the taking into custody or detaining of one who is suspected of committing a crime.

Prosecution and Pretrial Activities

Prosecution: the bringing of charges against an individual, based on probable cause, so as to bring the matter before a court.

Next we enter the **prosecution** and pretrial services phase of the process—and the realm of the powerful individuals who "control the floodgates" of the courts process. After an arrest, police present information concerning the case and the accused (typically in the form of an official offense/arrest report) to the prosecutor, who will decide—at his or her discretion—if formal charges will be filed with the court. If no charges are filed, the accused must be released. The prosecutor can also elect, after initially filing charges, to drop charges (*nolle prosequi*) if he or she determines that the probable cause and/or evidence in the matter is weak. (Probable cause, discussed more fully in Chapter 6, is a legal term that basically refers to information that would lead a reasonable person to believe that a person has committed, is committing, or is about to commit a crime.) Furthermore, in some jurisdictions, defendants, often those without a prior criminal record, may be eligible for diversion from prosecution subject to the completion of specific conditions such as drug treatment. Successful completion of the conditions may result in charges

Figure 1.2 /// The Sequence of Events in the Criminal Justice System

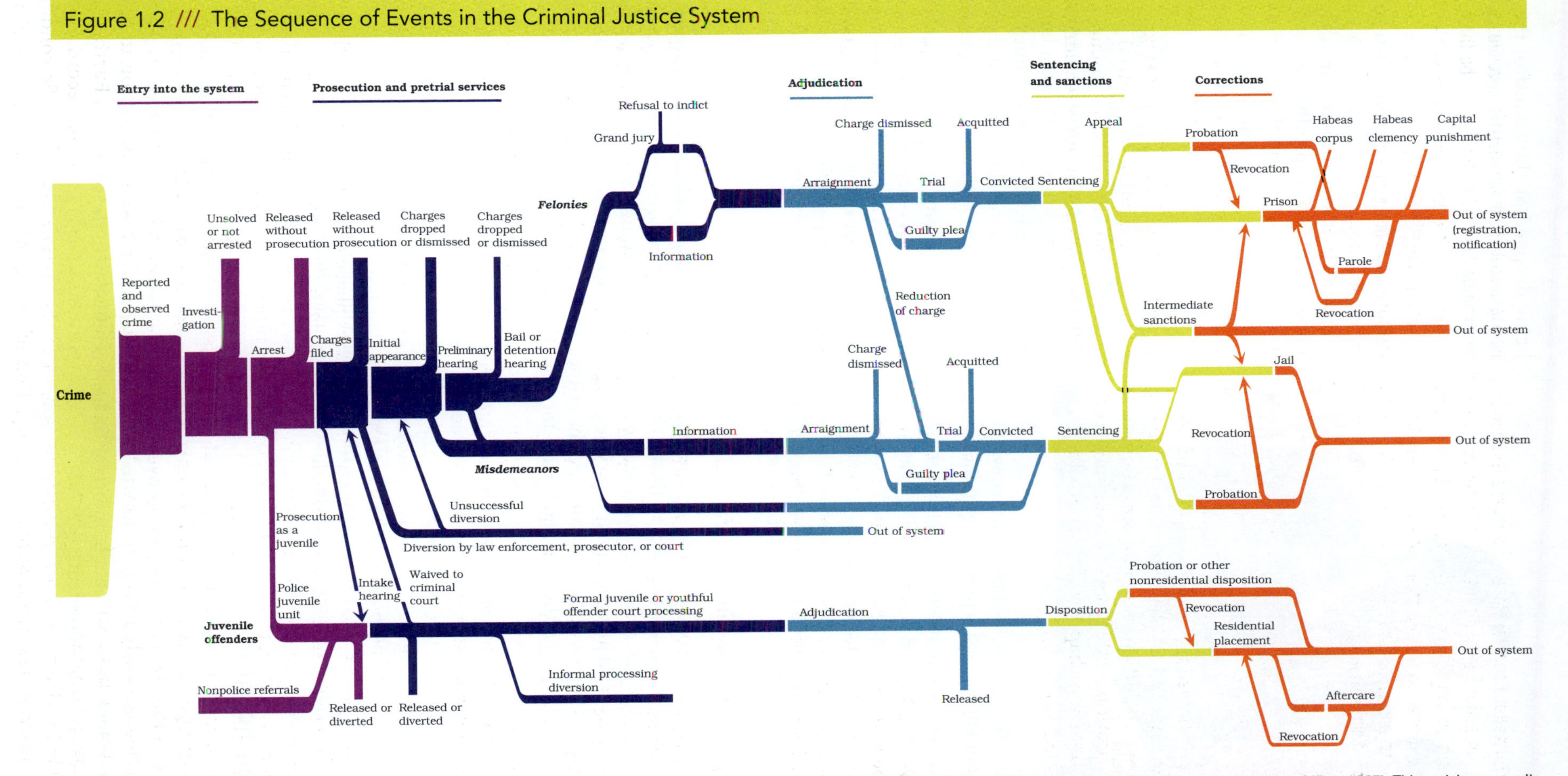

Source: Adapted from President's Commission on Law Enforcement and Administration of Justice, *The Challenge of Crime in a Free Society* (Washington, D.C.: U.S. Government Printing Office, 1967). This revision, a result of the Symposium on the 30th Anniversary of the President's Commission, was prepared by the Bureau of Justice Statistics in 1997.

Note: This chart gives a simplified view of case-flow through the criminal justice system. Procedures vary among jurisdictions. The weights of lines are not intended to show actual size of caseloads.

Scott Dekraai (L), sits next to his attorney, Assistant Public Defender Scott Sanders, at Orange County Superior Court in Santa Ana, California. The former tugboat worker pleaded guilty to first-degree murder for killing eight people in a Southern California hair salon where his ex-wife worked. Under the United States' system of justice, all criminal defendants prosecuted for serious crimes have the right to be represented by an attorney.

being dropped or the record of the crime being expunged (meaning to legally strike or erase).

Initial Appearance

Persons charged with a crime must be taken for an initial appearance before a judge or magistrate without unnecessary delay (the amount of time is typically specified in the state's statutes or in municipal ordinances). There, the judge will inform the accused of the charges and decide whether there was probable cause for the police to make an arrest. If the offense is not very serious, the determination of guilt and assessment of a penalty may also occur at this stage.

Often, a defense attorney is also assigned at the initial appearance. All defendants who are prosecuted for serious crimes have a right to be represented by an attorney. If the court determines the defendant is indigent and cannot afford such representation, the court will assign counsel at the public's expense.

A decision about whether to release the defendant on bail or some other conditional release may also be made at the initial appearance. The court considers factors such as the seriousness of the charge; whether the defendant is a flight risk; and if he or she has a permanent residence, a job, and family ties. If the accused is likely to appear at trial, the court may decide that he or she should be released on recognizance (often termed "ROR," meaning that the defendant is released without having to provide bail, upon promising to appear and answer the criminal charge) or into the custody of a third party after the posting of a financial bond.

Preliminary Hearing or Grand Jury

The next step is to determine whether there is probable cause to believe the accused committed the crime and whether he or she should be tried. Depending on the jurisdiction and the case, this determination is made in one of two ways: through a preliminary hearing or through a grand jury. In a preliminary hearing, a judge determines if there is probable cause to believe that the accused committed the crime. If so, the case moves forward to trial, also known as "binding the defendant over" for trial. If the judge does not find probable cause, the case is dismissed.

In other jurisdictions and cases, the prosecutor presents evidence to a grand jury, which decides if there is sufficient evidence to bring the accused to trial. If the grand jury finds sufficient evidence, it submits to the court an indictment, a written statement of the essential facts of the offense charged against the accused. Misdemeanor cases and some felony cases proceed by the issuance of an "information," which is a formal, written accusation submitted to the court by a prosecutor (rather than an indictment from the grand jury). In some jurisdictions, indictments may be required in felony cases. Grand juries are discussed more in Chapter 7.

Adjudication

Next, in the middle of the flowchart shown in Figure 1.2, is the **adjudication** process. The adjudication process allows the defendant to respond to the charges brought against him or her, requires the government to prove its case (if the defendant claims that he or she is not guilty), and allows a judge or jury to decide whether the defendant is legally guilty or not guilty. Once an indictment or information has been filed with the trial court, the accused is scheduled for arraignment.

Adjudication: the legal resolution of a dispute—for example, when one is declared guilty or not guilty—by a judge or jury.

Practitioner's Perspective

Police Officer

Name: Kevin Wilmott
Position: Police Officer
Location: Southern California

What advice would you give to someone either wishing to study, or now studying, criminal justice and wanting to become a practitioner in this position? My biggest advice for students who are pursuing a career in law enforcement is to start looking into going on ride-alongs with agencies that they're interested in working for. A lot of times, agencies will offer civilian ride-alongs. Just call the department, and ask if you can go on a ride-along. Because that's the best way to really see if this is something that you're interested in. You can watch it on TV, you can read about it in a book, but until you actually go out in the field and shadow an officer and see what his or her "day in the life" is like, you don't really have a good idea. So I think the biggest thing is to go on ride-alongs and experience them first-hand.

Also, if you're not quite the age yet—typically the age is 21 for most agencies to be a police officer—a really good idea that I recommend is to look at civilian jobs in a police department. I started as a police dispatcher for two years. A lot of times there are positions for cadets, traffic enforcement or parking enforcement, dispatchers, records, whatever it may be. There are definitely civilian jobs where a lot of times the minimum age requirement is 18. So between 18 and 21 you can do that, and that actually gets you a foot in the door, as you get to see what it's like to work at a police department, and you can get to know the department and the community that you want to eventually work for as a police officer. It helps build your name and reputation with the department.

In general, what does a *typical* day look like for a practitioner in this position? Granted, our days are never quite routine, and every day is different. But typically, a day in the life for me, for the most part, would start with going to briefing. In our briefing we go over what's going on in our community, what to look out for, if there have been any recent crimes or crime trends that we need to be aware of. We also get our assignments, which area of the city we'll be working, what car we're going to be in. After briefing, we get our vehicle ready, and that means putting all of our equipment in the patrol car, and then we go out and begin our shift. I currently work nights and weekends, so I personally am looking for impaired drivers. Some officers look out for drug users or traffic enforcement, and some are interested in working with quality-of-life issues with transciency, loitering, and things of that nature. But I personally am interested in DUI enforcement. So a lot of times, when I'm not responding to a call for service, I'm out on the roads looking for people that may be impaired. And every so often we get dispatch for calls for service. Then at the conclusion of my shift, I gas up the vehicle, put it away, and take all the gear out of it. I also make sure I'm caught up on all my reports, because there's definitely a lot of report writing involved. The quieter hours are around three to five in the morning, so those are the times I try to catch up on all my reports.

To learn more about Kevin Wilmott's experiences as a police officer, watch the Practitioner's Perspective video in SAGE Vantage.

Arraignment

At the arraignment, the accused is informed of the charges, advised of the rights of criminal defendants, and asked to enter a plea to the charges. Generally, defendants enter a plea of guilty, not guilty, or *nolo contendere* (no contest).

©REUTERS/Reuters/POOL New

Gary Leon Ridgway and his attorneys look over the plea agreement allowing him to escape the death penalty by pleading guilty to 48 counts of aggravated first-degree murder in the Green River serial killing cases.

If the accused pleads guilty or *nolo contendere* (accepts penalty without admitting guilt), the judge may accept or reject the plea. If the plea is accepted, the defendant has in effect given up his or her constitutional right to a trial, no trial is held, and sentencing occurs at this proceeding or at a later date. But contrary to popular media depictions, not guilty pleas and trials are very rare; approximately 95 percent of criminal defendants plead guilty as a result of plea bargaining between the prosecutor and the defendant.

Trial

If the accused pleads not guilty or not guilty by reason of insanity, he or she is basically forcing the government to prove its case—to prove the defendant's guilt beyond a reasonable doubt. A person accused of a serious crime is guaranteed a trial by jury but may request a bench trial where the judge alone, rather than a jury, will hear both sides of the case. In both instances the prosecution and the defense are permitted to present physical evidence and question witnesses, while the judge decides on issues of law. The trial results in an **acquittal** (not guilty) or a **conviction** (guilty) on the original charges or on lesser included offenses.

Acquittal: a court or jury's judgment or verdict of not guilty of the offenses charged.

Conviction: the legal finding, by a jury or judge, or through a guilty plea, that a criminal defendant is guilty.

Sentencing and Sanctions, Generally

After a conviction, a sentence is imposed. With the exception of capital cases where the death penalty is being sought and the jury decides the punishment, the judge determines the sentence.

In arriving at an appropriate sentence, a sentencing hearing may be held at which time evidence of **aggravating** or **mitigating circumstances** is considered (aggravators are elements that tend to increase the offender's blame, such as use of torture; mitigators tend to reduce blame, such as youthfulness and lack of prior criminal record; these are discussed more in Chapter 9). Here the court may rely on presentence investigations by probation agencies and consider victim impact statements (a written or oral statement by the victim concerning the pain, anguish, and financial devastation the crime has caused).

Aggravating circumstances: elements of a crime that enhance its seriousness, such as the infliction of torture, killing of a police or corrections officer, and so on.

Mitigating circumstances: circumstances that would tend to lessen the severity of the sentence, such as one's youthfulness, mental instability, not having a prior criminal record, and so on.

The sentencing choices that may be available to judges and juries include one or more of the following:

- Death penalty (only in first-degree murder cases and only in certain states)
- Incarceration in a prison (for sentences of a year or longer), a jail (for sentences of up to a year), or another confinement facility
- Probation—allowing the convicted person to remain at liberty but subject to certain conditions and restrictions such as drug testing or drug treatment
- Fine—applied primarily as penalties in minor offenses
- Restitution—requiring the offender to pay compensation to the victim
- Intermediate **sanction** (used in some jurisdictions)—an alternative to incarceration that is considered more severe than straight probation but less severe than a prison term (e.g., boot camps, intense supervision often with drug treatment and testing, house arrest and electronic monitoring, and community service)

Sanction: a penalty or punishment.

Sentences and punishment are discussed in Chapter 9, whereas intermediate sanctions, probation, and parole are examined in Chapter 11.

Scott Olson/Getty Images News/Getty Images

The U.S. Supreme Court has held that trial juries may hear and consider victim impact statements (concerning such factors as the pain, anguish, and suffering the defendant's crime has caused) when making sentencing decisions.

Appellate Review

Following trial and sentencing, a defendant may appeal his or her conviction or sentence by requesting that a higher court review the arrest and trial (a process known as appellate review). The appellate process provides checks on the criminal justice system by ensuring that errors at trial (except for those considered to be "harmless") did not adversely affect the fairness of trial processes and the defendant's constitutional rights. In death penalty cases, appeals of convictions are automatic. In other cases, the appellate court has sole discretion over whether to review the case.

Corrections

The next phase into which the offender enters is corrections, as shown in Figure 1.2. Offenders sentenced to incarceration usually serve time in a local jail or a state prison. Offenders sentenced to less than one year generally go to jail; those sentenced to more than one year go to prison.

A prisoner may become eligible for parole after serving a portion of his or her **indeterminate sentence** (a range, such as 5–10 years). **Parole** is the conditional release of a prisoner before the prisoner's full sentence has been served. The decision to grant parole is made by an authority such as a parole board, which has power to grant or revoke parole (i.e., return the parolee to prison) or to discharge a parolee altogether. In some jurisdictions, offenders serving what is called a **determinate sentence**—a fixed number of years in prison—will not come before a paroling authority, because each offender is required to serve out the full sentence prior to release, less any earned "good time credits" (a reduction in the time served in jail or prison due to good behavior, participation in programs, and other activities).

Indeterminate sentence: a scheme whereby one is sentenced for a flexible time period (e.g., 5–10 years) so as to be released when rehabilitated or when the opportunity for rehabilitation is presented.

Parole: early release from prison, with conditions attached and under supervision of a parole agency.

Determinate sentence: a specific, fixed-period sentence ordered by a court.

If released by a parole board or through mandatory release, the parolee will be under the supervision of a parole officer in the community for the balance of his or her unexpired sentence. This supervision is governed by specific conditions of release, and the parolee may be returned to prison ("parole revocation") for violations of such conditions.

Once a person who is suspected of committing a crime is released from the jurisdiction of a criminal justice agency, he or she may commit a new crime (recidivate) and thus need to be processed again through the criminal justice system. Studies show that individuals with prior criminal histories are more likely to be rearrested than those without a prior history.

The Juvenile Justice System

Juvenile courts usually have jurisdiction over matters concerning children, including delinquency, neglect, and adoption. They also handle **status offenses** such as truancy and running away, which are not applicable to adults. State statutes define which persons are under the original jurisdiction of the juvenile court. The maximum age of original juvenile court jurisdiction in delinquency matters is 17 in most states.[31] The "Juvenile Justice Journal" boxes throughout this book examine the juvenile justice system.

Status offense: a crime committed by a juvenile that would not be a crime if committed by an adult; examples would be purchasing alcohol and tobacco products, truancy, and violating curfew.

Juvenile Justice Journal: A Brief History

Our society has long been concerned with its minors (usually defined as those persons between the ages of 10 and 18) who violate the law; such persons—commonly termed *juvenile delinquents*—have often been faced with adverse and traumatic home situations. As result of this concern, a U.S. juvenile justice system has developed that functions quite differently from the system that addresses crimes committed by adults. Therefore, as mentioned in the preface to this book, in several chapters we briefly discuss (in boxed features like this one) different facets of this system.

History of Juvenile Justice

In the early part of the 19th century, to the chagrin of prosecutors and many citizens, juries were acquitting children who were charged with crimes—not wishing to see children incarcerated with adults in ramshackle facilities. Quakers in New York City sought to establish a balance between those two camps—people wanting to see justice done with child offenders, and those not wanting them to be incarcerated—and founded the first house of refuge in 1825. The children worked an eight-hour day at various trades in addition to attending school for another four hours.

At about the middle of the 19th century, the house of refuge movement evolved into the slightly more punitive reform school, or reformatory, approach, which served to segregate young offenders from adult criminals; imprison the young and remove them from adverse home environments until the youths were reformed; help youth avoid idleness through military drills, physical exercise, and supervision; focus on education—preferably vocational and religious; and teach sobriety, thrift, industry, and prudence.

In 1899, the Illinois legislature enacted the Illinois Juvenile Court Act, creating the first separate juvenile court. At that time in the United States, juveniles were tried along with adults in criminal courts and sometimes sentenced to prison and occasionally to death. Prior to 1900, at least 10 children were executed in the United States for crimes committed before their 14th birthdays. Other children died in adult prisons. These deaths shocked the public conscience. Accordingly, Americans in the 20th century sought more pervasive reform than the infancy defense (a defense used when a person is too young to be held liable for his or her actions) to address the distinctive nature of children and youth.

Although the Illinois act did not fundamentally change procedures in the courts that were then sitting as juvenile courts to adjudicate cases involving children, it did emphasize the *parens patriae* ("state is the ultimate parent") philosophy (discussed in Chapter 2) to govern such cases. The act gave the courts jurisdiction over children charged with crimes, as well as children who lived in unsafe or harmful conditions. For example, courts were given jurisdiction over children who were homeless, lacked guardianship, needed public support, lived with disreputable persons or within unfit living conditions, habitually begged public for assistance, or, if under the age of 8, sold items or performed on the street. The act thus defined a rehabilitative rather than punishment purpose for that court, established the confidentiality of juveniles' court records to minimize stigma, required that juveniles be separated from adults when placed in the same institution in addition to barring altogether the detention of children under age 12 in jails, and provided for the informality of procedures within the court.

Status Offenses

The post–World War II period witnessed further developments, as the status offense became a separate category, covering acts that would not be criminal if committed by an adult (e.g., purchasing alcohol and tobacco products, truancy, and violating curfew). New York created a new jurisdictional category for persons in need of supervision (PINS): runaways, truants, and other youths who committed status offenses. Other states followed New York's lead. Then came the enactment of the very powerful and far-reaching Juvenile Justice and Delinquency Prevention Act of 1974, which removed status offenders from secure detention and correctional facilities, and more significant, perhaps, prevented the placement of any juveniles in any institutions where they would have regular contact with adults convicted of criminal charges.

THE WEDDING CAKE MODEL OF CRIMINAL JUSTICE

The criminal justice system flowchart shown in Figure 1.2 makes it easy to see the steps through which the offender moves through the process horizontally. It is also helpful to see how the system treats cases differently by viewing it vertically, as shown in the **wedding cake model of criminal justice**, developed by Samuel Walker (see Figure 1.3).[32]

Wedding cake model of criminal justice: a model of the criminal justice process whereby a four-tiered hierarchy exists, with a few celebrated cases at the top, and lower tiers increasing in size as the seriousness of cases declines and informal processes (use of discretion) become more likely to occur.

This approach begins with the premise that not all criminal cases are viewed or handled in the same manner—by either the police or the judiciary. The type of treatment given to a particular case, including its outcome, is determined mostly by factors such as the seriousness of the charge, current policies and political influences, and the defendant's status and resources. Some cases are run-of-the-mill and are treated as such, but some involve high-profile crimes and/or criminals and command much more attention.

As shown in Figure 1.3, the wedding cake model divides criminal justice system proceedings into four different categories: celebrated cases, serious felonies, lesser felonies, and misdemeanors. This partitioning of cases allows for a closer analysis of how the criminal justice system deals with them.

Layer 1: Celebrated Cases

The top layer of the wedding cake model includes the "celebrated cases." These cases command a great deal of media attention because the crimes are unusual (such as when James Holmes killed 12 people and injured 70 others in a mass shooting at a movie theater in Aurora, Colorado; or when Dzhokhar Tsarnaev, along with his brother, detonated two bombs at the Boston Marathon, killed a police officer, kidnapped a man, and engaged in a shootout with police) or because the defendants are celebrities or high-ranking officials (consider singer R. Kelly's criminal sexual abuse trial; O. J. Simpson, the celebrated athlete and actor whose "trial of the century" has been the topic of numerous books and documentaries; and Aaron Hernandez, the former NFL player who was convicted of murder). The legal process for these types of cases is not different from that of the "usual" case, but because of their complexity or high-profile nature, many more resources are devoted in the form of forensic tests, use of expert witnesses, jury sequestering (seclusion), cameras in the courtroom, and crowd control. At the same time, due to the widespread media and public

Figure 1.3 /// The Wedding Cake Model of Criminal Justice

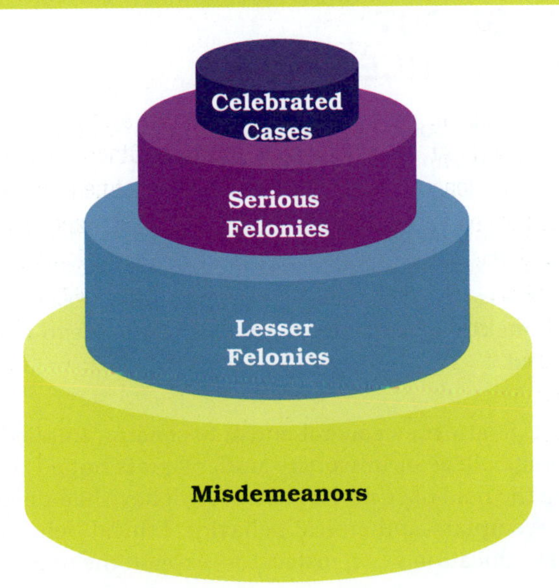

©iStockphoto.com/FrankvandenBergh

Convicted offenders who are to be incarcerated will serve time in either a local jail (typically for misdemeanants serving less than one year) or a federal or state prison (for felons, and usually for more than one year).

attention given to the cases, extra care is taken to ensure that defendants' rights are protected and that the accused are not given preferential treatment based on their status.

Layer 2: Serious Felonies

The second layer of the wedding cake includes serious felonies, which are violent crimes committed by people with lengthy criminal records and who often prey on people they do not know. These are viewed by the police and prosecutors as the cases that are most deserving of "heavy" treatment and punishment, and there is not as great a chance that the defendant will be allowed to enter into a plea agreement before trial.

Layer 3: Lesser Felonies

On the third layer of the wedding cake are the lesser felonies, which tend to be nonviolent and typically viewed as less important than the felonies in Layer 2. These offenders might have no criminal record; might have had a prior relationship with the victim; and might be charged with drug-related, financial, or other such crimes. A good portion of these cases will be filtered out of the system prior to trial and end in plea agreements.

Layer 4: Misdemeanors

Layer 4 consists of misdemeanor cases, which make up about 90 percent of all criminal matters. They include less serious and public order crimes: public drunkenness, minor theft, disturbing the peace, and so on. Police are more likely to deal with these cases informally and use their discretion to determine whether an arrest is necessary. When arrests are made, they will be handled by the lower courts—where the large number of cases handled by these courts require quick case processing, making trials rare. Many misdemeanor cases are resolved with plea agreements and penalties that involve fines, probation, or short-term jail sentences.[33]

ETHICS THROUGHOUT THE CRIMINAL JUSTICE SYSTEM

Robert F. Kennedy, in his 1960 book, *The Enemy Within: The McClellan Committee's Crusade Against Jimmy Hoffa and Corrupt Labor Unions,* stated that

> in the fall of 1959 I spoke at one of the country's most respected law schools. The professor in charge of teaching ethics told me the big question up for discussion among his students was whether, as a lawyer, you could lie to a judge. I told the professor . . . that I thought we had all been taught the answer to that question when we were six years old.[34]

As Kennedy, the late U.S. attorney general and U.S. senator, implied, by the time people reach the point of being college or university students, it is hoped that everyone—in particular, those studying the field of criminal justice—will have had deeply ingrained in them the need to practice exemplary and ethical behavior. Ethical behavior is often reemphasized in postsecondary education when instructors explain the need for academic honesty.

The importance of ethics is often reinforced in other social settings, for example, in work environments (employees are entrusted with property and other responsibilities), at home (parents and other caregivers explain the importance of treating others with respect), in sports (coaches emphasize the importance of fair play), and at church (treating others as you want to be treated is often taught as a guiding principle).

Character in the criminal justice arena is of utmost importance. Without it, nothing else matters. Character, it might be said, is who we are when no one is watching. Having character means that people do not betray their fellow human beings or violate oaths of office or public trust. Unfortunately, character of mind and actions cannot be taught solely in a college or university classroom, nor can it be implanted in a doctor's office, administered intravenously, or ingested as a pill.

Prior to commencing your journey into the field of criminal justice (and, later, reading Chapter 3, concerning ethics), you might do well to first ask yourself these questions: Should police officers receive free coffee from restaurants and quick-stop establishments? Free or half-priced meals? What about judges? Prison wardens? If judges, wardens, or other criminal justice professionals are not "rewarded" with such "freebies," then should police officers be able to accept such "gifts" while on- or off-duty? On what grounds do many police officers expect such favored treatment? And can this lead to ethical problems with respect to their work?

At its root, the field of criminal justice is about people and their activities; and in the end, the primary responsibilities of people engaged in this field are to ensure that they be of the highest ethical character and treat everyone with dignity and respect. Therefore, as indicated earlier, this textbook, unlike most or all others of its kind, devotes an entire chapter to the subject of **ethics**—or what essentially constitutes "correct" behavior in criminal justice.

Ethics: a set of rules or values that spell out appropriate human conduct.

You Be the . . .

Judge

Four male juveniles were burglarizing a home when Baltimore County Police Officer Amy Caprio was called to the scene. Officer Caprio approached a stolen vehicle the teens drove to the neighborhood. When she reached the vehicle, she encountered one of the juveniles behind the wheel. She drew her weapon and fired once after the vehicle accelerated toward her. The juvenile struck her with the vehicle and fled the scene. Officer Caprio, who had served on the department for 3 years and 10 months, was rushed to the hospital where she was pronounced dead. Further investigation revealed that the juvenile who killed the officer was waiting in the driver's seat of the stolen vehicle while his three associates carried out the burglary. All four teens were eventually charged with felony murder.

1. The juveniles were tried as adults for this crime. Does this decision by the prosecutor mostly closely align with the crime control or due process model of justice?
2. Should the prosecutor have had the discretion to try these juveniles as adults? Would it change your mind if they had been 12-year-olds rather than 17-year-olds?
3. Do you believe politics played a part in the decision to charge all four with murder?
4. What aggravating or mitigating circumstances of this case might be discussed at trial?
5. Where would this type of case fit on the wedding cake model of criminal justice?

Source: For more information, see Jessica Anderson, "Jury sees video showing moments before Baltimore County officer Amy Caprio's death," *Baltimore Sun*, April 23, 2019, https://www.baltimoresun.com/news/crime/bs-md-co-caprio-trial-openings-20190422-story.html.

/// IN A NUTSHELL

- It is important to study the criminal justice system because it serves to shape our government and culture, and because all of us are potential victims, witnesses, and taxpaying supporters of our justice system. Furthermore, we need to understand the rights afforded to us by our Constitution.
- The consensus theory of justice argues that laws reflect the values and beliefs shared by most people in society, whereas the conflict theory of justice maintains that laws are created to protect the interests of the most powerful people in society.
- The due process model of criminal justice holds that criminal defendants should be presumed innocent and constitutional rights of the accused should be emphasized over conviction of the guilty. Conversely, the crime control model, likened to an assembly line, emphasizes deterring criminal conduct and protecting society; eliminating legal loopholes, swiftly punishing offenders, and granting a high degree of discretion to police and prosecutors. Neither of these models, however, completely dominates national or local crime policy.
- Discretion is exercised throughout the criminal justice system, because violations of laws vary in their seriousness, and there are not enough human and financial resources to enforce all laws equally. Therefore, persons charged with enforcing laws, adjudicating cases, and punishing offenders exercise considerable judgment in terms of deciding whether to take action, which official response is appropriate, and to what extent the community's attitude toward specific types of criminal acts should affect such decisions.
- Although the offender's path through the criminal justice process may be viewed as horizontal in nature, there are many points through which an offender can take a vertical pathway out of the system.
- The wedding cake model of criminal justice argues that not all criminal cases are viewed or handled in the same manner by either the police or the courts—some are treated with more discretion, while others are subjected to more formal processes. The processing of cases by the criminal justice system is divided into four categories: celebrated cases, serious felonies, lesser felonies, and misdemeanors. The type of treatment given to a particular case is determined by factors such as the seriousness of the charge, current policies and political influences, and the defendant's status and resources.
- Criminal justice officials must behave ethically. People engaged in this field must be of the highest ethical character and treat everyone with dignity and respect.

/// KEY TERMS & CONCEPTS

Review key terms with eFlashcards at **edge.sagepub.com/peakbrief**.

Acquittal 16
Adjudication 14
Aggravating circumstances 16
Arrest 12
Conflict theory of justice 9
Consensus theory of justice 7
Conviction 16
Crime control model 9
Criminal justice flow and process 12
Determinate sentence 17
Discretion 11
Due process model 9
Ethics 21
Indeterminate sentence 17
Mitigating circumstances 16
Parole 17
Prosecution 12
Sanction 16
Status offenses 17
Three-strikes law 5
Wedding cake model of criminal justice 19

/// REVIEW QUESTIONS

Test your understanding of chapter content. Take the practice quiz at **edge.sagepub.com/peakbrief**.

1. Having read the chapter, do you believe it is important for you to study the structure and function of our criminal justice system? Why or why not?
2. How and why are laws created according to the consensus model of criminal justice? According to the conflict model of criminal justice?

3. How would you describe the crime control and due process models of criminal justice? What indicators might be present in your local community that would help you determine the most dominant model?
4. What are the major points at which an offender is dealt with in the criminal justice process, as he or she moves through the police, courts, and corrections components?
5. What are the four tiers of the wedding cake model of criminal justice, and how is discretion by criminal justice officials used differently at each stage?
6. How would you characterize the importance of discretion and ethics throughout the justice system?

/// LEARN BY DOING

As indicated in this textbook's preface, this "Learn by Doing" section, as well as those at the end of subsequent chapters, is an outgrowth of teachings by famed educator John Dewey, who advocated the "learning by doing" or problem-based approach to education. It also follows the popular learning method espoused by Benjamin Bloom in 1956, known as Bloom's Taxonomy, in which he called for "higher-order thinking skills"—critical and creative thinking that involves analysis, synthesis, and evaluation.[35]

The following scenarios and activities will shift your attention from textbook-centered instruction and move the emphasis to student-centered projects. By being placed in these hypothetical situations, you can thus learn—and apply—some of the concepts covered in this chapter, develop skills in communication and self-management, at times become a problem solver, and learn about and address current community issues.

1. Assume that you are an officer in your campus criminal justice honor society and are invited to speak at the society's monthly meeting concerning your view of how crime is perceived and dealt with in your community. You opt to approach the question from Packer's crime control and due process perspectives. Given what you know about crime and criminal justice in your community, what will you say in your presentation?
2. As a member of your campus criminal justice honor society, you are asked to speak at a meeting of your local police department's Citizens' Police Academy, focusing on the general need for citizens to "become involved" in addressing crime. What will you say?
3. Your criminal justice professor asks you to prepare your own succinct diagram of the criminal justice process, including brief descriptions of each of the major stages (arrest, initial appearance, and so on) as a case flows through the process. What will your final product look like?
4. As part of a class group project concerning the nature of crime and punishment, you are asked by your fellow group members to develop a 10-minute presentation on the wedding cake model of criminal justice. How will you describe it?

/// STUDY SITE

Get the tools you need to sharpen your study skills. SAGE edge offers a robust online environment featuring an impressive array of free tools and resources.

Access practice quizzes, eFlashcards, video, and multimedia at **edge.sagepub.com/peakbrief**

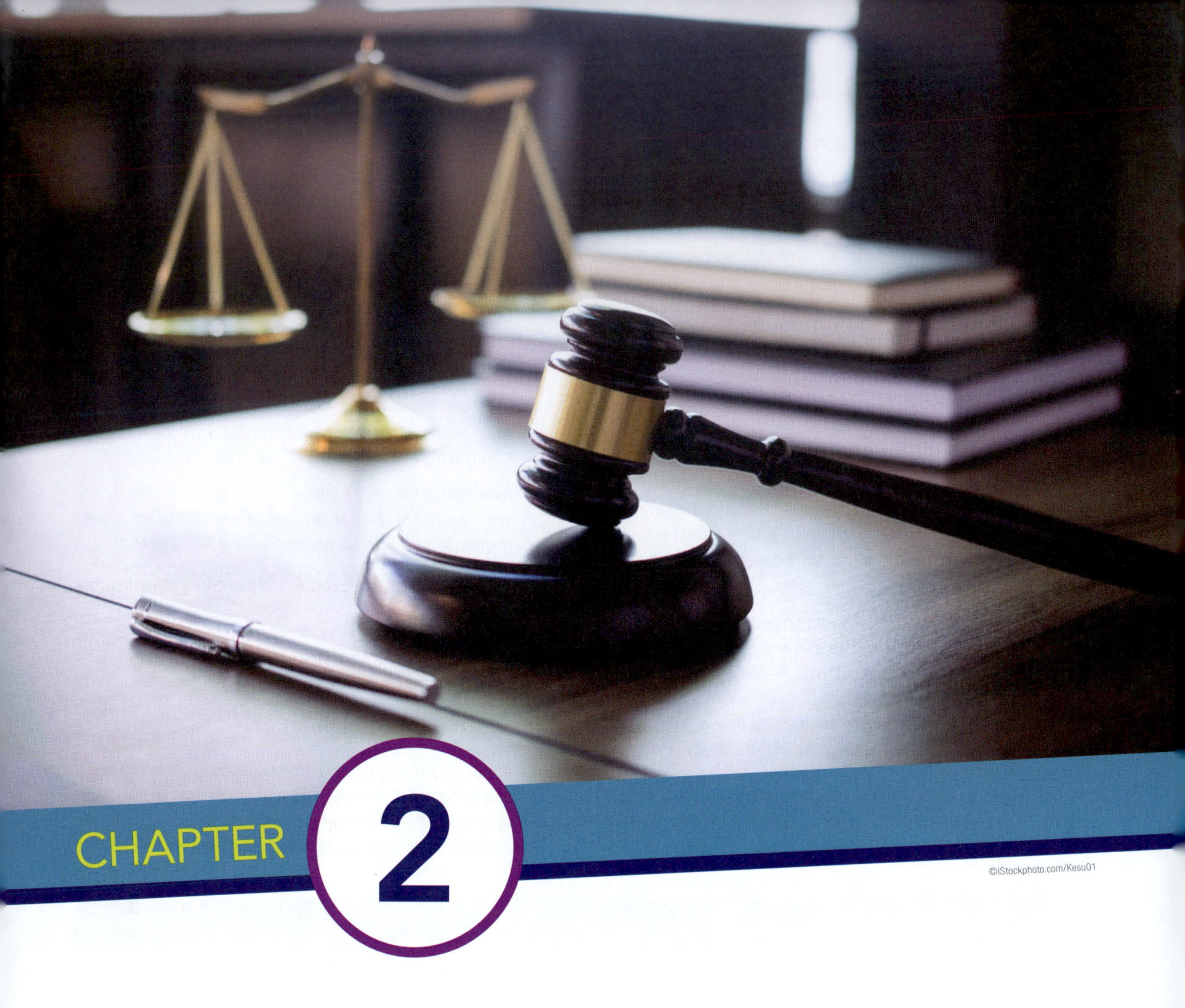

CHAPTER 2

©iStockphoto.com/Kesu01

FOUNDATIONS OF LAW AND CRIME

Nature, Elements, and Measurement

The Constitution of the United States was made not merely for the generation that then existed, but for posterity—unlimited, undefined, endless, perpetual posterity.

—Henry Clay

LEARNING OBJECTIVES

As a result of reading this chapter, you will be able to

1. Explain how modern-day law evolved from English common law
2. Describe the three sources of law in the U.S. legal system
3. Identify the differences between criminal and civil law
4. Explain the difference between substantive and procedural law
5. Review two critical elements of the criminal law—criminal intent (*mens rea*) and the physical commission of the criminal act (*actus reus*)
6. Describe how felonies and misdemeanors are placed into two separate broad categories
7. Delineate the definitions of, and distinctions between, crimes against persons and property; the different degrees of homicide; and offenses classified as public order, white-collar, and organized crimes
8. Explain the three primary methods for measuring crime and advantages and disadvantages of each

ASSESS YOUR AWARENESS

Test your knowledge of the foundations of law and crime by responding to the following nine true-false items; check your answers after reading this chapter.

1. The U.S. legal system is based on English common law.
2. A person who, upon returning home, discovers that her premises have been illegally entered and valuables removed, will correctly tell police, "I've been robbed."
3. Under current U.S. law, one can be charged only for those criminal acts that he or she actually performs, and not for failure to act or perform in some manner.
4. As opposed to the situation under English common law, today one may use force, even deadly force, if he or she reasonably believes that an attack against him or her is imminent.
5. No particular amount of time is necessary for one to legally premeditate a murder.
6. Rape and theft are often referred to as public order or victimless crimes.
7. The term *white-collar crime* was introduced in 1939, when a researcher discovered that many crimes were committed by persons of respectability and high social status in the course of their occupations.
8. If a burglar enters a premise and also commits a rape and a murder while inside, the hierarchy rule requires police to report to the FBI only the crime of murder.
9. Carjacking is now one of the FBI's eight Part I crimes.

<< Answers can be found on page 293.

As stated in this chapter, one's intent while committing a crime is not always easy to prove. For example, in a crime of homicide, there are a variety of possible outcomes, and the prosecutor will "look behind the act" to determine the killer's state of mind, the elements of the crime that were present, and so on. For example, in 2017, a 10-month-old girl in Methuen, Massachusetts, suffered cardiac arrest and was revived twice. While police worked to revive the child, the mother, who was a former drug addict, shuffled back and forth between her bedroom and bathroom. Police wondered whether this was nervous behavior, or whether she was flushing evidence of the drug that had harmed her baby. With no obvious signs of child endangerment in the residence, criminal justice officials found it difficult to identify the responsible party or determine intent. A prosecutor who reviews this case must ask and attempt to answer several questions:

- Was this a crime or an unfortunate accident?
- Did someone purposefully expose this child to a lethal dose of fentanyl?
- If not purposeful, did someone recklessly expose the child to this substance, or did the child ingest residue unintentionally left on an object?

As you will learn in this chapter, different crimes contain different elements and correspond with various levels of punishment. Think about why different facts should lead to a different criminal charge and punishment. What facts are important in defining crimes and why? Why should these facts matter? What other facts might help you to decide more easily what really happened and what type of crime might have been committed against this infant?

The answers to these questions are at the very heart of the criminal law and of our system of justice, which holds people criminally responsible only when the state can prove beyond a reasonable doubt that the accused committed each and every element of a crime without a valid defense.

INTRODUCTION

John Adams famously said we have a "government of laws and not of men," meaning that our democracy is not ruled at the whim of kings or rulers or demagogues. As explained in Chapter 1, Americans willingly give up some of their rights to their elected governments to create laws and to receive protection for their persons and property. It thus becomes important that Americans possess fundamental knowledge of our laws: how they are created and to what types of behaviors they apply.

Confusion sometimes surrounds the definition of specific crimes and criminal justice system responses. We might hear someone screaming, "I've been robbed!" on television or in the movies after returning home and discovering that his house has been entered and ransacked (he was not robbed; he was burglarized). Or, someone discovers a dead body and exclaims, "He's been murdered!" A person may have indeed been killed, but many killings are not criminal in nature; only a judge or jury can decide if a murder has taken place. News reporters commonly mistake jail for prison and might erroneously state, "John Jones,

The Code of Hammurabi, written in about 1780 B.C.E., set out crimes and punishments based on *lex talionis*—"an eye for an eye, a tooth for a tooth." It is the earliest-known example of a ruler's setting forth a body of laws arranged in orderly groupings.

a convicted murderer, was sentenced to 10 years in jail today" (if sentenced to 10 years, Jones will serve his time in a prison, not a jail). As French philosopher Voltaire wrote, "If you wish to converse with me, define your terms." That statement represents the major purpose of this chapter.

The reader should bear in mind, however, that several important components of the rule of law are not discussed here, primarily because they go beyond the reach of an introductory textbook. Persons wishing to inquire more deeply into the law—for example, in such areas as conspiracies, attempted crimes, omissions, and causation—would be wise to enroll in criminal law and procedure courses.

COMMON LAW AND ITS PROGENY

Rules were laid out in ancient societies. The use of societal laws can be traced back to the reign of Hammurabi (1792–1750 B.C.E.), the sixth king of the ancient empire of Babylon. The Code of Hammurabi set out crimes and punishments based on ***lex talionis***—"an eye for an eye, a tooth for a tooth." A more recent source of law is found in the Mosaic Code of the Israelites (1200 B.C.E.), in which, according to tradition, Moses—acting as an intermediary for God—passed on the law to the tribes of Israel. The Mosaic Code includes the Bible's Ten Commandments and contains descriptions of forbidden behaviors that are still defined as criminal acts in modern society (e.g., thou shall not kill nor steal).

Lex talionis: Latin for "an eye for an eye, a tooth for a tooth"; retaliation or revenge that dates back to the Bible and the Middle Ages.

But the system of law as we know it today (except for Louisiana's, which is based on the French civil code) is based on common law: collections of rules, customs, and traditions of medieval England, created during the reign of Henry II (1154–1189 C.E.), who began the process of unifying the law. Henry established a permanent body of professional judges who traveled a "circuit" and sat on tribunals in shires throughout the Crown's realm. These judges eventually gave the Crown jurisdiction over all major crimes and sowed the seeds of the trial by jury and the doctrine of *stare decisis.*[1]

The ***stare decisis*** (Latin for "to stand by things settled") doctrine is perhaps the most distinctive aspect of Anglo-American common law. According to *Black's Law Dictionary, stare decisis* is the doctrine stating that, when a court has once laid down a principle of law as applicable to a certain state of facts, it will adhere to that principle—and apply it in the same manner to all future cases where the facts are substantially the same.[2] This doctrine binds courts of equal or lower status or levels in a given jurisdiction to the principles established by the higher appellate courts within the same jurisdiction.

Stare decisis: Latin for "to stand by a decision"—a doctrine referring to court precedent, whereby lower courts must follow (and render the same) decisions of higher courts when the same legal issues and questions come before them, thereby not disturbing settled points of law.

For example, the federal courts have a three-tier structure (see Figure 2.1). According to the *stare decisis* doctrine, a federal district court in Maryland is required to follow the decisions of the Fourth Circuit Court of Appeals, of which it is a part, as well as those of the U.S. Supreme Court. However, it is not bound by the decisions of other district courts.[3] Note, however, that our system of law still must remain flexible and capable of change, and thus courts can also revisit earlier decisions and set new precedents.

MODERN-DAY SOURCES AND HIERARCHY OF LAW

In addition to the two primary types of law—criminal and civil (discussed in the next section)—the U.S. legal system features different sources of law and jurisdictions where those laws are enforced and administered. Our system of government is based on the concept of **federalism**, which divides government powers between the federal government and other government entities, including state and local governments. We have federal and state laws with corresponding federal and state courts to preside over such cases. Within those two legal arenas, laws—criminal and civil—are organized hierarchically, with different priorities and legal effects depending on their source and application.

Federalism: a type of government that divides powers between a national (federal) government and governments of smaller geographic territories, including states, counties, and cities.

Figure 2.1 /// Three-Tier Structure of the U.S. Federal Court System

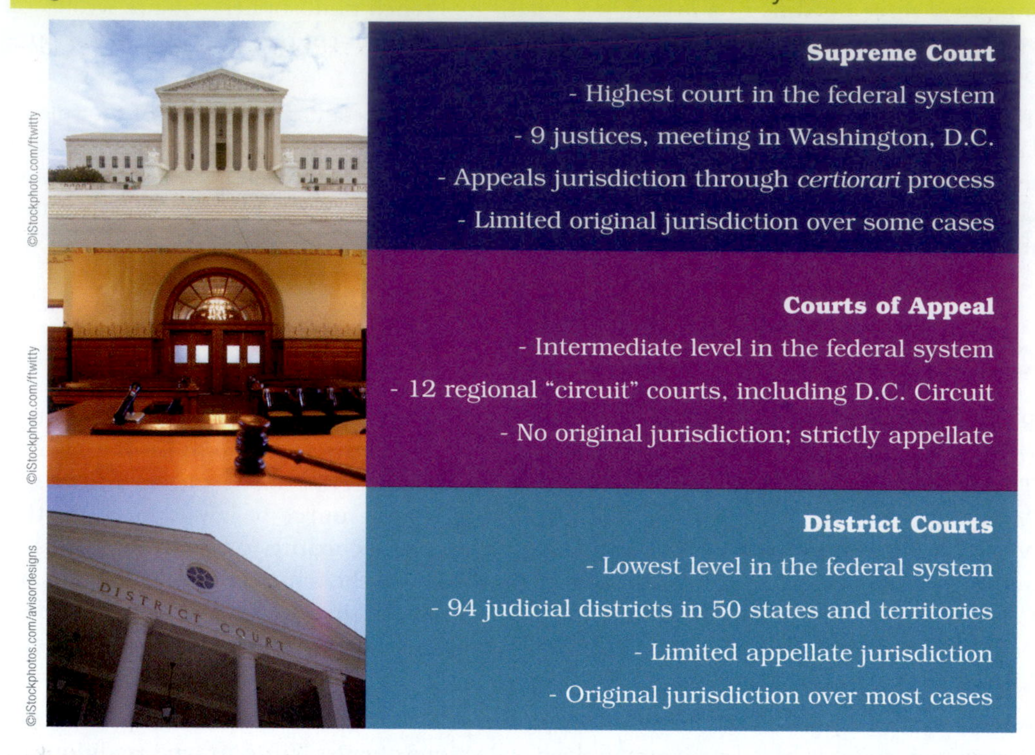

Supreme Court
- Highest court in the federal system
- 9 justices, meeting in Washington, D.C.
- Appeals jurisdiction through certiorari process
- Limited original jurisdiction over some cases
Courts of Appeal
- Intermediate level in the federal system
- 12 regional "circuit" courts, including D.C. Circuit
- No original jurisdiction; strictly appellate
District Courts
- Lowest level in the federal system
- 94 judicial districts in 50 states and territories
- Limited appellate jurisdiction
- Original jurisdiction over most cases
DISTRICT COURT

Again, terminology is critical to understanding law. Generally speaking, a *statute* is a law enacted by Congress (a federal law) or by a state legislature (a state law). Statutes are also known as *statutory law.* A *code* or *ordinance* typically refers to a law enacted by a local law-making body—a county board or a city council, for example (municipal laws). These and other types of laws are prioritized as follows:

Federal Law

1. The U.S. Constitution—"the supreme law of the land," which takes precedence over state constitutions and law even if they conflict
2. Federal statutes—civil and criminal laws enacted by Congress
3. Administrative laws—orders, directives, and regulations for federal agencies, such as workplace laws promulgated by the Occupational Safety and Health Administration (OSHA)
4. Federal common law—published decisions from the U.S. Supreme Court and the U.S. Circuit Courts of Appeal, which, like the common law from England (discussed earlier), establish legal "precedence" and must be followed by lower courts in the federal and state systems (Chapter 7 describes how the Supreme Court decides to hear cases and render its decisions.)

State Law

1. State constitutional law—state constitutional rulings (from a state's highest court) that may give greater protection or rights than the federal constitution but may not give less, and that contain protections similar to the U.S. Constitution—civil rights and liberties, separation of powers, and checks and balances[4]

2. State statutes—laws enacted by state legislatures, including criminal laws like statutes prohibiting murder or robbery
3. State common law—precedent established in published opinions by state appellate judges when deciding civil or criminal cases

City/County Law

Municipal ordinances or codes govern many aspects of our daily lives, including the following:

- Building and construction standards
- Rent control
- Noise and nuisance regulations
- Public health and safety
- Business licenses
- Civil rights and antidiscrimination

Criminal law: the body of law that defines criminal offenses and prescribes punishments for their infractions.

Civil law: a generic term for all noncriminal law, usually related to settling disputes between private citizens, governmental, and/or business entities.

Plaintiff: the party bringing a lawsuit or initiating a legal action against someone else.

Defendant: a person against whom a criminal charge is pending; one charged with a crime.

Burden of proof: the requirement that the state must meet to introduce evidence or establish facts.

CRIMINAL AND CIVIL LAW

Simply put, **criminal law** applies to criminal matters, for example, when someone breaks a law prohibiting the commission of a robbery. **Civil law** applies to civil matters, for example, when two parties have a property dispute or want to get divorced. These two distinct types of law differ in several critical ways.

First, each type involves different parties. Criminal law represents what we as a people have decided is criminal behavior, and crimes are considered harmful to all of us in society. As such, when someone commits a crime, the state or government—through the prosecutor or district attorney—prosecutes the person on behalf of the people. Since the title of a case always first references the party that brought suit against the other (i.e., the government entity responsible for prosecution in a criminal case), a typical criminal case name is something like *U.S. v. Jones* or *State of Maine v. Smith.* In a civil matter, two individuals (or business entities like a corporation) are on either side. In a property dispute, such as when A erects a fence over the property line of B, one neighbor brings a lawsuit against the other so that the court can referee their dispute. Or, if a person is injured using a product like a hairdryer, that person can bring a civil case against the manufacturer, probably a corporation. A typical civil law case name is something like *Jones v. Smith* or *Jones v. ABC Corporation.* In the former example, Jones is the party bringing the suit—the **plaintiff**—and Smith is the party defending against the suit—the **defendant**. Not surprisingly, these two types of cases are heard in different courts, and most courthouses have separate criminal and civil "divisions," where judges are more experienced in either of these particular types of cases.

U.S. National Archives and Records Administration

Sources of federal law include the U.S. Constitution, the U.S. Supreme Court, and those laws enacted by Congress.

Another important difference involves how these cases are decided. In a criminal case, the prosecutor or the state has the burden of proving—the **burden of proof**—the defendant's guilt "beyond a reasonable doubt." This sometimes-heavy burden is designed to force the government—the prosecutor—to prove its case with the strongest possible evidence to avoid wrongful convictions of innocent persons. Underlying our rule of law in the United States is the all-important concept that one is innocent until proven guilty. Sometimes the burden shifts to the defense to prove something, as with self-defense claims (discussed in Chapter 7). In a civil case, the burden on the party seeking damages or a remedy is less than in a criminal matter—those representing this party must prove their case by a "preponderance of the evidence." The precise difference between the legal standards of "beyond a reasonable doubt" and "preponderance of the evidence" can be hard to distinguish, even for experienced lawyers.

Reasonable doubt: the standard used by jurors to arrive at a verdict—whether or not the government (prosecutor) has established guilt beyond a reasonable doubt.

Reasonable doubt can be difficult to explain; in fact, several courts prefer not to attempt to give the jury any explanation at all. However, in most cases reasonable doubt means that, after hearing all of the evidence, jurors do not possess an abiding conviction—to a moral certainty—that the charges brought against the defendant are true. This does not necessarily mean absolute certainty—there can still be some doubt, but only to the extent that it would *not* affect a "reasonable person's" belief that the defendant is guilty. By contrast, if such doubt *does* affect a reasonable person's belief that the defendant is guilty, then the prosecution has not met its burden of proof, and the judge or jury must acquit—find the defendant not guilty.

The civil standard of a preponderance of the evidence is a much less difficult burden to meet and is often referred to as the "50 percent plus a feather" test. It asks jurors to decide which way the evidence causes the scales of justice to tip, toward guilt or innocence, and to decide the case on that basis.

The major difference between civil and criminal matters is the penalty. In a criminal case, the state or prosecutor seeks to punish a defendant with prison or jail time, a monetary fine, or both (or perhaps a community-based punishment such as probation, discussed in Chapter 11). In a civil matter, one party is seeking "damages" (money or some legal remedy) from the other party rather than trying to send him or her to jail or prison. In the property dispute example, A would be required to tear down or move the fence and perhaps pay for the legal fees B incurred in bringing the suit. And in the hairdryer example, the corporation could be ordered to pay the consumer's medical expenses and legal fees.

Some conduct can give rise to both a criminal and a civil matter. A few examples might serve to explain the difference—and how the two types of cases can cause problems for the police and misunderstanding by the public.

Assume that Jane calls the police to her home and tells the officer that she and her husband, Bill, recently separated and he went to their home and removed some furniture and other goods that she does not believe he should have taken. The officer must inform Jane that there is nothing he can do—at this point, this is a *civil,* not a criminal, matter. Jane may be upset with the officer, but legally the police have no jurisdiction in such disputes. Assume, however, that Bill goes to the home and makes serious threats of injury toward Jane, and she then goes to court and obtains a temporary protection order (TPO) commanding Bill to avoid any

Table 2.1 /// Differences Between Civil and Criminal Law

	Civil	Criminal
Burden of proof	"Preponderance of evidence"	"Beyond a reasonable doubt"
Nature of crime	A private wrong	A public wrong
Parties	Case is filed by an individual party	Some level of government files charges against the individual
Punishment	Usually in the form of monetary compensation for damages caused; no incarceration	Jail, fine, prison, probation, possibly death
Examples	Landlord/tenant dispute, divorce proceeding, child custody proceeding, property dispute, auto accident	Person accused of committing some crime (or, perhaps, neglecting a duty to act)

form of contact with her; Bill then disregards the TPO and stalks Jane at her place of employment. Because he violated the court's order, this has now become a *criminal* matter, and the police may arrest Bill.

Or, look again at the example of someone erecting a fence that cuts across his neighbor's property. The neighbor refuses to move or tear down the fence, so the matter is taken to court. This is a civil cause of action. Both sides have a dispute that the courts will resolve if the parties cannot reach an agreement. Assume further that the neighbors cannot resolve the fence matter, and one day while one of them is working in his yard the other attacks him with a club. Now the attacking neighbor could face a criminal charge from the state (assault with a deadly weapon) and a civil lawsuit by the injured neighbor (for medical expenses and emotional trauma).

Table 2.1 shows the primary differences between civil and criminal law.

SUBSTANTIVE AND PROCEDURAL LAW

Substantive law is the written law that defines criminal acts—the very "substance" of the criminal law. Examples are laws that define and prohibit murder and robbery. **Procedural law** (discussed in detail in Chapter 6) sets forth the procedures and mechanisms for processing criminal cases. The following examples of procedural law stem from constitutional amendments: Police officers are required to obtain search and arrest warrants, except in certain situations (Fourth Amendment), officers must give the *Miranda* warning before a suspect is interrogated while in custody (Fifth Amendment), and defendants have the right to an attorney at key junctures during criminal justice system processing (Sixth Amendment). Note that these laws govern how police officers, lawyers, judges, corrections officers, and a host of others do their work in the justice system. Procedural law also prescribes rules concerning **legal jurisdiction**, jury selection, appeal, evidence presented to a jury, order of conducting a trial, and representation of counsel.

Substantive law: the body of law that spells out the elements of criminal acts.

Procedural law: rules that set forth how substantive laws are to be enforced, such as those covering arrest, search, and seizure.

Legal jurisdiction: the authority to make legal decisions and judgments, often based on geographic area (territory) or the type of case in question.

ESSENTIAL ELEMENTS: *MENS REA* AND *ACTUS REUS*

The Latin term for criminal intent is *mens rea,* or "guilty mind," and its importance cannot be overstated. Our entire legal system and criminal laws are designed to punish only those actors who *intend* to commit their acts. As will be seen later in the discussion of homicide, acts that are clearly intentional and premeditated (e.g., murder in the first degree) are punished most severely, while those acts that are less intentional and/or accidental are punished less harshly. For example, assume that Bill, while hunting deer, shoots another hunter while out in the woods. If the prosecutor has evidence to show the shooting was purely accidental in nature (the other hunter was out of position, or Bill's gun malfunctioned), the prosecutor will not charge Bill with unlawful killing. If, by contrast, the evidence shows that Bill intended to shoot the other hunter (again, considering evidence of the position of the two men, the number of shots, or proof of Bill's planning), the prosecutor will charge Bill with homicide.

Mens Rea: Intent Versus Motive to Commit Crime

There is an important distinction to be made between *intent* and *motive.* One's **specific intent** concerns what he or she is seeking to do and is connected to a purpose or goal; **motive** refers to one's reason for doing something. For example, when a poverty-stricken

Intent, specific: a purposeful act or state of mind to commit a crime.

Motive: the reason for committing a crime.

Juvenile Justice Journal

The Foundations of Philosophy and Treatment

As noted in Chapter 1, ideological changes began to occur in the early 19th century with respect to strategies and practices concerning why and how youthful offenders and those in need of services could be differentiated from their adult counterparts. Two major aspects of that differentiation are the doctrines of *parens patriae* and *in loco parentis.*

Parens patriae means that the "state is the ultimate parent" of the child. In effect, this means that as long as we as parents adequately care for and provide at least the basic amenities for our children as required under the law, they are ours to keep. But when our children are physically or emotionally neglected or abused, the juvenile court and police may intervene and remove the children from that environment. Then the doctrine of *in loco parentis* takes hold, meaning that the state will act in place of the parent.

For police officers and other criminal justice personnel, there is perhaps no greater or more somber duty than having to testify under subpoena in juvenile court that a parent is unfit (and that his or her parental ties should be legally severed). However, when a parent or guardian's actions indicate a pattern of neglect and/or abuse toward a child, it is clearly better that the state assume responsibility for the child's care and custody.

Table 2.2 shows what might be termed the **idealistic contrast** between the juvenile court process and the adult criminal justice process. Note that this is the *ideal* process for juveniles, in keeping with the more nurturing and forgiving juvenile justice philosophy. However, that philosophy can easily go away when a juvenile commits a crime that is so heinous that he or she is deemed not to be amenable to the more lenient philosophy and jurisdiction of the juvenile courts; in such cases, he or she may be remanded to the custody of the appropriate adult court of jurisdiction. Also note that much of the difference between the philosophies of juvenile and adult courts is found in the terminology used.

Table 2.2 /// The Idealistic Contrast Between Juvenile and Adult Criminal Justice Processes

Adult	Juvenile
Adversarial procedure	Relatively amiable procedure
Individual responsibility for crime	Societal/family factors involved
Punishment is typically the goal	Rehabilitation is the goal
Arrest process	Petition
Trial—public	Hearing—private
Guilt or innocence	Guilt not the sole issue
Public record	Confidential record
Verdict	Decision
Sentence	Disposition

Idealistic contrast: the differences between juvenile and adult criminal justice processes, to include treatment and terminology.

woman steals milk for her child, her intention is to steal, but her motive is to provide for her child. Therefore, motive (the "why" someone is stirred to perform an action) is grounded primarily in psychology; intent, conversely, is the result of one's motive, is grounded in law, and carries a higher degree of blameworthiness because a harmful act was committed. Crime movies often focus a great deal on offenders' possible motives for committing particular crimes. However, as one law professor put it, "As any first year law student will tell you, motive is irrelevant in determining criminal liability. Unlike in the television show . . . in the perceived real world of criminal liability, motive is just a bit player, appearing only in limited circumstances, usually as a consideration in certain defenses. Ordinarily, the only real questions at trial are (1) did the defendant commit the illegal act and (2) did she have the necessary mental state (*mens rea*, or intent)?"[5]

To determine the requisite *mens rea* for a specific crime, we look to the criminal statute that defines the offense. But you will not find the term "*mens rea*" as part of any criminal law. Instead, this element is expressed through a variety of terms that differ from state to state. For example, an intentional crime might be defined as an act committed "intentionally," "willfully," or "maliciously," while a less serious crime—an accidental crime—can be expressed as an act committed "negligently," "recklessly," or "without due caution." The terms used to express *mens rea* are not uniform, so criminal justice professionals must learn to identify these terms and understand how they operate within a criminal statute (see the discussion of homicide and related exercises later in this chapter).

One's intent while committing a crime is not always easy to prove. As is discussed later, regarding the crime of homicide, when A shoots B, there are a variety of possible outcomes, and the prosecutor will "look behind the act" to determine what was going on in the mind of the killer, and whether to reduce what appears to be a charge of murder in the first degree (an intentional killing) to one of manslaughter (an accidental killing).

Actus Reus: The Act

Another critical feature of the U.S. criminal justice system is that we do not punish people for merely thinking about committing criminal acts; rather, the law generally requires a voluntary, overt act—killing, injuring, threatening, or breaking and entering, for example. The intentional *failure* to act (an "omission") can also be criminal but only when there is a legal duty to do something, such as when a person willfully fails to pay income tax owed or a parent fails to feed a child or give him or her medical attention.

The rule for establishing criminal liability is to prove that the defendant committed the **actus reus** element (the criminal act) with the **mens rea** ("guilty mind") set forth in the particular criminal law. When these two critical elements are present, it is known as "concurrence." For example, a man in a ski mask breaking into a locked home might seem to be committing the *actus reus* element for burglary (see "Crimes Against Property," below). But what if he is doing so to save himself from a terrible snowstorm after his car broke down? The all-important *mens rea* element is missing, and there is no concurrence. Accordingly, the prosecutor must prove both elements beyond a reasonable doubt—a question of fact for the jury to decide.

Actus reus: Latin for "guilty deed"—an act that accompanies one's intent to commit a crime, such as pulling out a knife and then stabbing someone.

Mens rea: Latin for "guilty mind"—the purposeful intention to commit a criminal act.

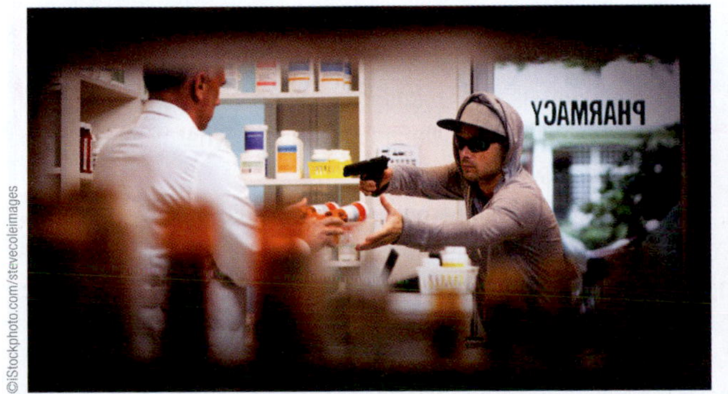

©iStockphoto.com/stevecoleimages

One's criminal intent is a major consideration in our legal system; for example, one who intentionally kills someone during a robbery may well be charged with premeditated murder.

©iStockphoto.com/Obradovic

A hunter who accidentally (without intent) shoots and kills another typically would not be charged with premeditated murder.

FELONIES AND MISDEMEANORS

Felony: a serious offense with a possible sentence of more than a year in prison.

Crimes are also classified into two broad categories based on the severity of the criminal act and the corresponding punishment. **Felonies** are offenses punishable by death or that have a possible sentence of more than one year of incarceration in prison. Many states further divide their felonies into different classes; for example, under Arizona's laws, first-degree murder is a Class 1 felony and is punishable by death or life imprisonment; sexual assault is a Class 2 felony (the number of years for which one may be sentenced to prison for this and other offenses will differ, depending on an offender's prior record); aggravated robbery (the offender has an accomplice) is a Class 3 felony; and forgery is generally a Class 4 felony.[6]

Misdemeanor: a lesser offense, typically punishable by a fine or up to one year in a local jail.

A **misdemeanor** is a less serious offense and is typically punishable by fines or incarceration for less than one year in a local jail. Like felonies, misdemeanors are often classified under state laws. In Arizona, shoplifting is a Class 1 misdemeanor (if the value of items taken is less than $1,000 and the item is not a firearm); reckless driving is a Class 2 misdemeanor; a vehicle driver who dumps trash on a highway or road is guilty of a Class 3 misdemeanor.[7]

You Be the . . .

Police Officer

The Nevada Revised Statutes[8] set forth the following statutory provisions related to driving. Read them carefully and then respond to the questions posed.

> NRS 484B.657. Vehicular manslaughter.
>
> A person who, while driving or in actual physical control of any vehicle, proximately causes the death of another person through an act or omission that constitutes simple negligence is guilty of vehicular manslaughter .
>
> NRS 484B.653. Reckless driving.
>
> It is unlawful for a person to: (a) Drive a vehicle in willful or wanton disregard of the safety of persons or property, (b) Drive a vehicle in an unauthorized speed contest on a public highway, or (c) Organize an unauthorized speed contest on a public highway.
>
> NRS 484B.165. Using handheld wireless communications device.
>
> Except as otherwise provided in this section, a person shall not, while operating a motor vehicle on a highway in this State: (a) Manually type or enter text into a cellular telephone or other handheld wireless communications device, or send or read data using any such device to access or search the Internet or to engage in nonvoice communications with another person, including, without limitation, texting, electronic messaging and instant messaging, or (b) Use a cellular telephone or other handheld wireless communications device to engage in voice communications with another person, unless the device is used with an accessory which allows the person to communicate without using his or her hands, other than to activate, deactivate or initiate a feature or function on the device.

1. What is the *mens rea* element under the vehicular manslaughter statute? What is the *actus reus* element?
2. Provide an example of what would qualify as reckless driving under section (a).
3. A woman who is texting while driving strikes and kills a pedestrian who is crossing the street. The woman was not speeding. Can she be charged for a violation of NRS 484B.165—Using handheld wireless communications device? Why or why not?

OFFENSE DEFINITIONS AND CATEGORIES

Crimes Against Persons

Crimes against persons are what most people consider "violent crime" or "street crime," such as certain homicides, sexual assaults, robberies, and aggravated assaults. As you will learn later in this chapter, the Federal Bureau of Investigation (FBI) defines these crimes as offenses that involve force or threat of force.

Crimes against persons: violent crimes, to include homicide, sexual assault, robbery, and aggravated assault.

Homicide

The taking of a human life—homicide—is obviously the most serious act that one can perpetrate against another person. But not every killing is criminal in nature, as the following examples demonstrate:

- Justifiable homicide—self-defense, legal state or federal executions, acts of war, or when a police officer uses lawful lethal force
- Excusable homicide—killings that are wholly accidental, such as when a person who is driving the speed limit and paying full attention hits a small child who darts into the street from behind a large RV, where no reasonable person could have known that such a risk was possible or preventable

Criminal homicides fall into two categories: murder (intentional killings) and manslaughter (accidental killings). Within these two categories, offenses are ranked in seriousness by degrees. As mentioned earlier, under our system of justice, the premise underlying homicide is that "when A shoots B, there are a variety of possible outcomes." The prosecutor must attempt to determine the shooter's intent, which can result in criminal charges ranging from murder in the first degree to involuntary manslaughter. A useful way to think of the universe of possible homicide charges is as a "ladder of offenses," going from the least serious up to the most serious (see Table 2.3).

Table 2.3 /// The "Ladder" of Homicide Crimes and the Required Mens Rea

Charge	*Mens Rea*	Example
Murder (Intentional Killings)		
First-Degree Murder	Premeditated, with malice aforethought	Planning to kill your wife's boyfriend by waiting at his home one evening and killing him when he arrives
Second-Degree Murder	Intentional but not premeditated	Unexpectedly seeing your wife's boyfriend several days after you discover their affair, then quickly grabbing a gun you keep in your car and killing him
Manslaughter (Accidental Killings)		
Voluntary Manslaughter	Committed in the "heat of passion"; being adequately provoked and not having time to "cool off" before the act	Arriving home and finding your wife and her boyfriend together, immediately grabbing a nearby weapon, and killing the boyfriend in a blind rage
Involuntary Manslaughter	Unintentional/accidental; causing a death while breaking the law (typically a misdemeanor criminal law) or while being negligent	Texting while driving and then accidentally hitting a child who darts out into the street
Felony Murder (Strict Liability)		
Felony/First-Degree Murder	Unintentionally causing a death while intentionally committing a dangerous felony; no *mens rea* required = strict liability	While burglarizing your boss's office to steal from the petty cash fund, she arrives unexpectedly, discovers you there, has a heart attack, and dies

AP Photo/Joshua Polson, The Greeley Tribune

Excusable killings include those that are accidental in nature, such as when a driver strikes and kills a toddler who darts out into the street; in such cases the driver will not be deemed culpable (blameworthy).

Felony-murder rule: the legal doctrine that says that, if a death occurs during the commission of a felony, the perpetrator of the crime may be charged with murder in the first degree, regardless of the absence of intent, premeditation and deliberation, or malice aforethought.

Murder. The term *murder* includes only intentional killings, which are categorized by degrees:

- *Murder in the first degree* (sometimes termed "murder one") is the unlawful, intentional killing of a human being with *premeditation and deliberation* (often termed "P&D") and *malice aforethought*. Federal and state statutes define murder and its elements differently, but generally they all require the elements of intent, P&D, and malice aforethought.

  P&D means the defendant thought about committing the act before doing so. Courts generally look at the following factors to determine P&D: evidence of planning, the manner of killing, and the prior relationship between the defendant and the victim.[9] Courts also look at the *time* the defendant may have spent contemplating or planning in order to determine whether premeditation existed, but courts differ on this issue. The federal courts have held that *no* minimum time period is necessary, and a jury can determine from the facts whether or not a defendant premeditated murder.[10] Malice aforethought is often said to be shown when someone acts with "a depraved heart," evidenced by one's conscious intent to shoot another person with a gun or stab someone with a knife.

  Dangerous conduct can also be prosecuted as first-degree murder under the **felony-murder rule**, which provides that if a death occurs during the commission of a felony, the defendant will be charged with murder in the first degree, regardless of his or her intent (a crime that does not require a *mens rea* element is known as a "strict liability" crime—see also the discussion of statutory rape, later). The classic example involves multiple defendants robbing a bank and during the robbery the bank security guard dies from a heart attack. In that case, all the defendants may be charged with first-degree murder under the felony-murder rule. Likewise, if one of the bank robbers panics and shoots a security guard, all of the defendants will be charged with first-degree murder even though the robbers intended only to rob, not to kill anyone. The felony-murder rule is designed to deter those who might otherwise commit dangerous felonies. If a would-be bank robber knows the penalty for an accidental death during his or her crime might be a charge of first-degree murder, perhaps the robber will think twice about committing the crime.

- *Murder in the second degree* ("murder two") is distinguished from first-degree murder in that it is also intentional—with malice—yet *impulsive*, without P&D. An example would be when two men get into an argument at a bar, and one pulls a knife and stabs the other to death. He intended to stab the other man, but the killing did not involve premeditation. Furthermore, although the intent to kill is an essential element of both first- and second-degree murder, a defendant can also be found guilty of second-degree murder if his or her actions show gross recklessness and a high disregard for human life, and there is extreme risk of death. Although it will depend on the jury's views, acts such as allowing a dangerous dog to run free and bite a child, intentionally shooting a gun into a crowd of people, throwing a heavy object off a roof onto a crowded street below, and playing Russian roulette (loading a gun and intentionally firing it at another person) have led to conviction for second-degree murder. As you will see later, these actions could also qualify as involuntary manslaughter.[11]

Manslaughter. The term *manslaughter* refers to accidental killings, categorized as voluntary (intentional but without malice) and involuntary (unintentional):

- *Voluntary manslaughter* is an intentional killing, but it involves (at least in the eyes of the law) no malice. Instead, there is "heat of passion" to a degree that a "reasonable person" might have been provoked into killing someone. The best example is when a person comes home early in the day and finds his or her significant other in the arms of another person, becomes enraged, grabs a gun, and kills one or both of them. The killer acted in the heat of passion rather than intentionally. A killing can be downgraded on the homicide ladder, from murder to voluntary manslaughter, only if the actor was adequately provoked (generally, words alone—as in an argument—are not enough to provoke, but seeing something like a cheating spouse is), and the actor must not have had time to "cool off." The person discovering his cheating spouse cannot leave, go to a bar and drink a few beers, and then return to the scene and kill the offending couple (this would be first-degree murder, as explained earlier). The "passion" that aroused the person to kill must have arisen contemporaneously—at the time of the provocation—and continued until the time of the criminal act.

  The circumstances of a killing might cause a prosecutor to look at both second-degree murder and voluntary manslaughter as possible charges. When a perpetrator clearly killed with intent but without P&D (e.g., two people fighting with weapons), a second-degree murder charge is appropriate. But when that same actor is fighting with weapons because the victim provoked her and she did not have time to "cool off" after that provocation, the charge could be downgraded to voluntary manslaughter. A prosecutor will consider all of the evidence to determine the right homicide charge, up or down the "ladder."

- *Involuntary manslaughter* is typically established in one of two ways: (1) through acts of negligence, such as when one is driving too fast on a slick road and kills a pedestrian, or (2) via the misdemeanor-manslaughter rule—similar to the felony-murder rule, but the crime involved is a misdemeanor. For example, a man enters a convenience store and shoplifts a six-pack of beer; the clerk chases him out the door but slips and falls, striking his head on the sidewalk and dying from the force of the impact. The shoplifter may be charged with involuntary manslaughter, as his actions caused the clerk's death.

The Methuen Fentanyl Case: Now that you have learned the different types of homicide and how the critical element of *mens rea* determines the degree and punishment, consider the scenario (from the beginning of the chapter) under which the jury could have convicted the mother from Methuen, Massachusetts, and how *mens rea* plays a part in each of the following:

- *First-degree murder:* The mother intentionally gave fentanyl to her child, and she planned the child's death.
- *Second-degree murder:* The mother gave fentanyl to her child intentionally but did not plan for the child to die.
- *Voluntary manslaughter:* The mother held her child while consuming fentanyl while experiencing severe withdrawal symptoms, knowing that the child would likely be exposed and harmed by the drug.
- *Involuntary manslaughter:* The mother used fentanyl, but never anticipated that the child would accidentally and unintentionally be exposed to the drug.

Further, consider the fact that drug paraphernalia was found in the mother's car during the subsequent investigation. But the police noted that the mother, father, and grandparents were highly cooperative during the investigation. How might those facts affect a case for homicide?

Sexual Assault

Sexual assault, referred to as "rape" or "forcible rape" under some older state laws, was historically defined as the carnal knowledge of a female forcibly and against her will, and

iStockphoto.com/GregorBister

Robbery involves taking, or attempting to take, anything of value from another person by force or threat of force or violence, where the victim is in fear of injury or death.

most criminal laws did not recognize rape of men, same-sex rape, or rape between spouses. Today, most states have adopted the modern definition of sexual assault—sexual contact without the victim's consent, but with no limitations on gender or relationship of perpetrator and victim. Most states have also recognized that sexual assault occurs when someone has sexual contact with a person who is incapable of consenting (someone who is intoxicated, under anesthesia, or mentally challenged). In such cases, the perpetrator knows or should know that the victim cannot consent to the act. Beginning in 2013, the FBI removed the term *forcible* from the agency's definition of rape and it is now defined as "penetration, no matter how slight, of the vagina or anus with any body part or object, or oral penetration by a sex organ of another person, without the consent of the victim." About 134,000 rapes are reported to the police annually in the United States.[12]

Like other crimes, sexual assault is categorized by degrees depending on the type of contact, from sexual penetration without the consent of the victim (first degree, the most serious) to unwanted sexual contact that does not result in physical injury (often third degree). Like homicide, sexual assault includes a "strict liability" crime with no *mens rea* element, also known as "statutory rape." This crime is defined as sexual contact between a "victim" of a certain age, typically between ages 12 and 16, and a "perpetrator" who is older, typically 19 and above.

Robbery

Robbery is the taking of or attempt to take anything of value from the care, custody, or control of a person or persons by force or threat of force or violence and/or by putting the victim in fear. About 319,000 robberies are reported to the police annually in the United States.[13] As stated earlier in this chapter, many times people who come home to find their houses have been broken into claim they have been "robbed" when they obviously have not been (they have been burglarized), given that robbery requires a face-to-face taking—a combination of theft and assault.

Aggravated Assault

Aggravated assault is an unlawful attack upon another for the purpose of inflicting severe or aggravated bodily injury. This offense is usually accompanied by the use of a weapon or by other means likely to produce death or great bodily harm. When aggravated assault (or even regular assault, as long as there is a threat) and larceny-theft (described in the next section) occur together, the offense falls under the category of robbery. Each year about 811,000 aggravated assaults are reported in the United States.[14]

As is the case with homicide crimes, criminal justice students often have difficulty understanding the "ladder" of assault crimes. A few examples will help to clarify the differences:

- First, the mere placing of someone in fear for his or her safety is an assault. If Joe yells at Jack threateningly, "I'm going to beat your brains out," this is an assault. An *assault*, then, occurs when one person makes threatening gestures that alarm someone and makes that person feel under attack; actual physical contact is not necessary.
- But if Joe intentionally strikes Jack on his cheek, the intentional physical contact intended to harm raises this conduct to the crime of *assault and battery*.

- Finally, if Joe gets a tire iron out of his car and strikes Jack with it several times, inflicting severe injury, Joe has now committed an *aggravated assault.*

Crimes Against Property

Crimes against property are offenses where no violence is involved, only the taking of property—crimes such as burglary, larceny-theft, motor vehicle theft, and arson.

Crimes against property: crimes during which no violence is perpetrated against a person, such as burglary, theft, and arson.

Burglary

Burglary is the unlawful entry of a structure to commit a felony or theft. To classify an offense as a burglary, the use of force to gain entry need not have occurred, nor does anything of value have to have been stolen. The FBI, in its *Uniform Crime Reports* (detailed later in this chapter), defines "structure" as an apartment, barn, house trailer or houseboat when used as a permanent dwelling, office, railroad car (but not an automobile), stable, or vessel (i.e., ship). About 1.4 million burglaries are reported annually in the United States.[15]

Larceny-Theft

Larceny-theft is the unlawful taking, carrying, leading, or riding away of property from the possession of another; it includes attempted thefts as well as thefts of bicycles, motor vehicle parts and accessories, shoplifting, pocket-picking, or the stealing of any property or article that is not taken by force and violence or by fraud. The value of the item stolen is significant, and in all states the monetary worth will determine whether the larceny-theft is a felony or a misdemeanor; each state's statutes will set forth its limits. For example, according to Massachusetts statutes, if the item is worth more than $250, it is a felony; Iowa, however, has several classifications: It is a "serious misdemeanor" if the item stolen is valued between $200 and $500, an "aggravated misdemeanor" if worth between $500 and $1,000, and a felony if worth more than $1,000.[16] Each year about 5.5 million larceny-thefts are reported to the police in the United States.[17]

Motor Vehicle Theft

Motor vehicle theft involves the theft or attempted theft of a land-based, self-propelled vehicle that does not run on rails and is not classified as farm equipment. Examples of motor vehicles include cars, buses, motorcycles, trucks, snowmobiles, and scooters. Approximately 773,000 motor vehicle thefts are reported annually in the United States. The estimated loss associated with these events is approximately $6 billion.[18]

Arson

Arson is any willful or malicious burning of or attempting to burn, with or without intent to defraud, a dwelling house, a public building, a motor vehicle or aircraft, personal property of another, and so forth. There are different types of arsonists, with very different motives for setting fires. Each year about 41,000 arsons are reported to police, with an average loss per event of about $15,600. About half (45 percent) of all arsons involve structures (residential, storage, public, and so on).[19]

Public Order Crimes

Public order crimes are offenses that violate general public values or the norms shared among most members of society. They are sometimes called "victimless" crimes, since many people argue that these crimes harm only the offenders. Public order crimes are also sometimes called "complaintless" crimes, since many involve consensual activities or do not involve clearly identifiable victims. Offenses classified as public order crimes include drug-related crimes, prostitution, public drunkenness, gambling, and various forms of disorderly conduct. Although public order crimes often involve consensual activities between offenders (e.g., a drug dealer and a drug user, a prostitute and a "John" who pays for sex), violence and

Public order crimes: offenses that violate a society's shared norms.

You Be the . . .

Prosecutor

A mother in New York asks her sister to babysit her infant son for two days while she goes to a job interview. The mother asks that either her sister or her sister's 17-year-old daughter give the boy a teaspoonful of medicine two times per day, but she fails to inform them of the proper dosages. The sister becomes distracted and forgets to medicate the infant. Later, realizing that the medicine has not yet been administered, the 17-year-old daughter gives the infant what turns out to be an overdose of the medicine, and the infant dies. A potentially relevant law in New York is as follows:

New York Penal Law § 125.25: A person is guilty of *murder in the second degree* when . . . under circumstances evincing a depraved indifference to human life, and being eighteen years old or more the defendant recklessly engages in conduct which creates a grave risk of serious physical injury or death to another person less than eleven years old and thereby causes the death of such person (penalty: 15 to 25-year prison term).

As the prosecutor reviewing this case, you must ask and attempt to answer several questions:

1. What crime, if any, was committed here?
2. If there was a crime committed, who bears responsibility (culpability) for it?
3. If any of the involved parties is charged, would you have enough evidence to prove the *actus reus* and *mens rea* elements for second-degree murder?

Note: See The New York State Senate, The Laws of New York, Section 125.25, Murder in the Second Degree, https://www.nysenate.gov/legislation/laws/PEN/125.25.

other serious criminal offenses often accompany these activities. For example, shootings and homicides are commonplace in open-air drug markets when dealers fight over market territory. Further, human trafficking is closely tied to prostitution. The United Nations Office on Drugs and Crime reports that sexual exploitation accounts for up to almost half of all human trafficking in North, Central, and South America.[20]

White-Collar Crime

White-collar crime: crimes committed by wealthy or powerful individuals in the course of their professions or occupations.

The concept of **white-collar crime**, sometimes called occupational or corporate crime, was introduced in 1939 in an address to the American Sociological Association by Edwin Sutherland, who defined it as "a crime committed by a person of respectability and high social status in the course of his occupation."[21] White-collar crime challenges many of the traditional assumptions concerning criminality—that it occurs in the streets and is mostly committed by criminals who are lower class and uneducated. Sutherland, after examining 40 years of records held by regulatory agencies, courts, and commissions, reported that of the 70 largest industrial and mercantile corporations, each had violated at least one law, averaged eight violations apiece, and had an adverse decision related to false advertising, patent abuse, wartime trade violations, price fixing, fraud, or intended manufacturing and sale of faulty goods.[22]

Although we frequently hear of white-collar criminals, the extent of white-collar crime is difficult to measure for several reasons. First, collecting accurate data is difficult since official statistics and victim surveys generally do not include much information concerning this type of crime. And, for obvious reasons, corporations zealously guard their public image and thus prefer to regulate themselves and maintain a code of silence. In addition, the police, social scientists, and members of the media are often inexperienced in the ways of corporate crime, much of which involves insider corporation- and industry-specific knowledge.

Furthermore, although the Sherman Antitrust Act (1890) and hundreds of laws and regulatory agencies exist to keep white-collar crime in check and to police corporations—through recalls, warnings, consent agreements, injunctions, fines, and criminal proceedings—guilty companies often maintain legions of attorneys and accountants who possess considerable expertise in seeing that their bosses are seldom if ever punished.[23] We do know, however, that white-collar crime is the largest and most costly type of crime in the United States. As criminologist Frank Hagan put it, "All the other forms of criminal behavior together do not equal the costs of occupational and organizational (corporate) crime."[24]

Several celebrity white-collar crime cases have made national news. Most people have heard of Bernie Madoff's $50 billion fraudulent Ponzi scheme (where one basically uses new investors' funds to pay early investors, rather than using profits earned by the individual or organization running the operation); Madoff was sentenced to 150 years in prison for his crimes.[25] Or they know that businesswoman/author/television personality Martha Stewart was convicted of committing and lying about insider trading, in which she sold stock that she knew was likely to plunge in price.[26] These celebrated cases are well known, but white-collar crime can take many more forms, including the following:

- A 2019 college admissions bribery scandal, called Operation Varsity Blues, exposed parents and university staff involved in bribery and other forms of fraud to facilitate applicant admissions to top American colleges and universities.
- A Swiss pharmaceutical company pleaded guilty and paid a record $500 million in fines for price fixing on vitamins.
- Medical quackery and unnecessary surgical procedures victimize more than two million Americans, cost $4 billion, and result in 12,000 deaths per year. Fee splitting and "ping-ponging" (in which doctors refer patients to other doctors in the same office) also occur.
- A major auto manufacturing company, aware that a defect (which would cost $11 to repair) in one of its vehicles could result in a fiery explosion during a rear-end collision (even at low speeds), decided it would be cheaper not to recall the vehicles and make repairs, but instead to pay drivers' injury and death claims.[27] Lawyers engage in "ambulance-chasing," file unnecessary lawsuits, and falsify evidence.
- Corporations dump or release toxic chemicals and hazardous materials into the environment to cut costs and avoid regulatory laws.
- Companies steal secrets and patents from one another and thus commit corporate espionage.
- Individuals steal the identities of others, with nearly 18 million persons in the United States being victimized by identity thieves in any given year.[28] Offenders may target victims of natural disasters or other types of victims in phony fund-raising schemes.
- Nigerian letter scams often involve emailed pleas to help transfer money from Nigeria or some other part of the world, with false promises of financial compensation if victims agree to provide their banking information.

TIMOTHY A. CLARY/AFP/Getty Images

Bernie Madoff, a former stockbroker and financial advisor, was convicted of operating a Ponzi scheme that is considered to be the largest of its kind in U.S. history.

An often-cited crime typology, developed by Herbert Edelhertz, distinguishes between two types of white-collar crime.[29] The first type involves crimes committed in the course of offenders' occupations by those operating inside business, government, or other establishments in violation of their duty of loyalty and fidelity to employer or client. These crimes could include commercial bribery and kickbacks, embezzlement, insider trading, expense account fraud, or any other offense committed by an offender in the course of carrying out his or her work duties. The second type of white-collar crime involves crimes incidental to and in furtherance of business operations (but not the central purpose of the business). For example, producing false company financial statements, using deceptive advertising, or committing antitrust violations can indirectly benefit the offender by benefiting his or her employment company.

Organized Crime

Organized crime: crimes committed by members of illegal organizations.

Organized crime is a term used to describe illegal acts committed by individuals involved in illegal organizations. White-collar criminals commit crimes as part of larger organizations, using their association or relationship with legitimate businesses and organizations to carry out their offenses. Organized crime offenders differ in that they use their association or relationship with illegal entities to carry out criminal activities. These illegal groups commonly engage in behaviors known as racketeering activities. The FBI defines these activities under the Racketeer Influenced and Corrupt Organizations (RICO) statute,[30] and they include the following: bribery (including sports), counterfeiting, embezzlement of union funds, mail and wire fraud, money laundering, obstruction of justice, murder for hire, drug trafficking, prostitution and sexual exploitation of children, alien smuggling, trafficking in counterfeit goods, theft from interstate shipment, and interstate transportation of stolen property. Organized crime may also include the following state crimes: murder, kidnapping, gambling, arson, robbery, bribery, extortion, and various drug offenses.

MEASURING CRIME AND VICTIMIZATION

Data about actual crime rates—incidents, offenders, location, and so on—help us more fully understand after the fact just how prevalent crime and offenders are. This information can then guide scholars and researchers to seek explanations for rates that are spiking or falling. Crime measurement attempts to answer the question of "how much" with respect to crime in the United States, arming both criminal justice professionals and policy makers with critically important information to run this complex, human system.

How Much Crime in the United States? Depends on Whom You Ask

Mark Twain once said, rather famously, that there are "lies, damn lies, and statistics."[31] Unfortunately, that assessment of statistics is somewhat (if not significantly) accurate with respect to U.S. reported crime figures. One example that will make the point is two different data sets reporting hate crimes. Specifically, in a recent year, the Federal Bureau of Investigation's (FBI's) *Uniform Crime Reports* (discussed in the next section) reported a total of 7,175 hate crimes (with 8,493 victims).[32] However, the National Crime Victimization Survey (NCVS, also discussed later in this chapter) reported an annual average of *250,000* violent hate crime victimizations. This significant discrepancy is largely due to the fact that more than half (54 percent) of victims in the NCVS did not report their crimes to the police. Furthermore, in the NCVS, hate-related victimizations are based on victims' *suspicion* of the offenders' motivation, rather than on the police's suspicion.[33] Thus, crime figures must be viewed with caution and with an understanding of the limitations associated with each data source.

Errors and variations notwithstanding, to comprehend the impact of crime on our society, criminologists, victimologists, sociologists, psychologists, and members of other related disciplines need such information for their research and for making planning and policy recommendations.

Knowing the nature and extent of crime also helps us understand the social forces that are driving those offenses and aggregated trends. Furthermore, measuring crime is one of the best ways to determine the effectiveness of our criminal justice agencies and policies. We can better examine the structure and functions of the police (by looking at reported crimes and the proportion of crimes solved by arrest or other means), the courts (to determine the types of punishment and treatment that offenders are receiving), and corrections organizations (to ascertain, among other things, whether or not offenders are being returned to prisons and jails).

FBI: UCR

Home

Uniform Crime Reporting

The Uniform Crime Reporting (UCR) Program has been the starting place for law enforcement executives, students of criminal justice, researchers, members of the media, and the public at large seeking information on crime in the nation. The program was conceived in 1929 by the International Association of Chiefs of Police to meet the need for reliable uniform crime statistics for the nation. In 1930, the FBI was tasked with collecting, publishing, and archiving those statistics.

Today, four annual publications, *Crime in the United States, National Incident-Based Reporting System, Law Enforcement Officers Killed and Assaulted*, and *Hate Crime Statistics* are produced from data received from over 18,000 city, university/college, county, state, tribal, and federal law enforcement agencies voluntarily participating in the program. The crime data are submitted either through a state UCR Program or directly to the FBI's UCR Program.

In addition to these reports, information is available on the Law Enforcement Officers Killed and Assaulted (LEOKA) Program and the Hate Crime Statistics Program, as well as the traditional Summary Reporting System (SRS) and the National Incident-Based Reporting System (NIBRS).

The FBI is undertaking a redesign of the system that has supported the FBI's UCR Program for more than 30 years. In support of this initiative, the FBI is currently managing the UCR Technical Refresh. **Details**

Explore Our Services

Publications
- Crime in the United States
- Law Enforcement Officers Killed and Assaulted (LEOKA)
- Hate Crime Statistics
- National Incident-Based Reporting System (NIBRS)
- Cargo Theft 2014 (pdf) | Cargo Theft 2013 (pdf)
- Human Trafficking (pdf)

NIBRS Resources
- NIBRS Overview

Recent Program Updates
- National Use-of-Force Data Collection
- Historical Rape Definition Retired (pdf)
- New Rape Definition (pdf)
- New Rape Definition Frequently Asked Questions (pdf)
- New Rape Fact Sheet (pdf)

Latest Releases

Law Enforcement Officers Killed and Assaulted 2016
The latest edition of Law Enforcement Officers Killed and Assaulted provides national statistics on lives lost in 2016.
Details

Crime Stats for 2016
Violent crime increased and property crimes decreased during 2016 when compared to 2015 data. Details

The Federal Bureau of Investigation's Uniform Crime Reporting Program publishes annual crime data for selected crimes as reported to the police in the United States.

Having crime data also allows for calculations of a national *crime rate* (discussed later), which in turn provides us with a "victim risk rate"—the odds that we (or relatives and friends) will become a victim of a serious crime.

Discussed first is the FBI's ***Uniform Crime Reports,*** followed by a review of a more in-depth method of capturing and analyzing reported crimes: the National Incident-Based Reporting System; finally, the National Crime Victimization Survey is examined. As you will learn, these three sources of crime information are very different in their approaches and findings.

Uniform Crime Reports: published annually by the FBI, each report describes the nature of crime as reported by law enforcement agencies; includes analyses of Part I crimes.

The FBI's *Uniform Crime Reports*

The Uniform Crime Reporting (UCR) Program was conceived in 1929 by the International Association of Chiefs of Police to meet a need for reliable, uniform crime statistics for the nation. In 1930, the FBI was tasked with collecting, publishing, and archiving those statistics. Today, several annual statistical publications, such as the comprehensive *Crime in the United States,* are produced from data provided by nearly 18,000 law enforcement agencies across the United States.

Other ancillary annual publications, such as *Hate Crime Statistics* and *Law Enforcement Officers Killed and Assaulted,* address specialized facets of crime such as hate crime or the murder and assault of law enforcement officers.[34] Special studies, reports, and monographs prepared using data mined from the UCR Program's large database are published each year as well. In addition to these reports, information about the National Incident-Based Reporting System (discussed later in this chapter), answers to general UCR questions, and answers to specific UCR questions are available on the program's website.

For the purposes of the *Uniform Crime Reports,* the FBI divides offenses into two groups, Part I and Part II crimes. Each month, contributing agencies voluntarily submit information to the FBI concerning the number of Part I offenses reported to them, as well as those offenses that were cleared by arrest, and the age, sex, and race of persons arrested for each of the offenses.

Crime Rate

A fundamental aspect of calculating and understanding crime concerns how the FBI calculates the **crime rate**. The formula used is not complicated in nature. It is as follows: number of crimes reported, divided by the population of the jurisdiction in question, then multiplied by 100,000; this renders the number of crimes reported for each 100,000 population. It is depicted as follows:

Crime rate: the number of reported crimes divided by the population of the jurisdiction, and multiplied by 100,000 persons; developed and used by the FBI Uniform Crime Reports.

$$\frac{\text{Number of offenses}}{\text{Population of the jurisdiction}} \times 100{,}000$$

As an example, assume, as in a recent year, that about 1,247,321 violent crimes were reported to the police in the United States; also, for that year the nation's reported population was 325,719,178. According to this formula, the resulting crime rate for that year was 382.9 violent crimes per 100,000 population.[35]

The crime rate formula can also be considered a "victim risk rate," or the chances of one's becoming a victim; therefore, the chances of one's being a victim of a violent Part I offense for that recent year were about 383 in 100,000.

Part I Offenses

Part I or "index" crimes are composed of eight serious felonies—murder, rape, robbery, aggravated assault, burglary, larceny-theft, motor vehicle theft, and arson. About 9 million such crimes are reported each year (about 1.3 million being violent and 7.7 million, property).[36] The first four of these eight offenses are deemed crimes against *persons* and are defined in the UCR Program as offenses that involve force or threat of force.

As seen in Table 2.4, crimes of murder had declined each year between 2012 and 2014. However, murders increased between 2014 and 2015 by about 12.1 percent and increased

Table 2.4 /// Rates of Murder Committed in the United States

Year	Population	Murder and Nonnegligent Manslaughter	Murder and Nonnegligent Manslaughter Rate
2010	309,330,219	14,722	4.8
2011	311,587,816	14,661	4.7
2012	313,873,685	14,856	4.7
2013	316,497,531	14,319	4.5
2014	318,907,401	14,164	4.4
2015	320,896,618	15,883	4.9
2016	323,405,935	17,413	5.4
2017	325,719,178	17,284	5.3

Source: Adapted from Federal Bureau of Investigation, "Table 1. Volume and Rate per 100,000 Inhabitants, 1998–2017," *Crime in the United States—2017* (Washington, D.C.: Uniform Crime Reporting Program).

Figure 2.2 /// FBI's UCR "Crime Clock"

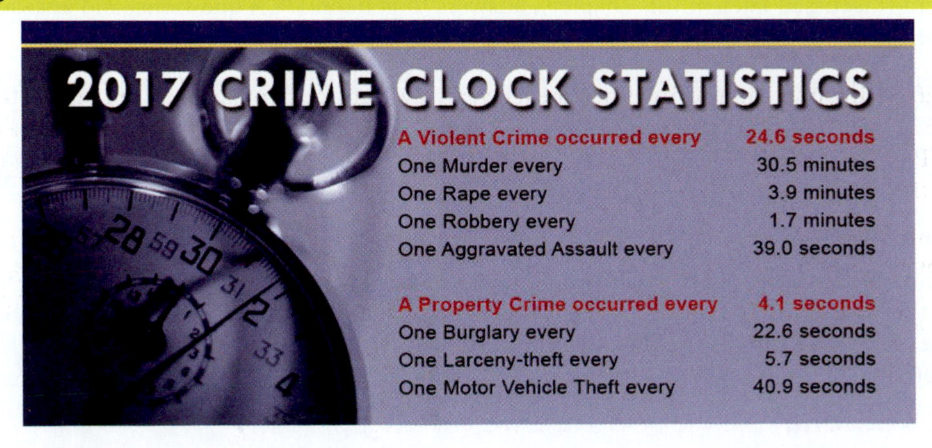

Source: Federal Bureau of Investigation

again by about 9.6 percent in 2016. Between 2016 and 2017 the number of homicides reported to the police in the United States remained relatively stable (less than a 1 percent decline).[37]

The FBI's annual *Uniform Crime Reports* contain a "Crime Clock" that depicts the average time intervals between Part I crimes; a sample Crime Clock is shown in Figure 2.2.

Part II Offenses

In addition to information concerning the eight Part I offenses, the FBI provides arrest-only data for about 20 Part II offenses—simple assaults, forgery, embezzlement, prostitution, vandalism, drug violations, and so forth.[38]

Cautions and Criticisms of UCR Data

Each year when the FBI's *Crime in the United States* is published, many people and groups with an interest in crime rush to use the crime figures in attempts to rank and compare crimes in cities and counties. The FBI is quick to point out each year in its *Uniform Crime Reports* that making such comparisons is ill advised, due to the variety of characteristics of different states, counties, and communities.

Many variables can contribute to the amount of crime occurring in a specific jurisdiction. First, it is important to understand a jurisdiction's industrial/economic base, its dependence on neighboring jurisdictions, its transportation system, its reliance on tourism and conventions, its proximity to military installations and correctional facilities, and so forth. Indeed, the strength and aggressiveness of the local police are also key factors in understanding the nature and extent of crime occurring in an area.

In addition to the crime theories discussed earlier, other factors known to affect the volume and type of crime are largely outside the control of the criminal justice system:

- Population density and degree of urbanization
- Variations in composition of the population, particularly youth concentration
- Stability of the population with respect to residents' mobility, commuting patterns, and transient factors
- Modes of transportation and highway system
- Economic conditions, including median income, poverty level, and job availability
- Cultural factors and educational, recreational, and religious characteristics[39]

Critics of UCR data—who include any person engaged in serious research into crime, criminology, victimology, and so on—commonly note the shortcomings of these data (see Table 2.5).[40]

The Hierarchy Rule: Definition and Application

In tabulating how many crimes occur each year, law enforcement agencies are instructed by the FBI to use what is known as the **hierarchy rule**, which basically says that when more than one Part I offense is committed during a criminal event, the law enforcement agency must identify and report only the offense that is highest on the hierarchy list.[41]

Put another way, the hierarchy rule requires counting only the most serious offense and ignoring all others. Note, however, that this rule applies only to the crime reporting process; it does *not* affect the number of charges for which the defendant may be prosecuted in the courts.

Hierarchy rule: in the FBI *Uniform Crime Reports* reporting scheme, the practice whereby only the most serious offense of several that are committed during a criminal act is reported by the police.

National Incident-Based Reporting System: a crime reporting system in which police describe each offense that occurs during a crime event as well as characteristics of the offender.

The National Incident-Based Reporting System

Being aware of the criticisms leveled over time against the UCR Program, the U.S. Department of Justice in 1988 launched the **National Incident-Based Reporting System**

Table 2.5 /// Limitations of the Uniform Crime Reports
Although the FBI's *Uniform Crime Reports* include several unique publications such as special reports providing statistics on hate crimes and law enforcement officers killed and assaulted, the data have some limitations.
UCR Limitations
• Offense data are available only for a small number (eight) of all crimes committed in the United States.
• The UCR data list only crimes reported to law enforcement agencies (not all crimes that occur are known to the police).
• Reporting of citizens' reports of crime by the police is voluntary; therefore, police may choose not to report or might report inaccurately (thus UCR data may be affected by the reporting practices of local law enforcement).
• The hierarchy rule (discussed in the text) is used, meaning that when a number of crimes are committed as part of a single criminal act, only the most serious offense of all of them is reported to the UCR.
• Attempted crimes are combined with completed crimes.
• When computing crime rates, the UCR counts incidents involving all kinds of targets (e.g., crimes of burglary are against businesses and residents, not against "populations").
• The UCR includes very little information concerning crime victims.

Source: A complete UCR handbook may be viewed at https://ucr.fbi.gov/additional-ucr-publications/ucr_handbook.pdf.

(NIBRS). Although the UCR Program collects and analyzes data for eight Part I offenses, the NIBRS furnishes crime data provided by more than 5,900 participating federal, state, and local law enforcement agencies for 49 specific crimes (called Group A offenses) grouped into 23 crime categories.[42] The FBI posts an interactive map, which depicts the locations of participating agencies and the number of offenses reported for each of the major crime categories. This map can be accessed at https://nibrs.fbi.gov/.

With NIBRS, legislators, municipal planners/administrators, academicians, sociologists, and the public have access to more comprehensive, detailed, accurate, and meaningful crime information than the traditional UCR system can provide. With such information, law enforcement can better make a case to acquire the resources needed to fight crime. The NIBRS also enables agencies to locate similarities in crime-fighting problems so that agencies can work together to develop solutions or discover strategies for addressing the issues. Several NIBRS manuals, studies, and papers are available on the UCR Program's website at https://www.fbi.gov/services/cjis/ucr.

There are important differences between the UCR and the NIBRS data sets. For example, the NIBRS does not apply the UCR's "hierarchy rule" (discussed earlier). If more than one crime was committed by the same person or group of persons during the same event, then all of the crimes would be reported in the NIBRS database. Further, in addition to having the UCR Program's two crime categories—crimes against persons (e.g., murder, rape, and aggravated assault) and crimes against property (e.g., robbery, burglary, and larceny-theft), the NIBRS includes a third crime category, crimes against society, to represent society's prohibitions against certain types of activities (e.g., drug or narcotic offenses). The more detailed NIBRS data are scheduled to become the national crime data standard by January 1, 2021.[43]

National Crime Victimization Survey: a random survey of U.S. households that measures crimes committed against victims; includes crimes not reported to police.

The National Crime Victimization Survey

The **National Crime Victimization Survey** (NCVS), created to address some of the shortcomings of the UCR Program, has been collecting data on personal and household

Practitioner's Perspective

Criminal Justice Program Director; Former Law Enforcement Officer and Drug Investigator

Name: Michael Verro
Position: Criminal Justice Program Director; Former Law Enforcement Officer and Drug Investigator
Location: Excelsior College, Albany, New York

What is your career story? Prior to working for Excelsior College, I was in law enforcement, starting when I was 18 years old. I started doing seasonal work as a peace officer, which is like private security, but they have police authority. When I got out of college, I realized that law enforcement was an area I wanted to go into as a profession. So for the next couple of years, I started like most people do in private security. I was a private investigator, a bodyguard, and worked my way up. At about age 27, I was hired as a police officer and I did that for 25 years. For 15 of those years, I worked as both a police officer and a detective. I was subsequently recruited to the DEA drug task force, and I worked for them for about three years.

I eventually went back to work as a police officer and then came back to New York and worked again as a peace officer. That really led into my world of academia; and now, I'm the program director for criminal justice at Excelsior.

What characteristics and skills are most helpful to succeed in this position? To me, skills are trainable. And there's a big distinction between training and education. That's one thing I think a lot of people have a misconception about. For education, you go to college. You learn theory, application, and critical thinking. Whereas training and some of the other skills are usually taught at an academy, things like driving, shooting, and defensive tactics.

But there are some psychological and mental skills people should possess for this career. And I consider those to be traits or characteristics of their broader personality. Things such as maturity, intelligence, wanting to help other people, and being level headed. Those are very important characteristics and skills, and some they should possess if they're going to go into this field.

Do you see any common trends in this position? When I went on the job almost 30 years ago, there wasn't a lot of technology. We carried a radio around on our hip that we used to call the brick, because it was about the size of a red building brick. It was very heavy and unreliable. Years ago, you had to call for dispatch and then wait and hope that they could go through their information, their databases or make phone calls as quickly as possible to obtain the information you wanted. Nowadays, every patrol car has a laptop with immediate access to information at your fingertips. Police officers wear body cameras, everything is digitally recorded, and they're using drones for surveillance and intelligence gathering. It used to be, again, when we looked to get a warrant—let's say a search warrant—it was basically searching a house. Now, there are digital search warrants if you want to ascertain information off computers or servers. So technology is growing—not only from the criminal justice perspective but also from the crime perspective. Every 90 days, there's some new form of digital crime being perpetrated, and the criminal justice system has to keep up with that.

What are some challenges and misconceptions you face in this position? One of the hardest things that we have to deal with is the fact that criminal justice education is often seen as more of a trainable thing, that it's not really taken seriously in academia as an actual degree, or not considered one of the real sciences. That's a misconception. It's kind of like psychology in that it wasn't really seen as a science until the late 1800s.

To learn more about Michael Verro's experiences as a Criminal Justice Program Director, Former Law Enforcement Officer, and Drug Investigator, watch the Practitioner's Perspective video in SAGE Vantage.

victimization since 1973. The NCVS claims on its website to be "the nation's primary source of information on criminal victimization." Each year, data are obtained from a nationally representative sample of about 95,000 households comprising nearly 240,000 persons on the frequency, characteristics, and consequences of criminal victimization in the United States. Each household is interviewed twice during the year. The survey enables the Bureau of Justice Statistics to estimate the likelihood of victimization by rape, sexual assault, robbery, assault, theft, household burglary, and motor vehicle theft for the population as a whole as well as for segments of the population such as women, the elderly, members of various racial groups, city dwellers, or other groups. The NCVS provides the largest national forum for victims to describe the impact of crime and characteristics of violent offenders.[44]

The survey—which asks respondents to report crime experiences occurring in the past six months—is administered by the U.S. Census Bureau (under the U.S. Department of Commerce) on behalf of the Bureau of Justice Statistics (under the U.S. Department of Justice).

The four primary objectives of the NCVS are to

- Develop detailed information about the victims and consequences of crime
- Estimate the number and types of crimes not reported to the police
- Provide uniform measures of selected types of crimes
- Permit comparisons over time and types of areas[45]

The NCVS categorizes crimes as "personal" or "property." Personal crimes cover rape and sexual attack, robbery, aggravated and simple assault, and purse-snatching/pocket-picking, whereas property crimes cover burglary, theft, motor vehicle theft, and vandalism.

A potential problem with the NCVS is that it estimates crime based on a representative sample; it would be too costly and time consuming to try to survey all U.S. residents. Also, since in relies on victim reports, there is the possibility that respondents might forget crimes, be unable to recall crime details, or provide inaccurate information to surveyors.[46]

/// IN A NUTSHELL

- Our legal system is based on the English common law: collections of rules, customs, and traditions. At its core is the doctrine of *stare decisis*, meaning that when a court has decided a case based on a set of facts, it will adhere to that principle, and other courts will apply that decision in the same manner to all future cases where facts are substantially the same.
- There are three sources of law in the U.S. legal system: federal law, state law, and city/county law.
- There are both civil and criminal laws at the federal, state, and local levels. Many more laws are enacted each year by the U.S. Congress, state legislatures, and city councils.
- Substantive law is the written law that defines or regulates our rights and duties. Procedural law sets forth the procedures and mechanisms for processing criminal cases.
- Two essential elements that underlie our system of law are *mens rea* (criminal intent) and *actus reus* (a criminal act or a failure to act where there is a legal duty).
- Criminal laws define crimes by specifying the *actus reus* (such as killing for homicide and the taking of property for robbery) and the *mens rea* (intentionally, recklessly, by force, unintentionally, etc.). Most crimes are categorized by degrees depending on the seriousness of the physical contact or threat (threatening versus physical contact) and the perpetrator's *mens rea* (intentional versus accidental), or based on the value of items stolen or damaged in crimes against property.
- Common crime categories include crimes against persons, crimes against property, public order crimes, white-collar crime, and organized crime.
- Crimes in the United States are measured and reported in three primary ways: the FBI's *Uniform Crime Reports*, the National Incident-Based Reporting System, and the National Crime Victimization Survey. Each has its unique approaches as well as advantages and disadvantages.

/// KEY TERMS & CONCEPTS

Review key terms with eFlashcards at **edge.sagepub.com/peakbrief**.

Actus reus 33
Burden of proof 29
Civil law 29
Crime rate 43
Crimes against persons 35
Crimes against property 39
Criminal law 29
Defendant 29
Federalism 27
Felony 34
Felony-murder rule 36
Hierarchy rule 45
Idealistic contrast 32
Intent, specific 31
Legal jurisdiction 31
Lex talionis 27
Mens rea 33
Misdemeanor 34
Motive 31
National Crime Victimization Survey 46
National Incident-Based Reporting System 45
Organized crime 42
Plaintiff 29
Procedural law 31
Public order crimes 39
Reasonable doubt 30
Stare decisis 27
Substantive law 31
Uniform Crime Reports 43
White-collar crime 40

/// REVIEW QUESTIONS

Test your understanding of chapter content. Take the practice quiz at **edge.sagepub.com/peakbrief**.

1. What are the differences between criminal law and civil law?
2. How would you define and explain the importance and contributions of *mens rea* and *actus rea* as they operate in our legal system?
3. What is the difference between one's motive and one's intent to commit a crime, and which is the most important in our legal system?
4. A Nebraska law states the following: "Any person who knowingly or intentionally causes or permits a child or vulnerable adult to ingest methamphetamine, a chemical substance used in manufacturing methamphetamine, or paraphernalia is guilty of a Class I misdemeanor." The *mens rea* for this crime is __________; the *actus reus* element of this crime is __________.
5. What is the difference between white-collar crime and organized crime?
6. What are the three primary methods of measuring the extent of crime, and what have been offered as disadvantages for some of them?

/// LEARN BY DOING

Below are two additional case studies, each of which is grounded in actual case facts and chapter materials (concerning sexual assault and homicide, respectively). While studying each scenario, first assume that you are the *prosecuting* attorney and decide which charge(s), if any, should be brought against the accused on the basis of the facts. Next, assume the role of the *defense* attorney and explain, given the facts, what defenses should legitimately be made against the crime(s) charged. Remember that the Sixth Amendment entitles every defendant to the "guiding hand" of effective counsel, whose job it is to ensure that all legal protections are afforded. Answers and/or outcomes for each case are provided in the Notes section. For purposes of discussion, however, approach all of them as if there are no absolute, totally correct answers.

A high school principal threatens a graduating senior that if she does not have sex with him, he will prevent her from graduating. She submits against her will and then reports the incident to police after she is able to graduate. The state prosecutes the man under a first-degree sexual assault statute that reads as follows:

> A person who knowingly has sexual intercourse without consent with a person of the opposite sex commits the offense of sexual intercourse without consent. "Without consent" shall mean the victim is compelled to submit by force or by threat of force.

1. What is the prosecution's best argument that this is precisely the type of situation the law seeks to prevent and that the

defendant is guilty of first-degree sexual assault? What facts would you want to know in your capacity as a prosecutor to bolster your case?

2. What is the defense attorney's best argument that this statute does not apply in this case? What other facts would you want to know to strengthen your argument?

3. You are a state lawmaker—how would you revise this statute to better address this case and ones like it? Be specific and rewrite the law as necessary.[47]

Peterson is relaxing at home when he hears noises in the alley behind his house; he looks in that direction and sees three men who are removing parts from his parked vehicle. Peterson approaches the men and tells them to stop what they are doing; he then runs inside his home, obtains a pistol, and returns to the alley. By now the three men are back in their vehicle and preparing to drive away. Peterson approaches them and tells them not to move, or he will shoot. The driver exits the car and, with a wrench in his hand, begins advancing toward Peterson. Peterson then warns the driver not to come any closer. Still carrying the wrench, the man continues to move toward Peterson, who then shoots and kills him.

4. What is the *primary* legal issue here?

5. Also consider the following questions:

   - What charge should the prosecutor bring against Peterson: Murder in the first degree? Murder in the second degree? Manslaughter?
   - Did Peterson act in self-defense?
   - What other options might have been available to Peterson, aside from obtaining a gun and returning to the alley? Besides shooting the driver?[48]

The following two activities will allow you to explore and discuss several of the crime data sources discussed in this chapter.

6. As part of a criminal justice honor society paper to be presented at a local conference, you are asked to use the Internet and determine how many crimes were reported for your city, your county, your state, and the United States during the past calendar year. How will you proceed?

7. Your criminal justice professor is working on a journal article concerning the dangers of police work, and she asks you to obtain some information on police deaths and assaults. Using the FBI's UCR website and its supplemental report, *Law Enforcement Officers Killed and Assaulted* (at https://ucr.fbi.gov/ucr-publications), what would you report?

/// STUDY SITE

Get the tools you need to sharpen your study skills. SAGE edge offers a robust online environment featuring an impressive array of free tools and resources.

Access practice quizzes, eFlashcards, video, and multimedia at **edge.sagepub.com/peakbrief**

CHAPTER 3

Associated Press/J. Scott Applewhite

ETHICAL ESSENTIALS

Doing Right When No One Is Watching

Ethics is knowing the difference between what you have a right to do and what is right to do.

—Justice Potter Stewart[1]

LEARNING OBJECTIVES

As a result of reading this chapter, you will be able to

1. Articulate legitimate ethical dilemmas that arise with police, courts, and corrections practitioners
2. Explain the philosophical foundations that underlie and mold modern ethical behavior
3. Discuss the need for, and application of, ethical standards as they concern police
4. Explain the importance of ethics in the court system
5. Delineate the unique ethical considerations and obligations that exist with federal employees
6. Explain the interplay of ethics with the subculture and job-related stress in corrections practitioners
7. Understand what ethical tests can help criminal justice students and practitioners decide how to address ethical dilemmas

ASSESS YOUR AWARENESS

Test your knowledge of ethics by responding to the following seven true-false items; check your answers after reading this chapter.

1. The term *ethics* is rooted in the ancient Greek idea of "character."
2. The "ends justify the means" philosophy is typically a good, safe philosophy for the police and judges to follow.
3. Communities sometimes seem to tolerate questionable police behavior, if it is carried out to benefit the greater public good (such as dealing with violent gang members).
4. During an oral interview, applicants for policing jobs should *never* indicate a willingness to "snitch" on another officer whom they observe doing something wrong.
5. The receipt of gratuities by criminal justice personnel is a universally accepted practice.
6. Whistleblowers who expose improper acts of their coworkers now have no legal protection.
7. Because of their constitutional obligations, prosecutors and defense attorneys are not bound to the same ethical standards as other criminal justice employees.

<< **Answers can be found on page 293**

In July 2017, a suspect attempting to flee from Utah state troopers ran his pickup truck head-on into a truck driven by 43-year-old William Gray. The suspect died in the crash. At the hospital, police attempted to obtain a blood sample from the unconscious, severely burned Gray. Nurse Alex Wubbels refused the police request for a blood draw, pointing out that Gray was not under arrest, that police did not have a warrant to obtain the blood, and that police could not obtain consent because the man was unconscious. The police believed they had implied consent by virtue of Gray's commercial driver's license (CDL). (This point soon became a legal controversy, as some attorneys argued that federal regulations regarding Gray's CDL allowed police to obtain the sample; other attorneys disagreed.) Video shows the officer saying, "I either go away with blood in vials or body in tow," then placing Wubbels under arrest, grabbing her, pulling her arms behind her back, handcuffing her, and dragging her out of the hospital as she screams. An internal investigation found the officer and his supervisor violated several department policies during Wubbels's arrest; a review by the city's civilian review board also found that both officers violated department policies. The arresting officer was terminated and his supervisor demoted; both appealed their punishments. Wubbels agreed to a $500,000 payment to settle the dispute.[2]

As you read this chapter, consider whether or not Wubbels should have been arrested and if there were other means available to police to address the situation. Also, consider how tenuous the relationship can be between police and citizens, how quickly matters can get out of hand, and how police use of force can easily be viewed as "crossing the line" and be deemed unethical in nature. Perhaps a good "test" in this particular case is whether or not the officer's actions were reasonable in light of the facts and circumstances facing him, without regard to his underlying intent or motivation.

INTRODUCTION

As regular practice is essential to being a renowned musician, and a perfect cake is essential to having a beautiful wedding, so too is ethics essential to being a criminal justice practitioner. A Latin term that might be used to describe this relationship is *sine qua non*—"without which, nothing." A fundamental knowledge of ethics, as well as some guideposts concerning what constitutes unethical behavior, is an important topic in today's society and for all criminal justice students—not only to guide their own behavior but also because unethical behavior at times appears to permeate contemporary U.S. politics/government, business, and sports.

What specific behaviors are clearly unethical for criminal justice employees? What criteria should guide these employees in their work? To what extent, if any, should the public allow criminal justice employees to violate citizens' rights in order to maintain public order?

This chapter attempts to address these questions and examines many types of ethical problems that can and do arise in police departments, courts, and corrections agencies. The focus is necessarily on the police, who find themselves in many more situations where corruption and brutality can occur than do judges and corrections personnel.

This is not a black-and-white area of study; in fact, there are definitely "shades of gray" for many people where ethics is called into question. Also problematic is that some people who are hired into criminal justice positions simply are not of good character. Furthermore, as addressed in Chapter 1, we cannot *train* people to have high ethical standards, nor can we infuse ethics intravenously. In sum, character and ethics are largely things that people either have or don't have.

GOOD EXAMPLES OF BAD EXAMPLES

To frame the concept of ethics and demonstrate how one's value system can easily be challenged in criminal justice work, consider the following scenarios, all of which are based on true events, and what you might deem to be an appropriate response and punishment (if any) for each:

- *Police*: A Chicago police officer was reprimanded for violating rules that prohibit officers from making political statements while on duty. Specifically, he had posted on social media a picture of himself, in uniform, holding an American flag and a homemade sign that read, "I stand for the anthem. I love the American flag. I support my president and the 2nd Amendment." But the reprimand only seemed to intensify his rhetoric, as he began posting inflammatory material about women, welfare recipients, and those who disagree with his politics. He also tangled with social media users, saying, "Keep listening for that knock on the door," and often boasted that he would continue to avoid serious punishment. His superiors twice tried to fire him, though he appealed those efforts and won. "The police dept didn't and CAN'T fire me," he wrote. The officer is one of the most disciplined officers in the department, having been suspended seven times for a total of 111 days.[3]
- *Courts*: For several weeks, a wealthy divorcée receives menacing telephone calls that demand social dates and sexual favors. The caller's voice is electronically disguised. The suspect also begins stalking the woman. After working several clues and surveillances at the victim's home, you, a federal agent, finally make contact with a suspect

and determine that he is the chief judge of the state's supreme court. Upon confronting him, you are told by the judge to "forget about it, or you'll be checking passports in a remote embassy."[4]

- *Corrections*: Following a number of high-profile deaths and cases of abuse, several civil rights organizations and advocates for the mentally ill demanded an investigation of a southern prison system where more than 12,000 inmates were alleged as being held in solitary confinement. Cases cited included an inmate who died after being gassed three times in a solitary confinement cell and one who was scalded to death while locked by corrections officers in a prison shower. It was also alleged that inmates were beaten and sexually assaulted, and that racial disparity exists: Black inmates made up 44 percent of the prison population but represented 57 percent of the inmates in solitary confinement. In addition, it was believed that nearly a quarter of all inmates with mental illnesses were housed in solitary confinement.[5]

PHILOSOPHICAL FOUNDATIONS

The term *ethics* is rooted in the ancient Greek idea of "character." Ethics involves doing what is right or correct, and the term is generally used to refer to how people should behave in a professional capacity. Many people would argue, however, that no difference should exist between one's professional and personal behavior. In other words, ethical rules of conduct should apply to everything a person does.

A central problem with understanding ethics concerns the questions of "whose ethics" and "which right." This becomes evident when one examines controversial issues such as the death penalty, abortion, use of deadly force, and gun control. How individuals view a particular controversy depends largely on their values, character, or ethics. Both sides of controversies such as these believe they are morally right. These issues demonstrate that to understand behavior, the most basic values must be examined and understood.

Another area for examination is **deontological ethics**, which does not consider consequences but instead examines one's duty to act. The word *deontology* comes from two Greek roots, *deos,* meaning "duty," and *logos,* meaning "study." Thus, deontology means the study of duty. When police officers observe a violation of law, they have a duty to act. Officers frequently use this duty as an excuse when they issue traffic citations that appear to have little utility and do not produce any great benefit for the rest of society. For example, when an officer writes a traffic citation for a prohibited left turn made at two o'clock in the morning when no traffic is around, the officer is fulfilling a departmental duty to enforce the law. From a utilitarian standpoint (where we judge an action by its consequences), however, little if any good was served. Here, duty, and not good consequences, was the primary motivator.

Deontological ethics: one's duty to act.

Pool/Getty Images News/Getty Images

One of the ethical responsibilities of police officers is to testify truthfully in court concerning what they saw, heard, and did during the performance of their duties.

Immanuel Kant, an 18th-century philosopher, expanded the ethics of duty by including the idea of "good will."[6] People's actions must be guided by good intent. In the previous example, the officer who wrote the traffic citation for an improper left turn would be acting unethically if the ticket was a response to a quota or some irrelevant motive. However, if the citation was issued because the officer truly believed that it would result in some good outcome, it would have been an ethical action.

Some people have expanded this argument even further. Richard Kania argued that police officers should be allowed to freely accept gratuities because such actions would constitute the building blocks of positive social relationships between the police and the public.[7] In this case, duty is used to justify what under normal circumstances would be considered unethical. Conversely, if officers take gratuities for self-gratification rather than to form positive community relationships, then the action would be considered unethical by many.

Types of Ethics

Ethics usually involves standards of fair and honest conduct; what we call conscience, the ability to recognize right from wrong; and actions that are good and proper. There are absolute ethics and relative ethics. **Absolute ethics** has only two sides—something is either good or bad, black or white. Some examples in police ethics would be unethical behaviors such as bribery, extortion, excessive force, and perjury, which nearly everyone would agree are unacceptable behaviors by the police.

Absolute ethics: the type of ethics where there are only two sides—good or bad, black or white; some examples would be unethical behaviors such as bribery, extortion, excessive force, and perjury, which nearly everyone would agree are unacceptable for criminal justice personnel.

Relative ethics: the gray area of ethics that is not so clear-cut, such as releasing a serious offender in order to use him later as an informant.

Relative ethics is more complicated and can have a multitude of sides with varying shades of gray. What one person considers to be ethical behavior may be deemed highly unethical by someone else. Not all ethical issues are clear-cut, however, and communities *do* seem willing at times to tolerate extralegal behavior if there is a greater public good, especially in dealing with problems such as gangs and the homeless. This willingness on the part of the community can be conveyed to the police. Ethical relativism can be said to form an essential part of the community policing movement, discussed more fully in Chapter 5.

A community's acceptance of relative ethics as part of criminal justice may send the wrong message: that few boundaries are placed on justice system employee behaviors and that, at times, "anything goes" in their fight against crime. As John Kleinig pointed out, giving false testimony to ensure that a public menace is "put away" or illegally wiretapping an organized crime figure's telephone might sometimes be viewed as "necessary" and "justified," though illegal.[8]

This is the essence of the crime control model of criminal justice (discussed in Chapter 1). Another example is that many police believe they are compelled to skirt along the edges of the law—or even violate it—in order to arrest drug traffickers. The ethical problem here is that even if the action could be justified as morally proper, it remains illegal. For many persons, however, the protection of society overrides other concerns.

This viewpoint—the "principle of double effect"—holds that when one commits an act to achieve a good end and an inevitable but intended effect is negative, then the act might be justified. A long-standing debate has occurred about balancing the rights of individuals against the community's interest in calm and order.

These special areas of ethics can become problematic and controversial when police officers use deadly force or lie and deceive others in their work. Police could justify a whole range of activities that others may deem unethical simply because the consequences resulted in the greatest good for the greatest number—the *utilitarian* approach (or **utilitarianism**). If the ends justified the means, perjury would be ethical when committed to prevent a serial killer from being set free to prey on society. In our democratic society, however, the means are just as important as, if not more important than, the desired end.

Utilitarianism: in ethics, as articulated by John Stuart Mill, a belief that the proper course of action is that which maximizes utility—usually defined as that which maximizes happiness and minimizes suffering.

As examples, citizens in some jurisdictions may not object to the police "hassling" suspected gang members—pulling them over in their cars, say, and doing a field interview—or telling homeless people who are loitering in front of a heavy tourism area or public park to "move along."

It is no less important today than in the past for criminal justice employees to appreciate and come to grips with ethical essentials. Indeed, ethical issues in policing have been affected by three critical factors:[9] (1) the growing level of temptation stemming from the illicit drug trade; (2) the potentially compromising nature of the organizational culture—a culture that can exalt loyalty over integrity, with a "code of silence" that protects unethical employees; and (3) the challenges posed by decentralization (flattening the organization and pushing officers' decision making downward) through the advent of community-oriented policing and problem solving.

Noble Cause Corruption

When the police practice relative ethics and the principle of double effect, described earlier, it is known as **noble cause corruption**—what Thomas Martinelli, perhaps gratuitously, defined as "corruption committed in the name of good ends, corruption that happens when police officers care too much about their work."[10] It basically holds that when an act is committed to achieve a good end (such as an illegal search) but its outcome is negative (the person who is searched eventually goes to prison), the act might still be justified.

Noble cause corruption: a situation in which one commits an unethical act but for the greater good; for example, a police officer violates the Constitution in order to capture a serious offender.

Although noble cause corruption can occur anywhere in the criminal justice system, we might look at the police for examples. Officers might bend the rules, such as not reading a drunk person his rights or performing a field sobriety test; planting evidence; issuing "sewer" tickets—writing a person a ticket but not giving it to her, resulting in a warrant issued for failure to appear in court; "testilying," or "using the magic pencil," whereby police officers write up an incident in a way that criminalizes a suspect—this is a powerful tool for punishment. Noble cause corruption carries with it a different way of thinking about the police relationship with the law. Here, officers operate on a standard that places personal morality above the law, becoming legislators *of* the law and acting as if they *are* the law.[11] Some officers rationalize such activities; as a Philadelphia police officer put it, "When you're shoveling society's garbage, you gotta be indulged a little bit."[12] Obviously the kinds of noble cause behaviors mentioned here often involve arrogance on the part of the police and ignore the basic constitutional guidelines the occupation demands. Administrators and middle managers must be careful to take a hardline view that their subordinates always tell the truth and follow the law. A supervisory philosophy of discipline based on due process, fairness, and equity, combined with intelligent, informed, and comprehensive decision making, is best for the department, its employees, and the community.[13]

ETHICS IN POLICING

Having defined the types of ethics and some dilemmas, next we discuss in greater detail some of the ethical issues faced by police leaders and their subordinates.

A Primer: The Oral Job Interview

During oral interviews for a position in policing, applicants are often placed in a hypothetical situation that tests their ethical beliefs and character. For example, they may be asked to assume the role of Officer Brown, who is checking on foot an office supplies retail store that was found to have an unlocked door during early morning hours. On leaving the building, Brown observes another officer, Smith, removing a $200 writing pen from a display case and placing it in his uniform pocket. What should Officer Brown do?

This kind of question commonly befuddles the applicant: "Should I rat on my fellow officer? Overlook the matter? Merely tell Smith never to do that again?" Unfortunately, applicants may do a lot of "how am I *supposed* to respond" soul searching and second-guessing with these kinds of questions.

Bear in mind that criminal justice agencies do not wish to hire someone who possesses ethical shortcomings; it is simply too potentially dangerous and expensive, from both legal and moral standpoints, to take the chance of bringing into an agency someone who is corrupt. That is the reason for such questioning and a thorough background investigation of applicants.

Before responding to a scenario like the one concerning Officers Brown and Smith, the applicant should consider the following issues: Is this likely to be the first time that Smith has stolen something? Don't the police arrest and jail people for this same kind of behavior?

In short, police administrators should *never* want an applicant to respond that it is acceptable for an officer to steal. Furthermore, it would be incorrect for an applicant to believe that police do not want an officer to "rat out" another officer. Applicants should never acknowledge that stealing or other such activities are to be overlooked.

Scott Olson/Getty Images News/Getty Images

Police executives typically address the public during critical incidents or investigations. Here, Chicago Police Chief Eddie Johnson comments on the allegation by actor and singer Jussie Smollett that he was the victim of a hate crime assault in February 2019. (Smollett was later indicted for paying two men to stage the crime against him.)

Police Corruption

"For as long as there have been police, there has been police corruption."[14] Thus observed police expert Lawrence Sherman about one of the oldest problems in U.S. policing. Indeed, the Knapp Commission investigated police corruption in the early 1970s, finding that there are two primary types of corrupt police officers: the "meat-eaters" and the "grass-eaters." Meat-eaters spend a good deal of their working hours aggressively seeking out situations that they can exploit for financial gain, including gambling, narcotics, and other lucrative enterprises.

Grass-eaters, the commission noted, constitute the overwhelming majority of those officers who accept payoffs; they are not aggressive but will accept gratuities from contractors, tow-truck operators, gamblers, and the like. Although such officers probably constitute a small percentage of the field, any such activity is to be identified and dealt with sternly.

Police corruption can be defined broadly, from major forms of police wrongdoing to the pettiest forms of improper behavior. Another definition is "the misuse of authority by a police officer in a manner designed to produce personal gain for the officer or for others."[15] Police corruption is not limited to monetary gain, however. Gains may be made through the acceptance of services received, status, influence, prestige, or future support for the officer or someone else.[16]

Police corruption: misconduct by police officers that can involve but is not limited to illegal activities for economic gain, gratuities, favors, and so on.

To Inform or Not to Inform: The Code of Silence

Let's continue with the earlier scenario. Remember that Officer Brown witnessed Officer Smith putting an expensive ink pen in his pocket after they found an unlocked office supplies retail business on the graveyard shift. If reported, the misconduct will ruin Smith, but if not reported, the behavior could eventually cause enormous harm. To outsiders, this is not a moral dilemma for Brown at all; the only proper path is for her to report the misconduct. However, arguments exist both for and against Brown's informing on her partner. Reasons for informing include the fact that the harm caused by a scandal would be outweighed by the public's knowing that the police department is free of corruption; also, individual episodes of corruption would be brought to a halt. Brown, moreover, has a sworn duty to uphold the law. Reasons against informing include the facts that, at least in Brown's mind, the other officer is a member of the "family" and a skilled police officer is a valuable asset whose social value far outweighs the damage done by moderate corruption.[17]

A person who is in charge of investigating police corruption would no doubt take a punitive view, because police are not supposed to steal, and they arrest people for the same kinds of acts every day. Still, the issue—and a common question during an oral interview when citizens are being tested for police positions—is whether or not Brown would come forth and inform on her fellow officer.

It is necessary to train police recruits on the need for a corruption-free department. The creation and maintenance of an internal affairs unit and the vigorous prosecution of lawbreaking police officers are also critical to maintaining the integrity of officers.

The Law Enforcement Code of Ethics and Oath of Honor

The Law Enforcement Code of Ethics (LECE) was adopted by the International Association of Chiefs of Police (IACP) in 1957 and has been revised several times since then. It is a powerful proclamation, and tens of thousands of police officers across the nation have sworn to uphold this code upon graduating from their academies. Unfortunately, the LECE is also quite lengthy, covering rather broadly the following topics as they relate to police officers: primary responsibilities, performance of one's duties, discretion, use of force, confidentiality, integrity, cooperation with other officers and agencies, personal/professional capabilities, and private life.

The IACP adopted a separate, shorter code that would be mutually supportive of the LECE but also easier for officers to remember and call to mind when they come face-to-face with an ethical dilemma. It is the Law Enforcement Oath of Honor, and the IACP is hoping the oath will be implemented in all police agencies and by all individual officers. It may be used at swearing-in ceremonies, graduation ceremonies, promotion ceremonies, beginnings of training sessions, police meetings and conferences, and so forth.[18] The Law Enforcement Oath of Honor is as follows:

On my honor, I will never
betray my badge, my integrity,
my character or the public trust.
I will always have the courage to hold
myself and others accountable for our actions.
I will always uphold
the constitution, my community and the
agency I serve.[19]

Accepted and Deviant Lying

In many cases, no clear line separates acceptable from unacceptable behavior in policing. The two are separated by an expansive gray area that comes under relative ethics. Some observers have referred to such illegal behavior as a "**slippery slope**," meaning that people tread on solid or legal ground but at some point slip beyond the acceptable into illegal or unacceptable behavior.

Slippery slope: the idea that a small first step can lead to more serious behaviors, such as the receipt of minor gratuities by police officers believed to eventually cause them to desire or demand receipt of items of greater value.

Criminal justice employees lie or deceive for different purposes and under varying circumstances. In some cases, their misrepresentations are accepted as an essential part of a criminal investigation, whereas in other cases they are viewed as violations of law. David Carter examined police lying and perjury and found a distinction between accepted lying and deviant lying.[20] **Accepted lying** includes police activities intended to apprehend or entrap suspects. This type of lying is generally considered to be trickery. **Deviant lying**, by contrast, refers to occasions when officers commit perjury to convict suspects or are deceptive about some activity that is illegal or unacceptable to the department or public in general.

Accepted lying: police activities intended to apprehend or entrap suspects. This type of lying is generally considered to be trickery.

Deviant lying: occasions when officers commit perjury to convict suspects or are deceptive about some activity that is illegal or unacceptable to the department or public in general.

Deception has long been practiced by the police to ensnare violators and suspects. For many years, it was the principal method used by detectives and police officers to secure confessions and convictions. Accepted lying is allowed by law, and to a great extent it is expected by the public. Gary Marx identified three methods police use to trick a suspect: (1) performing the illegal action as part of a larger, socially acceptable, and legal goal; (2) disguising the illegal action so that the suspect does not know it is illegal; and (3) morally weakening the suspect so that the suspect voluntarily becomes involved.[21] The courts have long accepted deception as an investigative tool. For example, the U.S. Supreme Court ruled in *Illinois v. Perkins* that, when investigating crimes, police undercover agents are not required to administer the *Miranda* warning to incarcerated inmates.[22] Lying, although acceptable by the courts and the public in certain circumstances, does result in an ethical dilemma. It is a dirty means to accomplishing a good end—the police using untruths to gain the truth relative to some event.

AP Photo/Drug Enforcement Agency

Lying and deception have long been used by the police to identify and arrest criminals; this undercover DEA agent is posing as a student as part of a drug investigation.

In their examination of lying, Thomas Barker and David Carter identified two types of deviant lying: lying that serves legitimate purposes and lying that conceals or promotes crimes or illegitimate ends.[23] Lying that serves legitimate goals occurs when officers lie to secure a conviction, obtain a search warrant, or conceal omissions during an investigation. Barker found that police officers believe that almost one-fourth of their agency would commit perjury to secure a conviction or to obtain a search warrant.[24] Lying becomes an effective, routine way to sidestep legal impediments. When left unchecked by supervisors, managers, and administrators, lying can become organizationally accepted as an effective means of nullifying legal entanglements and removing obstacles that stand in the way of convictions. Examples include using the services of nonexistent confidential informants to secure search warrants, concealing that an interrogator went too far, coercing a confession, or perjuring oneself to gain a conviction.

Lying to conceal or promote criminality is the most distressing form of deception. Examples range from when the police lie to conceal their use of excessive force when arresting a suspect to obscuring the commission of a criminal act.

Accepting Gratuities

Gratuities: the receipt of some benefit (a meal, gift, or some other favor) either for free or for a reduced price.

Many police officers commonly accept **gratuities** as a part of their job. Restaurants frequently give officers free or half-price meals and drinks, and other businesses routinely give officers discounts for services or merchandise. While some officers and their departments accept the receipt of such gratuities as a legitimate part of their job, other agencies prohibit such gifts and discounts but seldom attempt to enforce any relevant policy or regulation. Finally, some departments attempt to ensure that officers do not accept free or discounted services or merchandise and routinely enforce policies or regulations against such behavior.[25]

There are two basic arguments *against* police acceptance of gratuities. First is the slippery slope argument, discussed earlier, which proposes that gratuities are the first step in police corruption. This argument holds that once gratuities are received, police officers' ethics are subverted and they are open to additional breaches of their integrity. In addition, officers who accept minor gifts or gratuities are then obligated to provide the donors with some special service or accommodation. Furthermore, some critics propose that receiving a gratuity is wrong because officers are receiving rewards for services that they are obligated to provide as part of their employment. That is, officers have no legitimate right to accept compensation in the form of a gratuity. If the police ever hope to be accepted as members of a full-fledged profession, they must address whether the acceptance of gratuities is professional behavior.

Former New York police commissioner Patrick V. Murphy was one of those who believed that "except for your pay check, there is no such thing as a clean buck."[26] He also noted that judges, teachers, doctors, and other professionals do not accept special consideration from restaurants, convenience stores, movie theaters, and so on.

Here is an example of a policy developed by a sheriff's office concerning gratuities:

1. Without the express permission of the Sheriff, members shall not solicit or accept any gift, gratuity, loan, present, or fee where there is any direct or indirect connection between this solicitation or acceptance of such gift and their employment by this office.

2. Members shall not accept, either directly or indirectly, any gift, gratuity, loan, fee, or thing of value, the acceptance of which might tend to improperly influence their actions, or that of any other member, in any matter of police business, or which might tend to cast an adverse reflection on the Sheriff's Office.
3. Any unauthorized gift, gratuity, loan, fee, reward, or other thing falling into any of these categories coming into the possession of any member shall be forwarded to the member's commander, together with a written report explaining the circumstances connected therewith. The commander will decide the disposition of the gift.[27]

Greed and Temptation

Edward Tully underscored the vast amount of temptation that confronts today's police officers and what police leaders must do to combat it:

> Socrates, Mother Teresa, or other revered individuals in our society never had to face the constant stream of ethical problems of a busy cop on the beat. One of the roles of [police leaders] is to create an environment that will help the officer resist the temptations that may lead to misconduct, corruption, or abuse of power. The executive cannot construct a work environment that will completely insulate the officers from the forces which lead to misconduct. The ultimate responsibility for an officer's ethical and moral welfare rests squarely with the officer.[28]

Most citizens have no way of comprehending the amount of temptation that confronts today's police officers. They frequently find themselves alone inside retail businesses after normal business hours, clearing the building after finding an open door or window. A swing or graveyard shift officer could easily obtain considerable plunder during these occasions,

You Be the . . .

Police Officer

One of the authors was involved in the following case study. Consider how you would have handled this situation: you are a police officer for a small university community and it is summertime. You are driving your patrol car one evening when a vehicle suddenly pulls up beside you, and the woman driver yells, "Officer, my baby's dying!" You have her drive her car to the curb, and she then carries the child to your patrol car. You note that the infant is about eight months of age, her face has a blue tinge to it, her eyes are rolled back, and she does not appear to be breathing. You quickly check her airway, resolve the situation (the baby's airway had become obstructed by her tongue during a seizure) and convey them to the hospital. Two months later, learning from a friend where you live, the same woman comes to your home. Sobbing, she says, "Officer, my husband is in jail for disorderly conduct, I'm broke, and my kids have no food or milk. I need money to bail him out of jail so he can get back to work and buy groceries. You gave me help once—can you help me now?" Her husband, well known to your police agency for past crimes, does not enjoy a good reputation in town.

1. Will you do something to assist the woman financially? If not, explain.
2. If yes, will you loan her grocery money only, or enough so that she can also bail her husband out of jail?
3. Assume you recently received a large income tax refund and thus decide to give her enough money to bail her husband out of jail so that he can return to his job. Is this ethical? Why or why not? If you do, how might the other officers react if learning of your conduct?

Practitioner's Perspective

Police Internal Affairs Investigator

Name: Zack Thew
Position: Police Internal Affairs Investigator
Location: Reno, Nevada

What is your career story? I got a bachelor's degree in criminal justice. Shortly after that, I was employed by the Reno Police Department. I had the opportunity to work various assignments as an officer. I was a training officer, a lead negotiator, and a part of our mentoring program. After several years of doing that, I went into the detective division, where I worked in the fraud unit, the sex crimes unit, the robbery homicide unit, and the computer crimes unit before promoting to sergeant.

What misconceptions do you often hear about your position? A misconception for internal affairs is that investigators are on a witch hunt or that they're trying to discipline employees for the sake of discipline. Certainly, accountability is at the heart of what we do, but good internal affairs investigators are conscious of employees' rights. They work very closely with employees and the association, and they work to protect the rights of those employees and ensure that all the investigations are fact-based and fair, and that the conclusions are reasonable.

Another misconception held by some members of the public is that the unit exists to cover up or protect substandard or criminal activity by the officers. And that's simply not true. I think how you address both of those misconceptions is to be as transparent, fair, and consistent as possible.

What role does diversity play in your position?: Diversity plays a major role in internal affairs. Unfortunately, police often find themselves in situations that are grayer more than they are black-and-white. To be able to adapt to those situations and hold ourselves accountable under those circumstances, we need to understand the situation from as many perspectives as possible. We do that with a diverse approach. We look at it from the perspective of the involved citizens, the officers, the supervisors, the administrators, the media, the politicians, the community, and understand those perspectives so we can have an idea of the big picture with the course of action we ultimately take.

Do you see any common trends in this position? The current trend in internal affairs is a push for transparency. The public at large is very inquisitive, and the media and politicians are discerning. They want to know what's happening with their police department and how their organization is serving them. The fact of the matter is, we do not have as big of a segment of the population that blindly trusts the police as they existed before. So again, it's fostering that relationship, and holding our people accountable, and being transparent.

What advice would you give to someone either wishing to study, or now studying, criminal justice and wanting to become a practitioner in this position? My advice for students who want to pursue this career is to spend some time in internal affairs to understand the demands of the job. Be as varied as you possibly can be in your work and personal experience, and understand current events, both inside and outside of your organization. That's going to help you know what you need to do to make your organization successful and what the expectations are from the public. Set high standards and hold yourself accountable above all else.

To learn more about Zach Thew's experiences as a Police Internal Affairs Investigator, watch the Practitioner's Perspective video in SAGE Vantage.

acquiring everything from clothing to tires for a personal vehicle. At the other end of the spectrum is the potential for huge payoffs from drug traffickers or other big-money offenders who will gladly pay the officer to look away from their crimes. Some officers, like the general public, find this temptation impossible to overcome.

"Random Acts of Kindness:" The Hidden Side of Police Work

This chapter section has focused on efforts to evaluate, constrain, and discipline different forms of police behavior. Perhaps it is appropriate, therefore, to balance the issue by pointing out another, often overlooked side of the issue: police behaviors that are laudable and reveal a completely different face from that which the public perceives.

Each day police officers perform countless "random acts of kindness" that seldom come to the public's attention. Yet, as one of this book's authors can attest from personal experience, many are the times that officers do act in unique and unexpected ways in order to help someone in distress, such as passing the hat (literally) to collect funds to buy food for a group of hungry undocumented immigrants while they are being detained; or giving a hard-core alcoholic enough whiskey (from a bottle of bourbon confiscated long before) that he might sleep in his cell, keeping the hallucinations of the delirium tremens (DTs) at bay; lying on their bellies during a downpour to retrieve cash from a storm sewer so that an elderly couple (who, shortly before, had been bound, gagged, beaten, and robbed) might have at least some of their money returned (while fleeing their home, the robber had dropped a considerable number of bills in the yard and gutter); giving a woman money to feed herself and her baby while her husband is in the local jail.

The following deeds, however, *were* reported by the media:

- A Miami, Florida, officer caught a struggling single mother shoplifting food from a grocery store for her family. Rather than take her to jail, the officer purchased the food for the mother and showed her the local food banks and churches where she could get assistance until she got back on her feet.[29]
- Officers in New York City and Palatka, Florida, noticing that homeless men were virtually walking barefoot, purchased new shoes for them.[30]
- After a 20-year-old wheelchair-bound man was robbed of $4,000 he was saving for a new wheelchair, officers not only caught the robber and returned his money, but also created a fund so that he could purchase the very best of wheelchairs (his other one at times got stuck in the snow).[31]
- Officers in Benicia, California, and Phoenix, Arizona, learning that teens were walking long distances each day (one going nine miles to and from work, the other, two miles to school), purchased bicycles for them and arranged for them to obtain helmets and other necessities.
- When a Nicoma, Oklahoma, police officer learned that a bike had been stolen at knifepoint from a boy on his beat who had autism, he collected money from his department and his second job to buy the boy a new bike.[32]

ETHICS IN THE COURTS

Although the public tends to think of criminal justice ethics primarily in terms of the police, certainly other criminal justice professionals—including the court work group—have expectations in this regard as well. The ethical standards and expectations—and some examples of failings—of those individuals are discussed next.

Evolving Standards of Conduct

The first call during the 20th century for formalized standards of conduct in the legal profession came in 1906, with Roscoe Pound's speech "The Causes of Popular Dissatisfaction With the Administration of Justice."[33] However, the first canons of judicial ethics probably grew out of a professional baseball scandal in 1919, in which the World Series was "thrown" to the Chicago White Sox by the Cincinnati Reds. Baseball officials turned to the judiciary for leadership and hired U.S. District Court Judge Kenesaw Mountain Landis as baseball commissioner—a position for which Landis was paid $42,500, compared to his $7,500 earnings per year as a judge. This affair prompted the 1921 American Bar Association (ABA) convention to pass a resolution of censure against the judge and appoint a committee to propose standards of judicial ethics.[34]

Model Code of Judicial Conduct: adopted by the House of Delegates of the American Bar Association in 1990, it provides a set of ethical principles and guidelines for judges.

In 1924, the ABA approved the Canons of Judicial Ethics under the leadership of Chief Justice William Howard Taft, and in 1972, the ABA approved a new **Model Code of Judicial Conduct**; in 1990, the same body adopted a revised model code. Nearly all states and the District of Columbia have promulgated standards based on the code. In 1974, the United States Judicial Conference adopted a Code of Conduct for United States Judges, and Congress over the years enacted legislation regulating judicial conduct, including the Ethics Reform Act of 1989.

The Judge

Judges are discussed generally in Chapter 8; however, here the focus is on their ethical responsibilities. Ideally, our judges are flawless, not allowing emotion or personal biases to creep into their work, treating all cases and individual litigants with an even hand, and employing "justice tempered with mercy." The perfect judge has been described as follows:

> The good judge takes equal pains with every case no matter how humble; he knows that important cases and unimportant cases do not exist, for injustice is not one of those poisons which . . . when taken in small doses may produce a salutary effect. Injustice is a dangerous poison even in doses of homeopathic proportions.[35]

Not all judges, of course, can attain this lofty status and find themselves succumbing to temptation and human faults and foibles. Judges can become embroiled in improper conduct or overstep their bounds in many ways: abuse of judicial power (against attorneys or litigants); inappropriate sanctions and dispositions (including showing favoritism or bias); not meeting the standards of impartiality and competence (discourteous behavior, gender bias and harassment, incompetence); conflict of interest (bias; conflicting financial interests or business, social, or family relationships); and personal conduct (criminal or sexual misconduct, prejudice, statements of opinion).[36]

Following are some actual examples of true-to-life ethical dilemmas involving the courts:[37]

- A judge persuades jailers to release his son on a nonbondable offense.
- A judge is indicted on charges that he used his office for a racketeering enterprise.
- Two judges attend the governor's $500-per-person inaugural ball.
- A judge is accused of acting with bias in giving a convicted murderer a less severe sentence because the victims were homosexual.
- A judge whose car bears the bumper sticker "I am a pro-life Democrat" acquits six pro-life demonstrators of trespassing at an abortion clinic on the grounds of necessity to protect human life.

Such incidents certainly do little to bolster public confidence in the justice system. People expect more from judges, who are "the most highly visible symbol of justice."[38]

Unfortunately, codes of ethical conduct have not eradicated these problems or allayed concerns about judges' behavior. Indeed, one judge who teaches judicial ethics at the National Judicial College in Reno, Nevada, stated that most judges attending the college admit never having read the Model Code of Judicial Conduct before seeking judicial office.[39]

According to the American Judicature Society, during one year 25 judges were suspended from office, and more than 80 judges resigned or retired either before or after formal charges were filed against them; 120 judges also received private censure, admonition, or reprimand.[40]

The key to judicial ethics is to identify the troublesome issues and to create an "ethical alarm system" that responds.[41] Perhaps the most important tenet in the code and the one that is most difficult to apply is that judges should avoid the *appearance* of impropriety—in other words, it is not enough that judges *do* what is just; they must also avoid conduct that would create in the public's mind a perception that their ability to carry out responsibilities with integrity, impartiality, and competence is impaired.

Ethical requirements for the federal judiciary and other federal employees are discussed later in this chapter.

Prosecutors

Given their power and authority to decide which cases are to be prosecuted, prosecuting attorneys must closely guard their ethical behavior. It was decided in 1935 (in *Berger v. United States*) that the primary duty of a prosecutor is "not that he shall win a case, but that justice shall be done."[42]

Instances of prosecutorial misconduct were reported as early as 1897[43] and are still reported today. One of the leading examples of unethical conduct by a prosecutor is *Miller v. Pate* (1967), in which the prosecutor concealed from the jury in a murder trial the fact that a pair of undershorts with red stains contained not blood but red paint.[44]

According to Elliot Cohen, misconduct works: Oral advocacy is important in the courtroom and can have a powerful effect. Another significant reason for such conduct is the harmless error doctrine, in which an appellate court can affirm a conviction despite the presence of serious misconduct during the trial. Only when appellate courts take a stricter, more consistent approach to this problem will it end.[45]

You Be the . . .

Ethics Committee Member

A U.S. district court judge in Alabama was controversial even before he was arrested on allegations of beating his wife. He was criticized for hearing government cases while his aviation company was getting hundreds of thousands of dollars in government business. He was also infamous for having an extramarital affair with his courtroom assistant, as well as for his messy public divorce.

In May 2015, Judge Mark Fuller resigned because of a fight he had with the same former courtroom assistant—now his wife—who is heard yelling, "He's beating on me! Please help me" to a police dispatcher.

A five-judge review panel investigated Fuller's behavior, and a U.S. House of Representatives committee considered conducting impeachment hearings against him.

However, because of his lifetime appointment to the bench, the judge could not be forced off the bench; he could only be reprimanded and asked to resign, and the congressional committee could have recommended impeachment.

1. Should judges with lifetime appointments receive such protections while sitting on the bench?
2. If not, what kind of system would you propose?

Source: See Brad Freeman, "Wife-Beating Judge Loses It: Why Mark Fuller's New Defense Is Beyond the Pale," *Salon,* March 31, 2015, http://www.salon.com/2015/03/31/wife_beating_judge_loses_it_why_mark_fullers_new_defense_is_beyond_the_pale/.

Defense Attorneys

Defense attorneys, too, must be legally and morally bound to ethical principles as agents of the courts. Cohen suggested the following moral principles for defense attorneys:[46]

- Treat others as ends in themselves and not as mere means to winning cases.
- Treat clients and other professional relations in a similar fashion.
- Do not deliberately engage in behavior apt to deceive the court as to truth.
- Be willing, if necessary, to make reasonable personal sacrifices of time, money, and popularity for what you believe to be a morally good cause.
- Do not give money to, or accept money from, clients for wrongful purposes or in wrongful amounts.
- Avoid harming others in the course of representing your client.
- Be loyal to your client, and do not betray his or her confidence.

Other Court Employees

Other court employees have ethical responsibilities as well. For example, an appellate court judge's secretary is asked by a good friend who is a lawyer whether the judge will be writing the opinion in a certain case. The lawyer may be wishing to attempt to influence the judge through his secretary, renegotiate with an opposing party, or engage in some other improper activity designed to alter the case outcome.[47] Bailiffs, court administrators, court reporters, courtroom clerks, and law clerks all fit into this category. It would be improper, say, for a bailiff who is accompanying jurors back from a break in a criminal trial to mention that the judge "sure seems annoyed at the defense attorney" or for a law clerk to tell an attorney friend that the judge she works for prefers reading short bench memos.[48]

ETHICAL CONDUCT OF FEDERAL EMPLOYEES

The laws governing the ethical conduct of federal employees are contained in a variety of statutes, the two major sources of which are Title 18 of the U.S. Code and the Ethics in Government Act of 1978 (enacted following the Watergate scandal of the early 1970s to promote public confidence in government). The latter act has been amended a number of times, with its most significant revision occurring in the Ethics Reform Act of 1989 (Public Law 101-194). A brief general description of that law, as well as expectations of the federal judiciary, is provided next.

The Ethics Reform Act and the Whistleblower Protection Act

The Ethics Reform Act addresses a number of areas of ethical concern, including the receipt of gifts, financial conflicts involving employees' positions, personal conflicts that may affect their impartiality, misuse of position (for private gain), and outside activities or employment that conflicts with their federal duties (such as serving as an expert witness or receiving payment for speaking, writing, and teaching).

Whistleblower Protection Act: a federal law prohibiting reprisal against employees who reveal information concerning a violation of law, rules, or regulations; gross mismanagement or waste of funds; an abuse of authority; and so on.

In 1989, the **Whistleblower Protection Act** (Public Law 101-12) strengthened the protections provided in the Ethics Reform Act. These whistleblower protection laws prohibit reprisal against federal employees who reasonably believe that their disclosures show "a violation of law, rule, or regulation, gross mismanagement, a gross waste of funds, an abuse of authority, or a specific and substantial danger to public health and safety."

Federal judges have the authority to resolve significant public and private disputes. Occasionally, however, a matter assigned to them may involve them or their families personally, or affect individuals or organizations with which they have associations outside of their official duties. In these situations, if their impartiality might be compromised, they must disqualify (or recuse) themselves from the proceeding.

Disqualification is required under Canon 3C(1) of the ABA's Code of Conduct for United States Judges, if the judge

- Has personal knowledge of disputed facts
- Was employed in a law firm that handled the same matter while he or she was there
- Has a close relative who is a party or an attorney
- Personally owns, or has an immediate family member who owns, a financial interest in a party
- As a government official, served as a counsel in the case

The Federal Judiciary

The Code of Conduct for United States Judges was initially adopted by the Judicial Conference on April 5, 1973. This code applies to U.S. circuit judges, district judges, Court of International Trade judges, Court of Federal Claims judges, bankruptcy judges, and magistrate judges. Following are the code's canons of ethical behavior[49]:

Canon 1: A judge should uphold the integrity and independence of the judiciary.

Canon 2: A judge should avoid impropriety and the appearance of impropriety in all activities.

Canon 3: A judge should perform the duties of the office impartially and diligently.

Canon 4: A judge may engage in extrajudicial activities that are consistent with the obligations of judicial office.

Canon 5: A judge should refrain from political activity.

Implicit in these canons are restrictions on judges' soliciting or accepting gifts; outside employment; and payment for appearances, speeches, or written articles.

Boston Globe/Getty Images

Like police officers, correctional officers in jails and prisons at times must exercise force—which raises ethical considerations and concerns for them as well.

ETHICS IN CORRECTIONS

Most correctional officers—like police officers and judges—who work in jails and prisons are dedicated, honest, and law-abiding in nature. Occasionally, however, correctional officers are found to have engaged in inappropriate behaviors. Some of those behaviors involve a variety of inappropriate relationships that can develop between inmates and staff members, to include such activities as bringing in contraband (such as drugs or tobacco) or physical or sexual abuse (generally involving male officers and female inmates).[50]

The strength of the corrections subculture can also contribute to ethical problems in correctional facilities; it correlates with the security level of a correctional facility and is strongest in maximum-security institutions. Powerful forces within the correctional system have a

You Be the . . .

Correctional Officer

Correctional Officer Ben Jones has worked for one year in a medium-security housing unit in a state prison and has gotten on friendly terms with an inmate, Stevens. Known to have been violent, manipulative, and associating with a similarly rough crowd while on the outside, now Stevens appears to be a model inmate; in fact, Officer Jones relies heavily on Stevens to keep him informed of the goings-on in the unit as well as to maintain its overall cleanliness and general appearance. Over time, the two address each other on a first-name basis and increasingly discuss personal matters; Jones occasionally allows Stevens to get by with minor infractions of prison rules (e.g., being in an unauthorized area or entering another inmate's cell). Today Stevens mentions that he is having problems with his fiancée—specifically, that he has received a "Dear John" letter from her, stating that she is dating other men and is "moving on." Upon arriving home from work that evening, Jones finds a case of wine on his porch. No card was left with the case of wine, but at work the next morning Stevens winks at Jones and asks if he "ventured into the vineyard last night."

1. Should Officer Jones report the incident?
2. Has Jones's behavior thus far violated any standards of ethics for correctional officers? If so, what form of punishment (if any) would be appropriate?
3. What should be the relationship between Jones and Stevens in the future?
4. What could Jones have done differently, if anything?

stronger influence over the behavior of correctional officers than do the administrators of the institution, legislative decrees, or agency policies.[51] Indeed, it has been known for several decades that the exposure to external danger in the workplace creates a remarkable increase in group solidarity.[52] Some of the job-related stressors for correctional officers are similar to those the police face: the ever-present potential for physical danger, hostility directed at officers by inmates and even by the public, unreasonable role demands, a tedious and unrewarding work environment, and dependence on one another to effectively and safely work in their environment.[53] For these reasons, several norms of corrections work have been identified: Always go to the aid of an officer in distress, do not "rat," never make another officer look bad in front of inmates, always support an officer in a dispute with an inmate, always support officer sanctions against inmates, and do not wear a "white hat" (participate in behavior that suggests sympathy or identification with inmates).[54] Security issues and the way in which individual correctional officers have to rely on each other for their safety make loyalty to one another a key norm.

The proscription against ratting out a colleague is strong. Cases have been documented in which officers who reported inappropriate activities were labeled by colleagues as "rats" and "no-goods" and sometimes received death threats; even though they were transferred to other institutions, the labels traveled with them. In this regard, the American Correctional Association (ACA) Code of Ethics states that "[m]embers shall report to appropriate authorities any corrupt or unethical behaviors in which there is sufficient evidence to justify review.[55]

In another case, a female correctional officer at a medium-security institution reported some of her colleagues for sleeping on the night shift. She had first approached them and expressed concern for her safety when they were asleep and told them that if they did not refrain from sleeping, she would have to report them to the superintendent. They continued sleeping, and she reported them. The consequences were severe: Graffiti was written about her on the walls, she received harassing phone calls and letters, her car was vandalized, and some bricks were thrown through the windows of her home.[56] Obviously, she deserved better, both in terms of protection during these acts, and with the investigation and prosecution of the parties involved.

It would be grossly unfair to suggest that the kinds of behaviors depicted here reflect the behavior of correctional officers in all places and at all times. The case studies do demonstrate, however, the power and loyalty of the group, and correctional administrators must be cognizant of that power. It is also noteworthy that the corrections subculture, like its police counterpart, provides several positive qualities, particularly in crisis situations, including mutual support and protection, which is essential to the emotional and psychological health of officers involved; there is always the "family" to support you.

ETHICS TESTS FOR THE CRIMINAL JUSTICE STUDENT

Following are some tests to help guide you, the criminal justice student, to decide what is and is not ethical behavior:[57]

- *Test of common sense*: Does the act make sense, or would someone look askance at it?
- *Test of publicity*: Would you be willing to see what you did highlighted on the front page of the local newspaper?
- *Test of one's best self*: Will the act fit the concept of oneself at one's best?
- *Test of one's most admired personality*: What would one's parents or minister do in this situation?
- *Test of hurting someone else*: Will it cause pain for someone?
- *Test of foresight*: What is the long-term likely result?

Other questions that a criminal justice practitioner might ask are "Is it worth my job and career?" and "Is my decision legal?"

Another tool is that of "the bell, the book, and the candle." Ask yourself these questions: Do bells or warning buzzers go off as I consider my choice of actions? Does it violate any laws or codes in the statute or ordinance books? Will my decision withstand the light of day or spotlight of publicity (the candle)?[58] In sum, all we can do is seek to make the best decisions we can and be a good person and a good justice system employee, one who is consistent and fair. We need to apply the law, the policy, the guidelines, or whatever it is we dispense in our occupation without bias or fear and to the best of our ability, being mindful along the way that others around us may have lost their moral compass and attempt to drag us down with them. To paraphrase Franklin Delano Roosevelt, "Be the best you can, wherever you are, with what you have."

In closing, it might be good to mention that ethics is important to all criminal justice students and practitioners, not only because of the moral/ethical issues and dilemmas they confront each day but also because they have a lot of discretion with the people with whom they are involved—such as the discretion to arrest or not arrest, to charge or not charge, to punish or not punish, and even to shoot or not shoot.

©iStockphoto.com/terrymorris, ©iStockphoto.com/DNY59, ©iStockphoto.com/maerapaso

"The bell, the book, and the candle" test can be used as a guide against making unethical decisions.

/// IN A NUTSHELL

- Ethics involves doing what is right or correct in a professional capacity. Deontological ethics considers one's duty to act. Immanuel Kant expanded the ethics of duty by including the idea of "good will." People's actions must be guided by good intent.
- There are two types of ethics: absolute and relative. Absolute ethics has only two sides—something is either good or bad. Relative ethics is more complicated and can have varying shades of gray. What is considered ethical behavior by one person may be deemed highly unethical by someone else.
- The "principle of double effect"—also known as noble cause—refers to the commission of an unethical act in order to achieve a good outcome.
- The Knapp Commission identified two types of corrupt police officers: the "meat-eaters" and the "grass-eaters." The branch of the department and the type of assignment affect opportunities for corruption. Officers' code of silence can interfere with the efforts of police leadership to uncover police corruption.
- The Law Enforcement Code of Ethics (LECE) was adopted in 1957; more recently a separate, shorter code was adopted, which is easier for officers to remember when they come face-to-face with an ethical dilemma; it is the Law Enforcement Oath of Honor.
- Accepted lying includes police activities intended to apprehend or entrap suspects. This type of lying is generally considered to be trickery. Deviant lying occurs when officers commit perjury to convict suspects or are deceptive about some activity that is illegal or unacceptable.
- There are two basic arguments *against* police acceptance of gratuities. First is the slippery slope argument, which proposes that gratuities are the first step in police corruption. In addition, when officers accept minor gifts or gratuities, they may then be obligated to provide the donors with some special service or accommodation.
- The first call for formalized standards of conduct in the legal profession came in 1906, with Roscoe Pound's speech "The Causes of Popular Dissatisfaction with the Administration of Justice." In 1924, the ABA approved the Canons of Judicial Ethics, and in 1972, the ABA approved a new Model Code of Judicial Conduct; in 1990, the same body adopted a revised model code. Nearly all states and the District of Columbia have established standards based on the code.
- In 1974, the United States Judicial Conference adopted a Code of Conduct for federal judges, and Congress over the years has enacted legislation regulating judicial conduct.
- Judges can engage in several types of abuses of judicial power, such as showing favoritism or bias, not being impartial, engaging in conflicts of interest, and being unethical in their personal conduct.
- Prosecutors and defense attorneys, too, must be legally and morally bound to ethical principles as agents of the courts.
- Federal employees are governed by the Ethics Reform Act of 1989, which addresses the receipt of gifts, financial and personal conflicts, and outside activities or employment that conflicts with their federal duties.
- Corrections personnel confront many of the same ethical dilemmas as police personnel.
- The strength of the corrections subculture correlates with the security level of a correctional facility and is strongest in maximum-security institutions.
- Some simple ethical tests can guide criminal justice students and practitioners in working through ethical quandaries.

/// KEY TERMS & CONCEPTS

Review key terms with eFlashcards at **edge.sagepub.com/peakbrief**.

Absolute ethics 56
Accepted lying 59
Deontological ethics 55
Deviant lying 59
Gratuities 60
Model Code of Judicial Conduct 64
Noble cause corruption 57
Police corruption 58
Relative ethics 56
Slippery slope 59
Utilitarianism 56
Whistleblower Protection Act 66

/// REVIEW QUESTIONS

Test your understanding of chapter content. Take the practice quiz at **edge.sagepub.com/peakbrief**.

1. How would you define *ethics?*
2. What are examples of relative and absolute ethics?
3. What specific examples of legitimate ethical dilemmas arise with police, courts, and corrections practitioners in the course of their work?

4. How would you describe the codes and canons of ethics that exist in police departments, courts, and corrections agencies? What elements do they have in common?
5. How does the principle of double effect pose problems for criminal justice and society?
6. Why was the Law Enforcement Oath of Honor developed, and how does it differ from the Code of Ethics?
7. What constitutes police corruption? What are its types, and what are the most difficult ethical dilemmas presented in this chapter? Consider the issues presented in each dilemma.
8. Do you believe criminal justice employees should be allowed to accept minor gratuities? Explain your response as well as pros and cons for doing so.
9. In what ways can judges, defense attorneys, and prosecutors engage in unethical behaviors?
10. What forms of behavior by correctional officers in prisons or jails may be unethical?
11. What are some of the ethics "tests" for criminal justice students?

/// LEARN BY DOING

Following are several brief case studies (based on actual occurrences) involving criminal justice employees. Having read this chapter's materials, determine for each case the ethical dilemmas involved and what you believe is the appropriate outcome.

1. You are sitting next to a police officer in a restaurant. When the officer attempts to pay for the meal, the waiter says, "Your money is no good here. An officer just visited my son's school and made quite an impression. Plus, I feel safer having cops around." The officer again offers to pay, but the waiter refuses to accept payment. The police department has a policy prohibiting the acceptance of free meals or gifts.
2. A judge often makes inappropriate sidebar comments and uses sexist remarks or jokes in court. For example, a woman was assaulted by her husband who beat her with a telephone; from the bench the judge said, "What's wrong with that? You've got to keep her in line once in a while." He begins to address female lawyers in a demeaning manner, using such terms as *sweetie, little lady lawyer,* and *pretty eyes.*[59]
3. (a) An associate warden and "rising star" in the local prison system has just been stopped and arrested for driving while intoxicated in his personal vehicle and while off duty. There are no damages or injuries involved, he is very remorseful, and he has just been released from jail. His wife calls you, the warden, pleading for you to allow him to keep his job. (b) One week later, this same associate warden stops at a local convenience store after work; as he leaves the store, a clerk stops him and summons the police—the individual has just been caught shoplifting a package of cigarettes. You have just been informed of this latest arrest.

/// STUDY SITE

Get the tools you need to sharpen your study skills. SAGE edge offers a robust online environment featuring an impressive array of free tools and resources.

Access practice quizzes, eFlashcards, video, and multimedia at **edge.sagepub.com/peakbrief**

©iStockphoto.com/tillsonburg

THE POLICE

PART 2

This part consists of three chapters. **Chapter 4** discusses the organization and operation of law enforcement agencies at the federal, state, and local (city and county) levels. Included are discussions of their English and colonial roots, the three eras of U.S. policing, and brief considerations of INTERPOL and the field of private security.

Chapter 5 focuses on the kinds of work that police do. After looking at their recruitment and training, considered next are the roles, styles, and tasks of the police as well as patrol and investigative functions, generally. Also included are the dangers of the job, the use of police discretion, and the work of criminal investigators. We also broadly examine several policing issues that exist today, including calls for police reform regarding the use of force, greater diversity in police agencies, and body-worn cameras.

Chapter 6 examines the constitutional rights of the accused as well as limitations placed on the police under the Fourth, Fifth, and Sixth Amendments; the focus is on arrest, search and seizure, the right to remain silent, and the right to counsel.

©iStockphoto.com/Vilches

CHAPTER 4

POLICE ORGANIZATION

Structure and Functions

We are born in organizations, educated by organizations, and most of us spend much of our lives working for organizations. We spend much of our leisure time paying, playing, and praying in organizations. Most of us will die in an organization, and when the time comes for burial, the largest organization of all—the state—must grant official permission.

—Amitai Etzioni

LEARNING OBJECTIVES

As a result of reading this chapter, you will be able to

1. Describe how policing began in England, to include the four major police-related offices that evolved and came to America
2. Discuss the three eras of policing and August Vollmer's major contributions to advancing early policing
3. Explain the duties and functions of selected federal law enforcement agencies and the departments into which they are organized
4. Delineate the various types of specialized functions that are found in state law enforcement agencies
5. Distinguish between the functions of municipal police and county sheriff's agencies
6. Describe the basic qualifications and attributes for one to obtain a career in a law enforcement agency
7. Explain how and why private policing was developed, and the contemporary purposes and issues of the private police

ASSESS YOUR AWARENESS

Test your knowledge of police structure and functions by responding to the following nine true-false items; check your answers after reading this chapter's materials.

1. The four primary criminal justice officials of early England—sheriff, constable, coroner, and justice of the peace—remain in existence today.
2. The "architect" and "crib" of policing—the person and agency where most initial practices were developed—were J. Edgar Hoover, in the Philadelphia Police Department.
3. The creation of the Department of Homeland Security in 2002 (and the concurrent reorganization of several major federal law enforcement departments and agencies) was the most significant transformation of the U.S. government in over a half century.
4. Full-time professional policing, as it is generally known today, began in the early 1900s in New York City.
5. The patrol function may be said to represent the backbone of policing.
6. State law enforcement agencies typically perform only one function: patrolling state highways.
7. INTERPOL is the oldest, best-known, and probably only truly international crime-fighting organization.
8. Private police (security officers) greatly outnumber the public police, and while some of their duties are similar, the overall powers of private police are entirely different.
9. For many good reasons, there is a strong likelihood that, within the next 10 years, all municipal and county police agencies will be absorbed and consolidated into a single, national, centralized force.

<< **Answers can be found on page 293.**

Most European and many Asian countries have police organizations that are centralized into one national police force. In the United States, however—where the citizens have always been much more fearful and mistrusting of a strong, centralized government—there has always been a tradition of utilizing a fragmented, localized approach to policing. This view has resulted in the decentralized system of policing that is seen today in its more than 18,000 individual municipal and county police agencies, in addition to those at the federal and state levels. Police jurisdictions are guarded zealously; this means that federal law enforcement agencies in the United States have virtually no authority over local police departments or their investigations unless a federal offense is involved. This fragmented policing model is favored by many or most people for several reasons.

First and foremost, they argue that local policing allows greater knowledge and solutions to local problems and that it provides local accountability, with officers being answerable to the citizens of that locality rather than to some remote federal or state authority (indeed, municipal police chiefs typically serve at the pleasure of the governing boards and county sheriffs are generally elected by the citizens at large). This model is also felt to spur innovation: Almost all of the recent advances in community policing, problem solving, smart policing, and crime prevention were initiated at the local level. Those who favor partial or total consolidation of police services, however, point to its potential advantages, including likely cost savings achieved by combining expensive services (e.g., jail, communications, equipment) and economies of scale in purchasing larger quantities of vehicles, uniforms, radios, and other equipment; consistency across the jurisdiction in terms of standards for recruitment and training; uniformity of policies and procedures; direct lines of communication; possibly better training, investigative, and other functions; coordination of efforts; and ease of information and data sharing.[1]

As you read this chapter and, in particular, study our federal system of law enforcement, consider whether or not there should be a single, national police force as is found in many other countries. Would such a system work in the United States? Would there be political obstacles to its creation? Other impediments?

INTRODUCTION

How did the police come into being? How are they "designed" and organized to accomplish their mission? These are legitimate questions.

This chapter's primary aims are (1) to look at the early development of policing and (2) to inform the reader about the role and functions of contemporary federal, state, and local agencies, as well as the private police.

The chapter begins with a brief history of the evolution of four primary criminal justice officers—sheriff, constable, coroner, and justice of the peace—from their roots in early England to their coming to America. Following is an overview of the three eras of policing in the United States, and the events and shortcomings that were a part of the first two eras, thus leading to today's *community era*. Discussed next are the structures and functions of selected major federal law enforcement agencies and **organizations** that compose the Department of Homeland Security and Department of Justice. After a discussion of the primary duties of state-level law enforcement organizations, next is a review of local law enforcement—municipal police and county sheriff's agencies. The world's premier international crime-fighting organization, INTERPOL, is described briefly; and the chapter concludes with general discussions of how to pursue a law enforcement career as well as the field of private policing.

Organization: an entity of two or more people who cooperate to achieve one or more objectives.

Note that this chapter's descriptions of federal law enforcement agencies' roles, composition, and organization may be modified by the time of this textbook's publication, particularly if the Trump administration opts to revise some agencies' policies, priorities, and practices.

ENGLISH AND COLONIAL ROOTS: AN OVERVIEW

All four of the primary criminal justice officials of early England—sheriff, constable, coroner, and justice of the peace—either still exist or existed until recently in the United States. Accordingly, it is important to grasp a basic understanding of these offices, including their early functions in England and, later, in America.

Sheriff

Sheriff: the chief law enforcement officer of a county, typically elected and frequently operating the jail as well as law enforcement functions.

The word **sheriff** is derived from *shire-reeve—shire* meaning "county" and *reeve* meaning "agent of the king." The shire-reeve appeared in England before the Norman Conquest of 1066. His job was to maintain law and order in the tithings (groupings of ten households). At the present time in England, a sheriff's only duties are to act as an officer of the court, to summon juries, and to enforce civil judgments.[2]

The first sheriffs in America appeared in the early colonial period. Today, the American sheriff remains the basic source of rural crime control. Sheriff's offices are discussed more later.

Constable

Like the sheriff, the **constable** can be traced back to Anglo-Saxon times. The office began during the reign of Edward I, when every parish or township had a constable. As the county militia turned more and more to matters of defense, only the constable pursued felons. Hence the ancient custom of citizens raising a loud "hue and cry" and joining in pursuit of criminals lapsed into disuse. The constable had a variety of duties, including collecting taxes, supervising highways, and serving as magistrate. The office soon became subject to election and was conferred upon local men of prominence. The creation of the office of justice of the peace around 1200 quickly changed this trend forever; soon the constable was limited to making arrests only with warrants issued by a justice of the peace. As a result, the office, deprived of social and civic prestige, was no longer attractive. It carried no salary, and the duties were often dangerous.[3] In the American colonies, the position fell into disfavor largely because most constables were untrained and believed to be wholly inadequate as officials of the law.[4]

Constable: in England, favored noblemen who were forerunners of modern-day U.S. criminal justice functionaries; largely disappeared in the United States by the 1970s.

Coroner

The office of **coroner** has been used to fulfill many different roles throughout its history and has changed steadily over the centuries since it began functioning by the end of the 12th century. From the beginning, the coroner was elected; his duties included oversight of the interests of the Crown, including criminal matters. In felony cases, the coroner could conduct a preliminary hearing, and the sheriff often came to the coroner's court to preside over the coroner's jury. This "coroner's inquest" determined the cause of death and the party responsible for it. Initially, coroners were given no compensation, yet they were elected for life. Soon, however, they were given the right to charge fees for their work.[5] The office was slow in gaining recognition in America, as many of the coroners' duties were already being performed by the sheriffs and justices of the peace. By 1933, the coroner was recognized as a separate office in two-thirds of the states. By then, however, the office had been stripped of many of its original functions. Today, in many states, the coroner legally serves as sheriff when the elected sheriff is disabled or disqualified. However, since the early part of the 21st century, the coroner has basically performed a single function: determining the causes of all deaths by violence or under suspicious circumstances.[6]

Coroner: an early English court officer; today one (usually a physician) in the United States whose duty it is to determine cause of death.

Justice of the Peace

The **justice of the peace (JP)** can be traced back as far as 1195 in England. Early JPs were wealthy landholders. The duties of JPs eventually included the granting of bail to felons, which led to corruption and criticism as the justices bailed out people who clearly should not have been released into the community. By the 16th century, the office came under criticism again because of the caliber of the people holding it (wealthy landowners who bought their way into office).[7] By the early 20th century, England had abolished the property-holding requirement, and many of the medieval functions of the JP's office were removed. Thereafter, the office possessed strictly extensive criminal jurisdiction but no civil jurisdiction whatsoever. This contrasts with the American system, which gives JPs limited jurisdiction in both criminal and civil cases. By 1930, the office had constitutional status in all of the states. JPs have long been allowed to collect fees for their services. As in England, it is typically not necessary to hold a law degree or to have pursued legal studies in order to be a JP in the United States.[8]

Justice of the peace (JP): a minor justice official who oversees lesser criminal trials; one of the early English judicial functionaries.

Police Reform in England, 1829

In England, after the end of the Napoleonic Wars in 1815, workers protested against new machines, food riots, and an ongoing increase in crime. The British army, traditionally

Practitioner's Perspective

Police Lieutenant

Name: Brian D. Fitch, Ph.D.
Position: Police Lieutenant (retired), Los Angeles County Sheriff's Department
Location: Los Angeles, California

What are the primary duties and responsibilities of a practitioner in this position? In most law enforcement agencies, particularly municipal agency sheriff departments, the core function of what we do is patrol—what we call field operations. That's the officer or the deputy on the street, driving around in a marked vehicle, interacting with the public, enforcing the law, writing traffic citations, taking police reports, investigating crimes. That is, by far, the bulk of what we do. The public expects us to staff our shift with enough people to respond to their calls in a timely manner. They expect us to investigate a crime as soon as it's occurred. So the bulk of what officers do is devoted to the first responder kind of fieldwork.

Law enforcement officers can be involved in anything from—for example, where I'm currently assigned, we're constantly dealing with all kinds of things, from homicide, to a body dump, to investigating a traffic citation, to arresting a drunk driver. But most of those functions are what we call line functions. Only a small percentage of a given agency is assigned to do detective work or to follow-up work. And that percentage becomes even smaller when you look at specialties with the detective work, like homicide. Some cities have relatively small populations, or have a particularly small rate of crimes, and those cities may not have any homicides in a given year. So that doesn't lend itself towards having a whole team of homicide investigators.

Many smaller agencies will employ the resources of a larger agency here within Los Angeles County. The Los Angeles County Sheriff's Department homicide unit investigates homicides for a number of smaller municipal police agencies, some of whom have 80, 90, or 100 officers. They still defer to the sheriff's department because of the amount of time and resources involved. And some of these investigations get very lengthy. If you look at the idea of becoming a profiler, which has become popular on TV, you'll see that the LA County Sheriff's Department, an agency of 10,000 sworn people, has one profiler assigned. At the LAPD, we have one profiler assigned. So within the whole Southern California basin of about 10 million people, we have about two profilers. So landing a job like that is as much luck and timing as is anything else.

What are some challenges you face in this position? Going back historically, police officers were familiar with people on their beat, like shop owners, residents, they got to talk with people on a daily basis. They were able to develop relationships, and that was important because when people felt they needed to report a crime, they knew exactly where to go. So now, when we put our officers in cars, we're able to cover a lot of geographic area and we increase our overall visibility, but we lose that relationship with the community. Law enforcement can't do its job without the community. I believe it was Sir Robert Peele who said, "The police are the public and the public are the police." In order for us to do our job effectively, law enforcement can't be everywhere and we can't be all-knowing and all-doing. Law enforcement needs and must have both the support and involvement of the community. So this move toward community policing is a move toward involving the community in improving quality of life, helping law enforcement officers solve crimes, and just generally trying to rekindle that relationship between law enforcement and the people who we are sworn to serve and protect.

To learn more about Brian Fitch's experiences as a Police Lieutenant, watch the Practitioner's Perspective video in SAGE Vantage.

used to disperse rioters, was becoming less effective as people began resisting its commands. In 1822, England's ruling party, the Tories, moved to consider new alternatives. The prime minister appointed Sir Robert Peel to establish a police force to combat the problems. Peel, a wealthy member of Parliament,[9] finally succeeded in 1829 when Parliament passed the Metropolitan Police Act. The London police are nicknamed "bobbies" after Sir Robert Peel.[10] Peel's remark that "the police are the public, and the public are the police" emphasized his belief that the police are first and foremost members of the larger society.[11]

Hulton Archive/Stringer/Hulton Archive/Getty Images

Sir Robert Peel's "bobbies" were established in London in 1829.

POLICING COMES TO THE UNITED STATES

This section provides an overview of the three eras of U.S. policing—political, reform, and community—that were shaped and defined by the varying goals and philosophies of each over time.

The Political Era, 1840s–1930s

In 1844, the New York state legislature passed a law establishing a full-time preventive police force for New York City. However, this new body came into being in a very different form than in Europe. The American version, as begun in New York City, was deliberately placed under the control of the city government and city politicians. The American plan required that each ward in the city be a separate patrol district, unlike the European model, which divided the districts along the lines of criminal activity. The process for selecting officers was also different. The mayor chose the recruits from a list of names submitted by the aldermen and tax assessors of each ward; the mayor then submitted his choices to the city council for approval. This system resulted in most of the power over the police going to the ward aldermen, who were seldom concerned about selecting the best people for the job. Instead, the system allowed and even encouraged political patronage and rewards for friends.[12]

This was the **political era** of policing, from the 1840s to the 1930s. Politics were played to such an extent that even nonranking patrol officers used political backers to obtain promotions, desired assignments, and transfers.

Political era: from the 1840s to the 1930s, the period of time when police were tied closely to politics and politicians, dependent on them for being hired and promoted, and for assignments—all of which raised the potential for corruption.

Police corruption also surfaced at this time. Corrupt officers wanted beats close to the gamblers, saloonkeepers, madams, and pimps—people who could not operate if the officers were "untouchable" or "100 percent coppers."[13] Political pull for corrupt officers could work for or against them; the officer who incurred the wrath of his superiors could be transferred to the outposts, where he would have no chance for financial advancement.

Still, it did not take long for other cities to adopt the general model of the New York City police force. New Orleans and Cincinnati adopted plans for a new police force in 1852, Boston and Philadelphia followed in 1854, Chicago in 1855, and Baltimore and Newark in 1857.[14] By 1880, virtually every major American city had a police force based on Peel's model.

The Reform Era, 1930s–1980s

During the early 20th century, reformers sought to reject political involvement by the police, and civil service systems were created to eliminate patronage and ward influences in hiring and firing police officers. In some cities, officers were not permitted to live in the same beat they patrolled in order to isolate them as completely as possible from political influences.[15] However, policing also became a matter viewed as best left to the discretion of police executives. Any noncrime activities required of police were considered "social work." The **reform era** (also termed the professional era) of policing would soon be in full bloom.

Reform era: also the professional era, from the 1930s to 1980s, when police sought to extricate themselves from the shackles of politicians, and leading to the crime-fighter era—with greater emphases being placed on *numbers*—arrests, citations, response times, and so on.

Figure 4.1 /// The Crib of Modern Law Enforcement: August Vollmer and the Berkeley Police Department

1905	Vollmer is elected Berkeley town marshal.
1906	Trustees create detective rank. Vollmer initiates a red light signal system to reach beat officers from headquarters; telephones are installed in boxes. A police records system is created.
1908	Hulton-Deutsch/Corbis Historical/Getty Images Two motorcycles are added to the department. Vollmer begins a police school.
1909	Trustees approve the appointment of a Bertillon expert and the purchase of fingerprinting equipment. A modus operandi file is created, modeled on the British system.
1911	All patrol officers are using bicycles.
1914	Three privately owned autos are authorized for patrol use.
1915	A central office is established for police reports.
1916	Vollmer urges Congress to establish a national fingerprint bureau (later created by the FBI in Washington, D.C.), begins annual lectures on police procedures, and persuades a biochemist to install and direct a crime laboratory at police headquarters.
1917	Vintage Images/Alamy Vollmer guides the development of the first lie detector and begins developing radio communications between patrol cars, handwriting analysis, and use of business machine equipment (a Hollerith tabulator).
1918	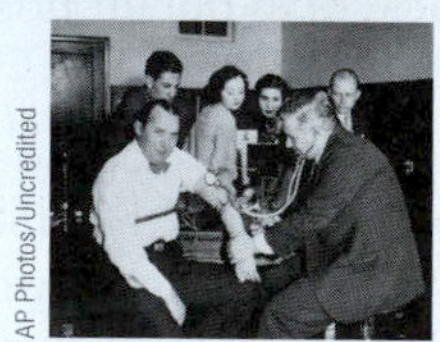 AP Photos/Uncredited Entrance examinations are initiated to measure the mental, physical, and emotional fitness of recruits; a part-time police psychiatrist is employed.
1919	Vollmer begins testing delinquents and using psychology to anticipate criminal behavior. He implements a juvenile program to reduce child delinquency.
1921	Vollmer has the first completely motorized force; officers furnish their own automobiles. Vollmer recruits college students for part-time police jobs.

The policing career of August Vollmer has been established as a major factor in the shaping and development of police professionalism, or the reform era (see Figure 4.1). In April 1905, at age 29, Vollmer became the town marshal in Berkeley, California; as indicated earlier, this was a time when police departments were notorious for their corruption and politics.

Vollmer commanded a force of only 3 deputies; his first act as town marshal was to request an increase in his force from 3 to 12 deputies in order to form day and night patrols.[16] Obtaining that, he soon won national publicity for being the first chief to order his men to patrol on bicycles (his research demonstrated that officers on bicycles would be able to respond three times more quickly to calls for service than men on foot). Vollmer then persuaded the Berkeley City Council to purchase a system of red lights. The lights, hung at each street intersection, served as an emergency notification system for police officers—the first such signal system in the country.[17]

Library of Congress, Prints & Photographs Division LC-USZ62-69954

Chief August Vollmer developed many "firsts" in the Berkeley, California, police department during the professional (reform) era of policing.

In 1906, Vollmer began to question the suspects he arrested, finding that nearly all criminals used their own peculiar method of operation, or modus operandi. In 1907, he sought the advice of a professor of biology at the University of California, becoming convinced of the value of scientific knowledge in criminal investigation.[18] Vollmer's most daring innovation, however, came in 1908 with the idea of a formal police school that drew on the expertise of university professors and included courses on police methods and procedures, fingerprinting, first aid, criminal law, anthropometry, photography, and public health. In 1916, he persuaded a professor of pharmacology and bacteriology to become a full-time criminalist supervising the department's criminal investigation laboratory. By 1917, Vollmer had his entire patrol force operating out of automobiles; it was the first completely mobile patrol force in the country.[19] In 1918, to improve the quality of police recruits in his department, Vollmer began to hire college students as part-time officers and to administer a set of intelligence, psychiatric, and neurological tests to all applicants (out of this group of "college cops" came several outstanding and influential police leaders across the United States). Finally, in 1921, in addition to experimenting with the lie detector, two of Vollmer's officers installed a crystal set and earphones in a Model T touring car, thus creating the first radio car.[20]

The crime-fighter image gained popularity under the reform model of policing, when officers were to remain in their "rolling fortresses," going from one call to the next with all due haste. Much police work was driven by *numbers*—numbers of arrests and calls for service, response time to calls for service, numbers of tickets written and miles driven by a patrol officer during a duty shift, and so on. For many people, like Los Angeles chief of police William Parker, the police were the "thin blue line," protecting society from barbarism. Parker viewed urban society as a jungle, needing the restraining hand of the police; the police had to enforce the law without fear or favor. Parker opposed any restrictions on police methods. The law, he believed, should give the police wide latitude to use wiretaps and to conduct search and seizure.[21]

The Community Era, 1980s–Present

Today, policing is in the **community era**, practicing community policing and problem solving. This strategy was born because of the problems that overwhelmed the reform era, beginning in the 1960s. During the reform era, officers had little long-term effect in dealing with crime and disorder; they were neither trained nor encouraged to consider the underlying causes of problems on their beats.[22]

Community era: beginning in about 1980, a time when the police retrained to work with the community to solve problems by looking at their underlying causes and developing tailored responses to them.

In addition, several studies struck at the very heart of traditional police methods. For example, it was learned that response time had very little to do with whether or not an arrest was made at the scene; detectives were greatly overrated in their ability to solve crimes;[23] less than 50 percent of an officer's time was committed to calls for service; and of those calls handled, more than 80 percent were noncriminal incidents.[24] As a result, a new "common wisdom" of policing came into being. The police were trained to work with the community to solve problems by looking at their underlying causes and developing tailored responses to them. Community policing and problem solving are discussed more thoroughly in Chapter 5.

FEDERAL LAW ENFORCEMENT

Federal law enforcement agency personnel make more than 150,000 arrests per year. More than half (46 percent) are for immigration crimes, most of which are in the five judicial districts along the U.S.-Mexican border. About one in seven (15 percent) of all federal arrests are for drug offenses, and another 15 percent are for supervision violations (failure to appear, probation, and so on).[25]

Federal law enforcement agencies: federal organizations that, for example, are charged with protecting the homeland (DHS); investigating crimes (FBI) and enforcing particular laws, such as those pertaining to drugs (DEA) or alcohol/tobacco/firearms/explosives (ATF); and guarding the courts and transporting prisoners (USMS).

Most **federal law enforcement agencies** are found within the Department of Homeland Security (DHS) and the Department of Justice (see Table 4.1), both of which are discussed in this section (note that the Central Intelligence Agency, or CIA, works as an independent agency).

Department of Homeland Security

Within one month of the terrorist attack on U.S. soil on September 11, 2001, President George W. Bush issued a proposal to create a new Department of Homeland Security (DHS)—which was established in November 2002 and became the most significant transformation of the U.S. government in over a half century. All or parts of 22 different federal departments and agencies were combined, beginning in January 2003, and 80,000 new federal employees were immediately put to work.[26] Congress committed $32 billion toward safeguarding the nation, developing vaccines to protect against biological or chemical threats, training and equipping first responders (local police, firefighters, and medical personnel), and funding science and technology projects to counter the use of biological weapons and assess vulnerabilities. In the 10 years after 9/11, more than $635 billion was appropriated by the federal government to support homeland security efforts.[27] Figure 4.2 shows the current organizational structure of the DHS and its 230,000 employees.

Following are brief descriptions of the major law enforcement agencies that are organizationally located within the DHS; all of them have full law enforcement authority (Note: a number of additional federal agencies have law enforcement authority and are not discussed here, including the Federal Bureau of Prisons, U.S. Forest Service, U.S. Fish and Wildlife Service, U.S. Capitol Police, U.S. Mint Police, and the Bureau of Indian Affairs.]

Table 4.1 /// Selected Federal Law Enforcement Agencies

Department of Homeland Security
Citizenship and Immigration Services
Customs and Border Protection (CBP)
Immigration and Customs Enforcement (ICE)
Secret Service
Transportation Security Administration
Coast Guard
Department of Justice
Federal Bureau of Investigation (FBI)
Bureau of Alcohol, Tobacco, Firearms and Explosives (ATF)
United States Marshals Service (USMS)
Drug Enforcement Administration (DEA)
Bureau of Prisons (BOP)

Figure 4.2 /// Department of Homeland Security Organizational Structure

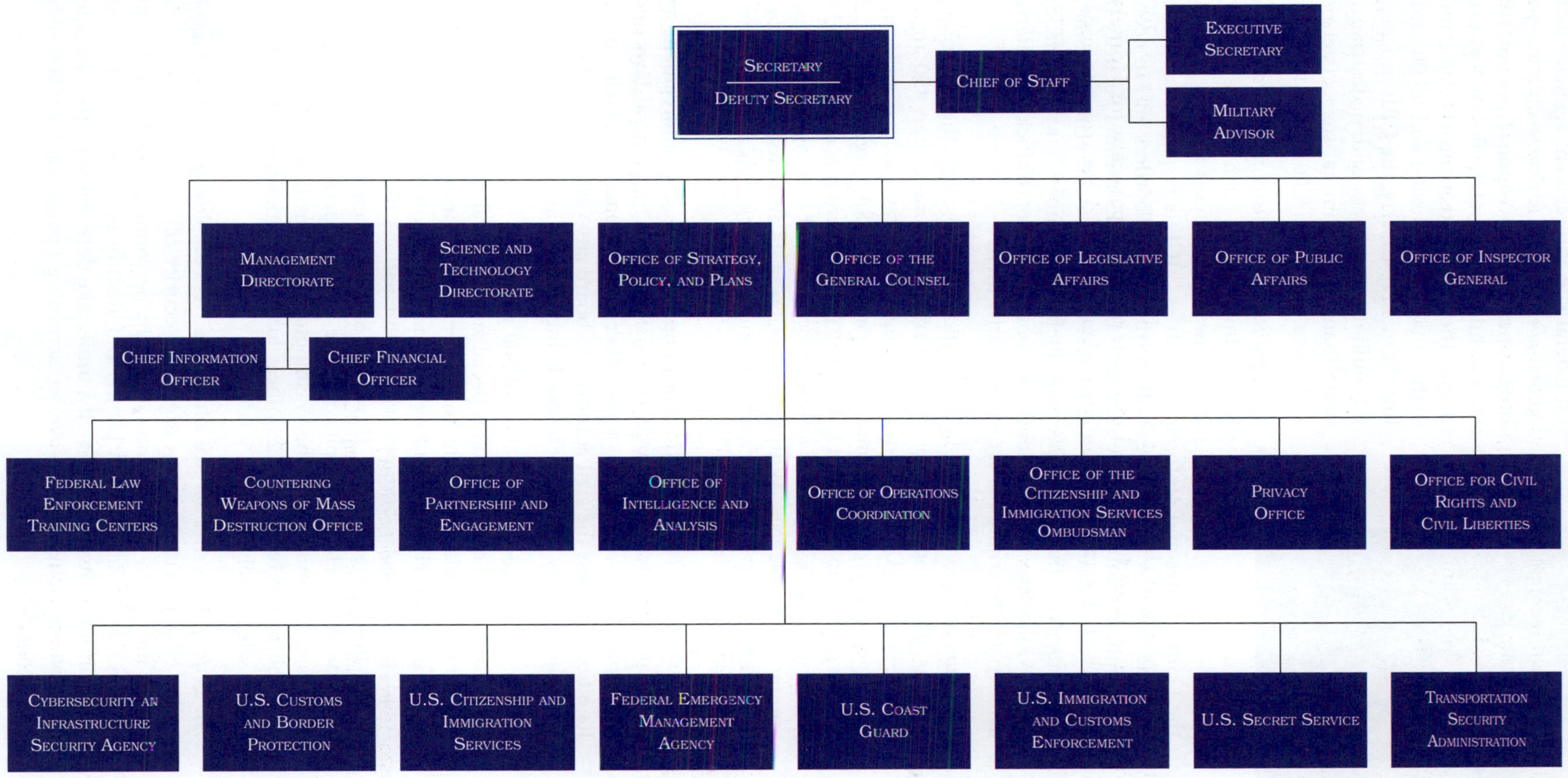

Secretary
Deputy Secretary

Chief of Staff

Executive Secretary

Military Advisor

Management Directorate

Science and Technology Directorate

Office of Strategy, Policy, and Plans

Office of the General Counsel

Office of Legislative Affairs

Office of Public Affairs

Office of Inspector General

Chief Information Officer

Chief Financial Officer

Federal Law Enforcement Training Centers

Countering Weapons of Mass Destruction Office

Office of Partnership and Engagement

Office of Intelligence and Analysis

Office of Operations Coordination

Office of the Citizenship and Immigration Services Ombudsman

Privacy Office

Office for Civil Rights and Civil Liberties

Cybersecurity an Infrastructure Security Agency

U.S. Customs and Border Protection

U.S. Citizenship and Immigration Services

Federal Emergency Management Agency

U.S. Coast Guard

U.S. Immigration and Customs Enforcement

U.S. Secret Service

Transportation Security Administration

Source: Department of Homeland Security

John Moore/Getty Images News/Getty Images

Homeland Security comprises multiple law enforcement agencies, including officers performing both customs and border protection duties.

- *Customs and Border Protection (CBP)* is one of the largest federal law enforcement agencies, with more than 60,000 agents. CBP is responsible for preventing terrorists and terrorist weapons from entering the United States while facilitating the flow of legitimate trade and travel. The CBP protects nearly 7,000 miles of border with Canada and Mexico and 95,000 miles of shoreline.[28] On a typical day, the CBP processes more than 1.1 million passengers and pedestrians, apprehends more than 1,100, criminal suspects at ports of entry, and seizes more than 9,000 pounds of narcotics.[29]
- *Immigration and Customs Enforcement (ICE)* is the largest investigative arm of the DHS, with about 20,000 sworn employees in more than 400 offices in the United States and 46 foreign countries.[30] ICE is responsible for identifying and shutting down vulnerabilities both in the nation's borders and in economic, transportation, and infrastructure security.[31] In 2010, a Homeland Security Investigations agency was formed within ICE to investigate financial crimes, money laundering and smuggling, intellectual property theft, cybercrimes; human trafficking and other rights violations, narcotics and weapons smuggling/trafficking, international gang activity, and international art and antiquity theft.[32]
- *The Transportation Security Administration (TSA)* protects the nation's transportation systems. TSA employs about 50,000 personnel at more than 450 airports and field offices. With the U.S. economy improving and air travel increasing, agents are screening approximately two million people each day. TSA's organization includes federal air marshals, who are armed federal law enforcement officers deployed on passenger flights worldwide to protect airline passengers and crew against the risk of criminal and terrorist violence.[33] With the exception of its air marshals, TSA personnel hold the title of "officer" in name only; they neither carry guns nor have law enforcement powers.[34]
- *The Coast Guard* is the nation's leading maritime law enforcement agency; it has broad police power to enforce, or assist in enforcing, federal laws and treaties on waters under U.S. jurisdiction and other international agreements on the high seas. Its officers possess the civil authority to board any vessel subject to U.S. jurisdiction; once aboard, they can inspect, search, inquire, and arrest.[35]
- *The Secret Service* employs about 6,500 personnel worldwide (3,200 special agents, 1,300 uniformed officers, 2,000 specialized/technical support personnel).[36] Its agents protect the president and other high-level officials and investigate counterfeiting and other financial crimes, including financial institution fraud, identity theft, computer fraud, and computer-based attacks on our nation's financial, banking, and telecommunications infrastructure. The Secret Service's domestic Uniformed Division protects the White House complex and the vice president's residence as well as foreign embassies and missions in the Washington, D.C., area. The Secret Service has agents assigned to approximately 125 offices located in cities throughout the United States and in select foreign cities.[37]
- *The Federal Protective Service* within the DHS employs about 1,000 law enforcement police officers, criminal investigators, security officers, and support personnel (as well as more than 10,000 contract security personnel); they focus on federal facilities (e.g., buildings, dams, power plants) and identify and assess threats, conduct surveillance, monitor suspicious activity, undertake deterrence patrols, engage in community policing activities, and oversee screening operations for 9,500 federal facilities worldwide.[38]

Because its roles and purpose are closely related to the protection of the United States against terrorism and other crimes, a brief discussion of INTERPOL is included later in this section.

INTERPOL: the only international crime-fighting organization; it collects intelligence information, issues alerts, and assists in capturing world criminals, and it has nearly 200 member countries.

Department of Justice

The Department of Justice (DOJ) is headed by the attorney general, who is appointed by the U.S. president and confirmed by the Senate. The president also appoints the attorney general's assistants and the U.S. attorneys for each of the judicial districts. The U.S. attorneys in each judicial district control and supervise all federal criminal prosecutions and represent the government in legal suits in which it is a party. These attorneys may appoint committees to investigate other governmental agencies or offices when questions of wrongdoing are raised or when possible violations of federal law are suspected or detected.

The DOJ is the official legal arm of the U.S. government. Within the Justice Department are several law enforcement organizations that investigate violations of federal laws; we discuss the Federal Bureau of Investigation; Bureau of Alcohol, Tobacco, Firearms and Explosives; Drug Enforcement Administration; and U.S. Marshals Service in the sections that follow. Figure 4.3 shows the organizational chart for the Department of Justice.

Federal Bureau of Investigation

The Federal Bureau of Investigation (FBI) was created and funded through the Department of Justice Appropriation Act of 1908. A new era was begun for the FBI in 1924 with the appointment of J. Edgar Hoover, who was determined to professionalize the organization, as director. Special agents were college graduates, preferably with degrees in law or accounting. During Hoover's tenure in office, many notorious criminals, such as Bonnie Parker, Clyde Barrow, and John Dillinger, were tracked and captured or killed.[39] The bureau's top four priority areas are to protect the United States:

1. From terrorist attacks
2. Against foreign intelligence operations and espionage
3. Against cyberattacks and high-tech crimes
4. Against public corruption at all levels

Other priorities are to: protect civil rights, combat transnational/national criminal organizations and enterprises, combat major white-collar crime, and combat significant violent crime.[40]

Today the FBI has 56 field offices, approximately 350 resident agencies, and more than 60 foreign liaison posts called legal attachés. About 30,000 agents, analysts, and support employees perform professional, administrative, technical, and other functions.[41]

To combat terrorism, the bureau can now monitor Internet sites, libraries, churches, and political organizations. In addition, under revamped guidelines, agents can attend public meetings for the purpose of preventing terrorism.[42] The FBI's laboratory examines blood, hair, firearms, paint, handwriting, typewriters, and other types of evidence—at no charge to state police and local police agencies. The FBI also operates the National Crime Information Center (NCIC), through which millions of records—including wanted persons as well as stolen vehicles and all kinds of property items or any object with an identifying number—are entered. The FBI also publishes the annual *Uniform Crime Reports*, discussed in Chapter 2.

Figure 4.3 /// Department of Justice Organizational Chart

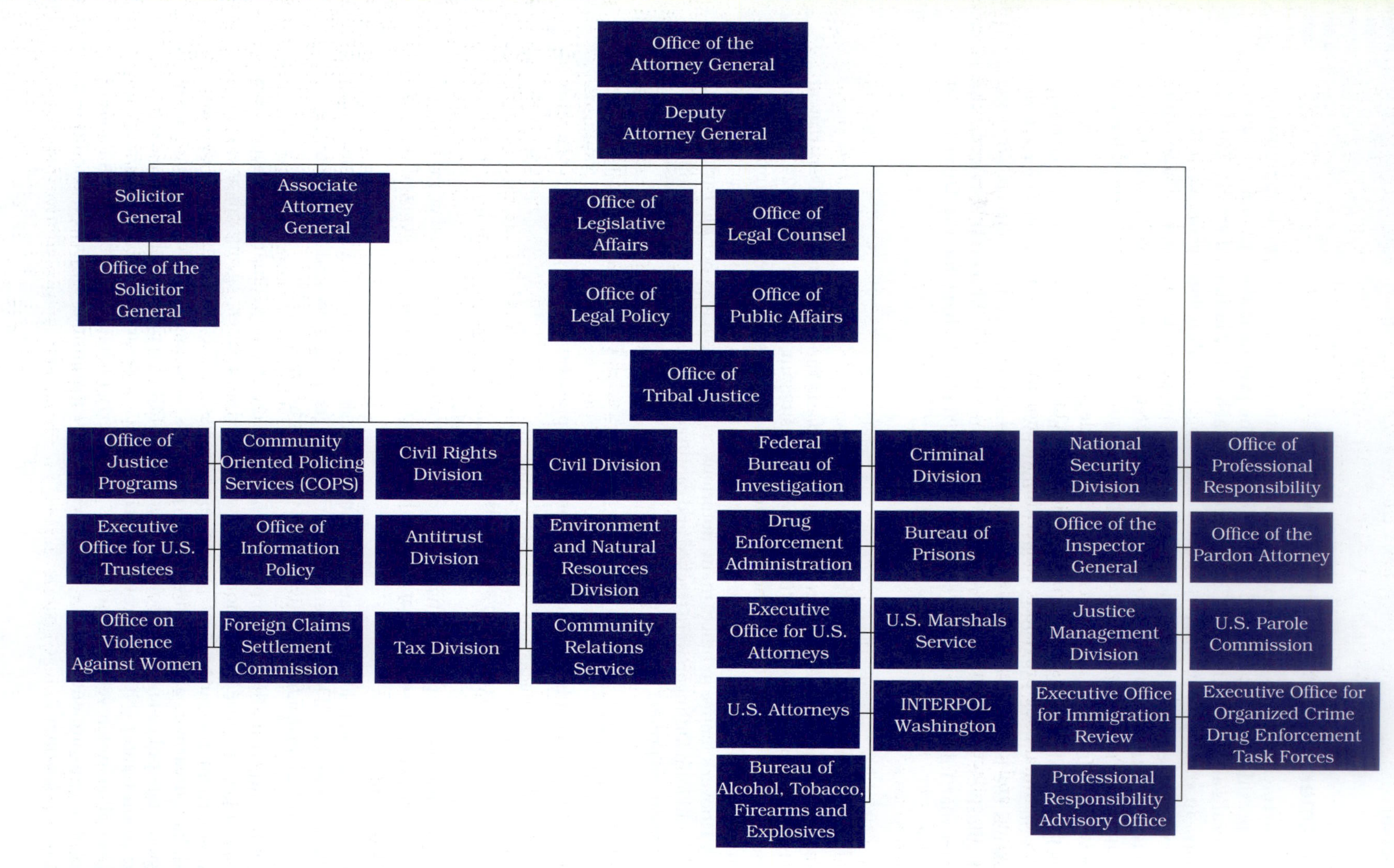

Source: Department of Justice

The FBI's Special Agent Selection System is described at https://www.fbijobs.gov/sites/default/files/how-to-apply.pdf (also see FBI job postings by clicking on the "Search Jobs" tab from the main https://www.fbijobs.gov page). Although simplistic in its appearance, the process involves a number of individual activities by applicants and FBI personnel at each phase and typically requires one year or longer to complete.

Library of Congress Prints and Photographs Division, Harris & Ewing

J. Edgar Hoover was the first director of the Federal Bureau of Investigation and served in that capacity from 1924 until his death in 1972, at age 77.

Bureau of Alcohol, Tobacco, Firearms and Explosives

The Bureau of Alcohol, Tobacco, Firearms and Explosives (ATF) originated as a unit within the Internal Revenue Service in 1862, when certain alcohol and tobacco tax statutes were created. Like the FBI and several other federal agencies, ATF has a rich and colorful history, much of which has involved capturing bootleggers and disposing of illegal whiskey stills during Prohibition. ATF administers the U.S. criminal code provisions concerning alcohol and tobacco smuggling and diversion. ATF also maintains a U.S. Bomb Data Center (to collect information on arson- and explosives-related incidents) and a Bomb and Arson Tracking System, which allows local, state, and other federal agencies to share information about bomb and arson cases.[43]

Today ATF has about 5,100 full-time employees, 2,600 of whom are special agents in 25 field divisions; agents work nearly 38,000 cases per year (approximately 93 percent of which are firearms cases, 4 percent are arson cases, 2 percent are explosives cases, and less than 1 percent are alcohol and tobacco cases).[44]

Drug Enforcement Administration

Today's Drug Enforcement Administration (DEA) had its origin with the passage of the Harrison Narcotics Tax Act, signed into law on December 17, 1914, by President Woodrow Wilson.[45] It was created as the DEA by President Richard Nixon in July 1973 in order to create a single federal agency to coordinate and enforce the federal drug laws. In 1982, the organization was given primary responsibility for drug and narcotics enforcement, sharing this jurisdiction with the FBI. Briefly, major responsibilities of the more than 5,000 DEA agents, under the U.S. Code, include the following: investigation of, and coordination with, major violators of controlled substance laws in domestic and international venues; management of a national drug intelligence program in cooperation with federal, state, local, and foreign officials; and seizure and forfeiture of assets derived from, traceable to, or intended to be used for illicit drug trafficking.[46] The DEA makes about 30,000 arrests per year, and recently its number of arrests for opioids has increased by 154 percent.[47]

U.S. Marshals Service

The U.S. Marshals Service (USMS) is one of the oldest federal law enforcement agencies, established under the Judiciary Act of 1789. Today, the USMS has 94 U.S. marshals, one for each federal court district. The USMS employs about 3,600 deputy U.S. marshals and criminal investigators, who arrest about 84,000 fugitives per year, transport federal prisoners, and protect about 2,600 federal judges and provide security for prosecutors and witnesses; they also conduct courthouse threat analyses and investigations (about 3,000 threats are received each year), and perform security, rescue, and recovery activities for natural disasters and civil disturbances.[48] Each district headquarters office is managed by a politically appointed U.S. marshal and a chief deputy U.S. marshal, who direct a staff of supervisors, investigators, deputy marshals, and administrative personnel.

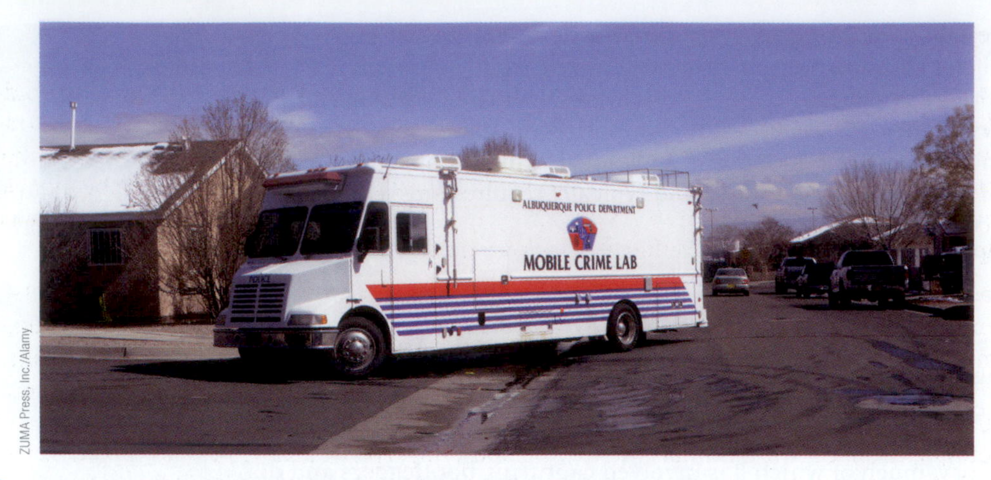

ZUMA Press, Inc./Alamy

Some agencies operate mobile forensic crime laboratories in addition to those that are fixed.

The USMS also operates the Witness Security Program. Federal witnesses are sometimes threatened by defendants or their associates; if certain criteria are met, the USMS will provide a complete change of identity for witnesses and their families, including new Social Security numbers, residences, and employment. In 1971, the USMS created the Special Operations Group (SOG), consisting of a well-trained elite group of deputy marshals that could respond to priority or dangerous missions anywhere within the United States within a few hours.[49]

Other Federal Agencies

Two other significant federal agencies outside of the Department of Justice with unique missions and contributions to the enforcement of the U.S. Code are the Central Intelligence Agency and the Internal Revenue Service.

Central Intelligence Agency

Although not a law enforcement agency, the Central Intelligence Agency (CIA) is of significance at the federal level to the nation's security and warrants a brief discussion. The National Security Act of 1947 established the National Security Council, which in 1949 created a subordinate organization, the CIA. Considered the most clandestine government service, the CIA participates in undercover and covert operations around the world for the purposes of managing crises and providing intelligence.[50] The agency offers career opportunities in five broad areas (analysis, clandestine, STEM, enterprise & support, and foreign language) as well as undergraduate student scholarship and internship opportunities.[51] Persons who are interested may also take a "Job Fit" test on the CIA website, at https://www.cia.gov/careers/opportunities/job-fit-tool.

Internal Revenue Service

The Internal Revenue Service (IRS) has as its main function the monitoring and collection of federal income taxes from American individuals and businesses. Since 1919, the IRS has had a Criminal Investigation (CI) division employing "accountants with a badge." While other federal agencies also have investigative jurisdiction for money laundering and some Bank Secrecy Act violations, the IRS is the only federal agency that can investigate potential criminal violations of the Internal Revenue Code.[52] The CI branch of the IRS comprises approximately 3,500 employees worldwide, about 2,500 of whom are special agents whose investigative jurisdiction includes tax, money laundering, and Bank Secrecy Act laws.[53]

INTERPOL

The International Criminal Police Organization, better known as INTERPOL, is the oldest, the best-known, and probably the only truly international crime-fighting organization for crimes committed on an international scale, such as drug trafficking, bank fraud, money laundering, and counterfeiting. INTERPOL agents do not patrol the globe, nor do they make arrests or engage in shootouts. They are basically intelligence gatherers who have helped many nations work together in attacking international crime since 1923.

Lyon, France, serves as the headquarters for INTERPOL's crime-fighting tasks and its 194 member countries. INTERPOL focuses on a number of crime areas that have transnational dimensions, including cybercrime, corruption, drugs, financial and high-tech crime, fugitives, maritime piracy, organized crime, terrorism (including bioterrorism), and trafficking in human beings. It also manages a range of databases with information on names and photographs of known criminals, wanted persons, fingerprints, DNA profiles, stolen or lost travel documents, stolen motor vehicles, child sex abuse images, and stolen works of art. INTERPOL also disseminates critical crime-related data through its system of international notices. There are eight kinds of notices, of which the most well known is the Red Notice, an international request for an individual's arrest.

INTERPOL has one cardinal rule: It deals only with common criminals; it does not become involved with political, racial, or religious matters. It has a basic three-step formula for offenses that all nations must follow for success: (1) Pass laws specifying the offense is a crime, (2) prosecute offenders and cooperate in other countries' prosecutions, and (3) furnish INTERPOL with and exchange information about crime and its perpetrators. This formula could reverse the trend that is forecast for the world at present: criminals' increasing capability for violence and destruction. The following crimes, because they are recognized as crimes by other countries, are covered by almost all U.S. treaties of extradition: murder, rape, bigamy, arson, robbery, burglary, forgery, counterfeiting, embezzlement, larceny, fraud, perjury, and kidnapping.[54]

STATE AGENCIES

As with federal law enforcement organizations, a variety of organizations, duties, and specializations can be found in the 50 states—although, generally, state troopers and highway patrol officers perform a lot of the same functions as their county and municipal counterparts: enforcing state statutes, investigating criminal and traffic offenses (and, by virtue of those roles, knowing and applying laws of arrest, search, and seizure), making arrests, testifying in court, communicating effectively in both oral and written contexts, using firearms and self-defense tactics proficiently, and effectively performing pursuit driving, self-defense, and lifesaving techniques until a patient can be transported to a hospital. Such agencies also maintain a wide array of special functions, including special weapons and tactics (SWAT) teams, drug units and task forces, marine and horse patrol, and so on.[55]

Patrol, Police, and Investigative Organizations

Perhaps the first distinction that might be made at the state level is between state police or highway patrol organizations and the state bureaus of investigation. Each state's statutes—and, often, the agency's name—will indicate the types of functions their agencies are authorized to perform. For example, a state's *highway patrol* division typically patrols and investigates crashes on the state highway system, while an agency that is officially named a **state police** organization typically performs a broader array of law enforcement functions in addition to those related to traffic. As an example, although the Missouri State Highway Patrol states on its website that its troopers provide assistance to motorists and "investigate highway traffic crashes and other roadway emergencies," it also states that "other responsibilities include assisting local peace officers upon request, investigating crimes, and enforcing criminal laws."[56]

State police: a state agency responsible for highway patrol and other duties as delineated in the state's statutes; some states require their police to investigate crimes against persons and property.

©John Spink/Atlanta Journal-Constitution via AP)

While many state troopers are, by statute, authorized to provide traffic control on their state's highways, others can conduct criminal investigations. Here, Georgia troopers investigate a shooting.

Some agencies are even designated as *public safety* organizations and often encompass several agencies or divisions. For example, the Hawaii Department of Public Safety, by statute, includes a law enforcement division (with general arrest duties, a narcotics division, a sheriff division, and an executive protection unit), a corrections division (handling inmate intake, incarceration, paroling authority, and industries), and a victim compensation commission.[57] **State bureaus of investigation** (SBIs), as their name implies, are investigative in nature and might be considered a state's equivalent to the FBI; they investigate all manner of cases assigned to them according to their state's laws and usually report to the state's attorney general. SBI investigators are plainclothes agents who usually investigate both criminal and civil cases involving the state and/or multiple jurisdictions. They also provide technical support to local agencies in the form of laboratory or record services and may be asked by city and county agencies to assist in investigating more serious crimes (e.g., homicide).

State bureau of investigation: a state agency that is responsible for investigating crimes involving state statutes; they may also be called in to assist police agencies in serious criminal matters, and often publish state crime reports.

Special-purpose state agencies: specially trained units for particular investigative needs, such as those for violations of alcoholic beverage laws, fish and game laws, organized crime, and so on.

Other Special-Purpose State Agencies

In addition to the traffic, investigative, and other units mentioned earlier, several other **special-purpose state agencies**, including police and other law enforcement organizations, have developed over time to meet particular needs. For example, many state attorney general's offices have units and investigators that investigate white-collar crimes; fraud against or by consumers, Medicare providers, and food stamp recipients; and crimes against children and seniors.[58]

States may also have limited-purpose units devoted to enforcing the following:

- Alcoholic beverage laws (regarding the distribution and sale of such beverages, monitoring bars and liquor stores, and so on)
- Fish and game laws (relating to hunting and fishing, to ensure that persons who engage in these activities have proper licenses and do not poach, hunt, or fish out of season, exceed their limit, and so on)
- State statutes and local ordinances on college and university campuses
- Agricultural laws, to include cattle brand inspection and enforcement
- Commercial vehicle laws, such as those federal and state laws pertaining to weights and permits of interstate carriers (i.e., tractor-trailer rigs) and ordinances applying to taxicabs
- Airport laws—in addition to TSA employees (discussed earlier), airport police provide support for the local city/county police by enforcing statutes and ordinances; patrolling (foot and vehicle); monitoring threats to people, property, and aircraft; and generally providing information and service to the traveling public.

LOCAL AGENCIES: MUNICIPAL POLICE DEPARTMENTS AND SHERIFF'S OFFICES

Today, Sir Robert Peel (discussed earlier in this chapter) would be amazed because, according to the federal Bureau of Justice Statistics, there are now about 18,000 federal, state, county, and municipal police agencies in the United States.[59] The municipal agencies comprise about 477,000 sworn full-time police officers,[60] and sheriff's offices employ about 190,000 sworn full-time deputies.[61] Next we focus on the organization and functions of these local agencies.

Basic Operations

Following is a brief overview of their employee composition, educational requirements, starting salaries, and some authorized equipment. **Municipal police departments** provide a range of enforcement, investigative, and order-maintenance functions; they employ an average of 2.1 full-time officers per 1,000 population; about one in eight of these sworn employees is a woman, and one in four is a member of a racial or ethnic minority. For educational requirements, about 15 percent of local police agencies require new officers to have some college experience, with 11 percent requiring at least a 2-year college degree. The average starting salary for an entry-level local officer is about $44,000. Significantly, about half of local police agencies employ fewer than ten sworn personnel. Almost three-fourths of these agencies require their officers to wear protective body armor at all times while on duty; 68 percent use video cameras in patrol cars, over 80 percent authorize the use of electronic control devices (such as a Taser), and more than 90 percent of agencies serving 25,000 or more residents use in-car computers. More information concerning municipal police agencies may be obtained from the Bureau of Justice Statistics.[62]

Municipal police department: a police force that enforces laws and maintains peace within a specified city or municipality.

In **county sheriff's offices**, about one in eight sworn employees is a woman, and 19 percent are members of a racial or ethnic minority. For educational requirements, about 10 percent of sheriff's offices require new deputies to have some college experience, with 7 percent requiring at least a 2-year college degree. The average starting salary for an entry-level deputy is about $31,000. About three-fifths of sheriff's departments employ fewer than 25 sworn personnel. Fifty-seven percent of these agencies require their officers to wear protective body armor at all times while on duty; two-thirds use video cameras in patrol cars, 66 percent authorize the use of electronic control devices, and over 85 percent of agencies serving 100,000 or more residents use in-car computers.[63]

County sheriff's office: a unit of county government with a sheriff (normally elected) and deputies whose duties vary but typically include policing unincorporated areas, maintaining county jails, providing security to courts in the county, and serving warrants and court papers.

Because of the diversity of sheriff's offices throughout the country, it is difficult to describe a "typical" office.[64] In addition to the aforementioned investigative, enforcement, and order-maintenance duties found in municipal police agencies, sheriff's offices are found to

- Maintain and operate county correctional institutions
- Serve civil processes (protective orders, liens, evictions, garnishments, and attachments) and perform other civil duties, such as extradition and transportation of prisoners
- Collect certain taxes and conduct real estate sales (usually for nonpayment of taxes) for the county
- Serve as bailiff of the courts

In Chapter 5 we will consider the kinds of screening methods that tend to be used in hiring new officers (see Figure 5.1), as well as typical topics covered in basic recruit academies (see Table 5.1). For more related discussion, see the "Interested in a Career?" section below.

Organization

Every police agency, no matter what its size, has an **organizational structure or chart**, even though it may not be prominently displayed for all to see in the agency's facility—or it may not even be written down at all. Even a community with only a town marshal has an organizational structure, although the structure will be very horizontal, with the marshal performing all of the functions displayed in Figure 4.4, the basic organizational chart for a small agency.

Organizational structure or chart: a diagram of the vertical and horizontal parts of an organization, showing its chain of command, lines of communication, division of labor, and so on.

Figure 4.4 /// A Basic Police Organizational Structure

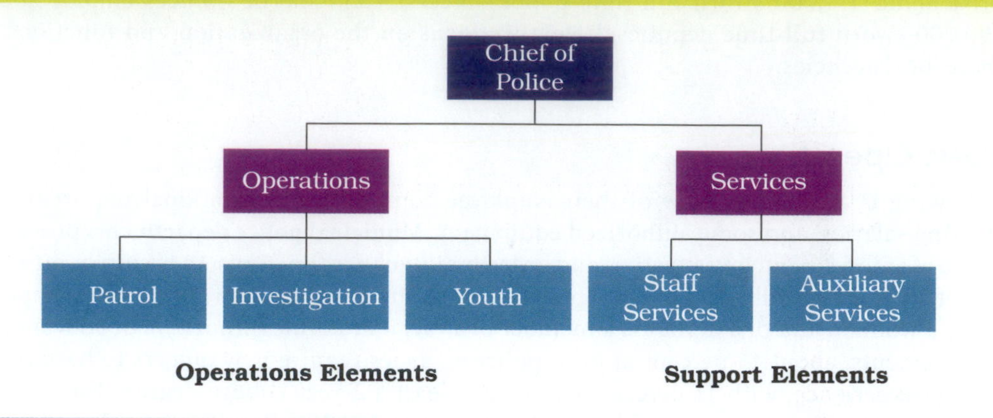

In Figure 4.4, agency operations, or line-element, personnel are engaged in active police functions in the field. They may be subdivided into primary and secondary operations elements. The patrol function—often called the backbone of policing—is the primary operational element because of its major responsibility for policing. (The patrol function is examined in Chapter 5.) In most small police agencies, patrol forces are responsible for all operational activities: providing routine patrols, conducting traffic and criminal investigations, making arrests, and functioning as generalists. The investigative and youth functions are the secondary operations elements. (The investigative function is discussed below.) The support (or nonline) functions and activities can become quite numerous, especially in a large agency. These functions fall within two broad categories: staff (or administrative) services and auxiliary (or technical) services. The staff services usually involve personnel and include such matters as recruitment, training, promotion, planning and research, community relations, and public information services. Auxiliary services are the kinds of functions that civilians rarely see. They include jail management, property and evidence, crime laboratory services, communications, and records and identification. Many career opportunities exist for those who are interested in police-related work but who cannot or do not want to be a field officer.

Greater specialization and variety of assignments may be seen in the organizational structure of a larger police organization, such as the Portland, Oregon, Police Bureau (PPB), shown in Figure 4.5. This structure not only shows the various components of the organization but also does the following:

- Apportions the workload among members and units according to a logical plan
- Ensures that lines of authority and responsibility are as definite and direct as possible
- Places responsibility and authority and, if responsibility is delegated, holds the delegator responsible
- Coordinates the efforts of members so that all will work harmoniously to accomplish the mission

Chain of command: vertical and horizontal power relations within an organization, showing how one position relates to others.

In sum, this structure establishes the **chain of command** and determines lines of communication and responsibility.

Obviously, the larger the agency, the greater the need for specialization and the more vertical the organizational chart will become. With greater specialization comes the need and opportunity for officers to be assigned to different tasks, often rotating from one assignment to another after a fixed interval. For example, in a medium-sized department serving a community of 100,000 or more, it would be possible for a police officer with 10 years of police experience to have been a dog handler, a motorcycle officer, a detective, and a traffic officer while simultaneously holding a slot on the special-weapons or hostage-negotiation team.

Figure 4.5 /// Portland Police Bureau Organizational Chart

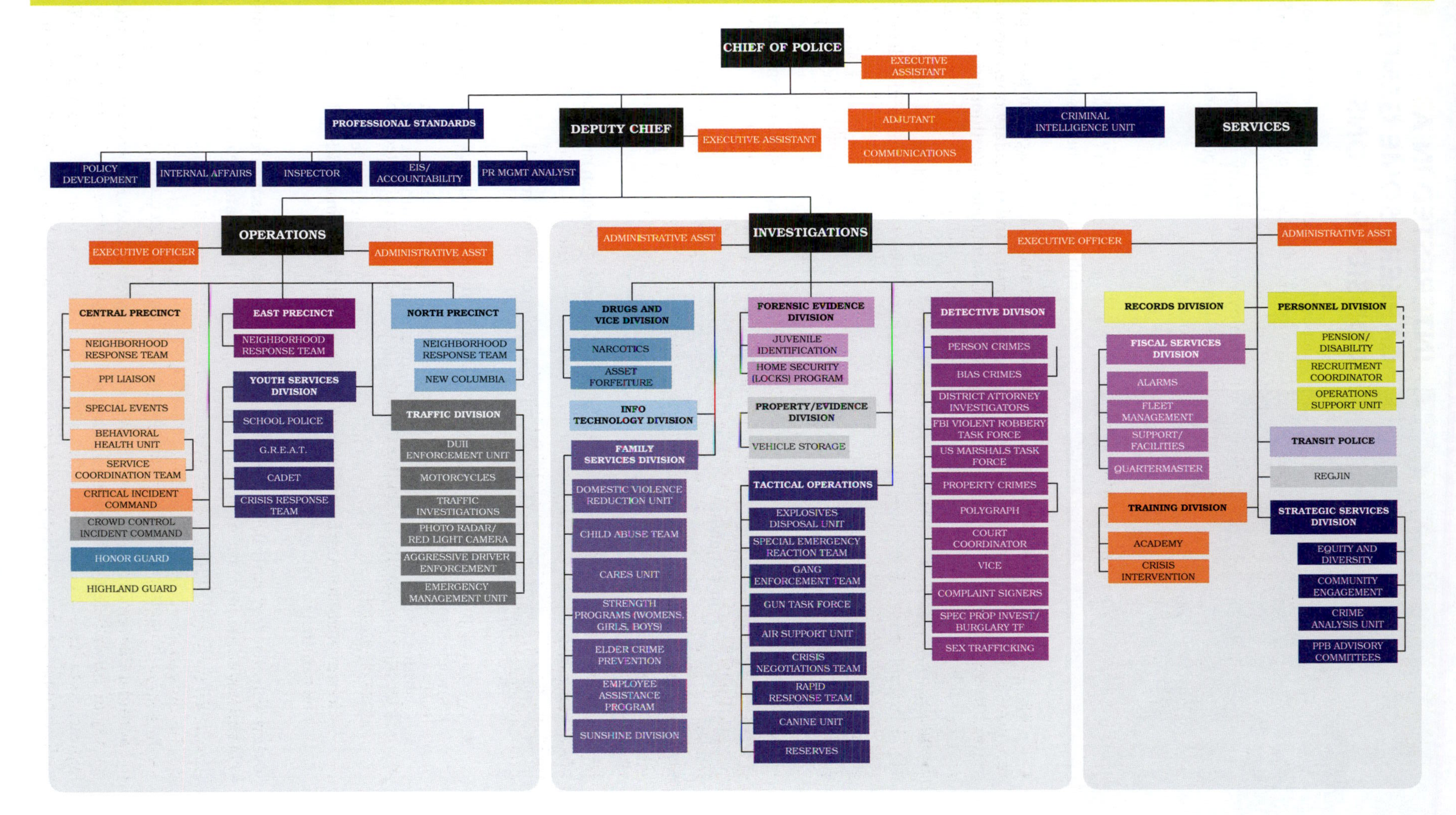

Source: Printed with the permission of Chief Danielle Outlaw, City of Portland, Oregon, Police Bureau.

Note: REGJIN = Regional Justice Information Network; DUII = driving under the influence of intoxicants; PPI = Portland Patrol Inc.; G.R.E.A.T. = Gang Resistance Education and Training; Sunshine Division is a program providing emergency food and clothing relief to families and individuals in need; https://www.portlandoregon.gov/police/article/250329.

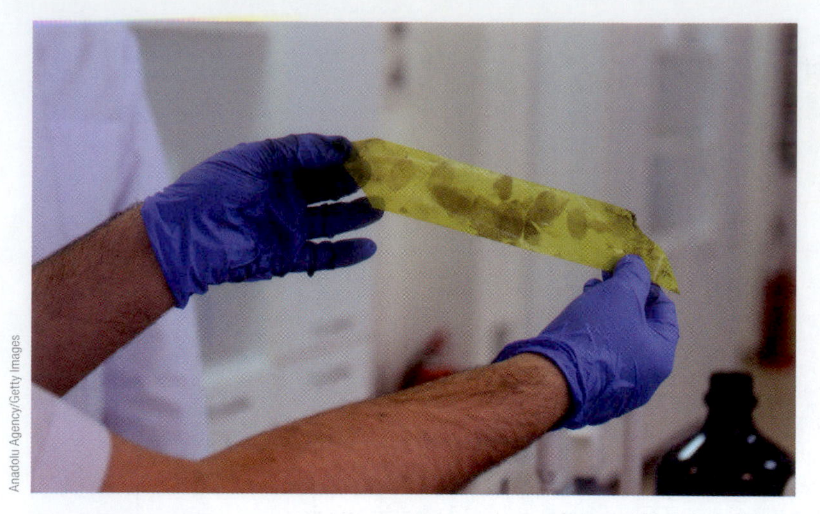
Anadolu Agency/Getty Images

An example of a police organization's auxiliary services is the work done by crime laboratory services.

INTERESTED IN A CAREER? SOME GENERAL CONSIDERATIONS

Generally speaking, there are a number of common elements—often termed KSAs (for knowledge, skills, and abilities)—of employment in many federal, state, and local law enforcement positions, such as the following:

- U.S. citizenship
- Age requirement (at the federal level, applicants must be less than 37 years of age)
- Written test
- Structured oral interview (typically consisting of situational questions posed by an oral board)
- Memory recall assessment (e.g., applicants might be provided with a person's photograph and identifying characteristics and then asked to recall as many details as possible)
- Medical exam (to test for any chronic disease or condition that could impair full performance of the job duties)
- Drug testing (typically, satisfactory completion of a drug test is a condition for placement)
- Background investigation[65]

Furthermore, the following minimum qualifications may be in effect prior to an offer of employment, depending on the agency:

- A basic training program (typically of 4–6 months' duration) must be completed successfully.
- A person convicted of a crime of domestic violence cannot lawfully possess a firearm or ammunition (see 18 U.S.C. Section 1001).
- Persons required to carry a firearm while performing their duties must satisfactorily complete firearms training.
- Positions (particularly federal and state) may require mobility and relocation, including travel relating to the duties of the job and in terms of assignment to a duty station; applicants must also sign a mobility agreement.[66] At the federal level, a law enforcement availability pay bonus is provided to agents that is an additional 25 percent of their annual base pay.

Several websites are invaluable for persons seeking law enforcement careers. The following websites include information on qualifications, physical fitness and health standards, and pretraining requirements:

- USAJOBS, jobsearch.usajobs.gov (search for "Law Enforcement")
- Federallawenforcement.org, "Careers with Federal Law Enforcement Agencies," https://www.federallawenforcement.org/careers/

- The Balance, "Federal Criminal Justice and Law Enforcement Careers," https://www.thebalancecareers.com/federal-law-enforcement-jobs-974533
- Federal Government Jobs, "FBI, CIA, & Homeland Security Jobs," http://www.federaljobs.net/law.htm
- Police Employment, The Police Job Board, "Federal Police Jobs," http://www.policeemployment.com/federal-police

More general information concerning federal employment may be obtained from the federal government's human resources office, the Office of Personnel Management, at www.opm.gov.

Although some agencies require a high school diploma, others require at least a bachelor's degree. Grade-point-average (GPA) requirements can vary; in some federal agencies, a superior academic achievement pay bonus is available to those applicants with higher GPAs; some agencies also provide a Foreign Language Award Program, which provides cash awards for persons qualifying. Some agencies prohibit employees from having visible body markings (tattoos, body art, branding) on the head, face, neck, and extremities.[67] Again, websites should be consulted for such requirements and proscriptions.

A Comment on Immigration and Law Enforcement at All Levels

Chapter 12 discusses immigration in detail as a national challenge and policy issue. This is an area in which law enforcement at all levels (federal, state, and local) is sometimes used as a political tool and may be caught in the middle as emotions and sometimes outright violence flare. Immigration has become a national talking point and political lightning rod. Some states have passed laws that undermine federal law, and some cities and even individual police agencies have passed laws and enacted policies dictating whether to limit their cooperation with federal law enforcement agencies. Sanctuary cities and protests have cropped up and are a source of governmental clashes, immigrant children have been separated from their parents at the U.S.-Mexican border, governments debate aid to undocumented immigrants, the federal government has issued threats to those jurisdictions violating federal laws, our nation squabbles with other nations, and so on. Meanwhile, detention centers have sprung up nationwide to hold undocumented immigrants (a number of whom have died), thousands of people have walked across the southern border seeking asylum, and a national debate is ongoing over the building of a border wall.[68] Again, law enforcement agencies at all levels are caught up in these political skirmishes and policy matters, while political entities debate and vacillate on matters such as whether or not to investigate, arrest, deport, detain, and house these individuals under a vast array of immigration laws.

ON GUARD: THE PRIVATE POLICE

Much has changed in society and the private security industry since 1851, when Allan Pinkerton initiated the Pinkerton National Detective Agency, specializing in railroad security. Pinkerton established the first private security contract operation in the United States. His motto was "We Never Sleep," and his logo, an open eye, was probably the genesis of the term "private eye."

Today, according to the late loss-prevention expert Saul Astor, "We are a nation of thieves"[69]—and, it might be added, a nation that needs to be protected against would-be active shooters, terrorists, rapists, robbers, and other dangerous people. According to the federal Bureau of

Justice Statistics, there are about 3.1 million violent-crime victimizations and 13.3 million property-crime victimizations each year in the United States.[70]

As a result, and especially given the recent spate of mass shootings (discussed in Chapter 12), the United States has become highly security minded concerning its computers, lotteries, celebrities, college campuses, casinos, nuclear plants, airports, shopping centers, mass transit systems, hospitals, and railroads. Such businesses, industries, and institutions have recognized the need to conscientiously protect their assets against threats of crime and other disasters—as well as the limited capabilities of the nation's full-time sworn officers and agents to protect them—and have increasingly turned to the "other police"—those of the private police or security—for protection. As examples, in Oakland, California, several neighborhoods have hired private security patrols, with each household paying $20 per month for unarmed patrols to operate 12 hours per day; Beverly Hills Public Schools contracts with a firm to provide armed safety personnel at a cost of about $1 million per year; and many retailers have created a Business Improvement District where each pays an additional "tax" for projects, some of which include security services. Closely related are the numerous police agencies and sheriff's offices using civilians to take reports (both in person and on the phone), enforce parking regulations, respond to some calls, and all manner of other duties that free sworn officers for higher priority calls.

Private police/security officers: all nonpublic officers, including guards, watchmen, private detectives, and investigators; they have limited powers and only the same arrest powers as regular citizens.

Proprietary services: in-house security services whose personnel are hired, trained, and supervised by the company or organization.

Contract services: a for-profit firm or individuals hired by an individual or company to provide security services.

According to the federal Bureau of Labor Statistics, today about 1.1 million **private police/security officers** provide such services (compared with 807,000 police officers and detectives); their median salary is about $27,000 per year.[71] Indeed, *Forbes* estimates there are now about 20 million private security officers working worldwide, in an industry worth about $180 billion—and growing.[72]In-house security services, directly hired and controlled by the company or organization, are called **proprietary services**. **Contract services** are those outside firms or individuals hired by the individual or company to provide security services for a fee. The most common security services provided include contract guards, alarm services, private investigators, locksmith services, armored-car services, and security consultants.

Although some of the duties of the security officer are similar to those of the public police officer, their overall arrest powers are entirely different. First, because security officers are not police officers, court decisions have stated that the security officer is not bound by the *Miranda* decision concerning suspects' rights. Furthermore, security officers generally possess only the same authority to effect an arrest as does the common citizen (the extent of a citizen's arrest power varies, however, depending on the type of crime, the jurisdiction, and the status of the citizen). In most states, warrantless arrests by private citizens are allowed when a felony has been committed and reasonable grounds exist for believing that the person arrested committed it. Most states also allow citizen's arrests for misdemeanors committed in the arrester's presence.

©iStockphoto.com/AndreyPopov

Our society has increasingly turned to the "other police"—those of the private sector—for protection.

The tasks of the private police are similar to those of their public counterparts: protecting executives and employees, tracking and forecasting security threats, monitoring alarms, preventing and detecting fraud, conducting investigations, providing crisis management and prevention, and responding to substance abuse.[73] Still, there are concerns about the field. First, studies have shown that security officer recruits often have minimal education and training; because the pay is usually quite low, the jobs often attract only those people who cannot find other jobs or who are seeking temporary work. Thus, much of the work is done by the young and the retired.[74]

Fortunately, over the past two decades, the private security field has seen

- Gains in certification, standards, and academic programs in colleges and universities
- Encouragement for greater professionalism through conferences, research and publications, and associations such as ASIS International, the International Association of Chiefs of Police, and the National Sheriffs' Association
- Improved background screening for hiring security personnel (mandated by the Private Security Officer Employment Authorization Act of 2004, which authorized a fingerprint-based criminal history check of state and national criminal history records to screen prospective and current private security officers)
- Improved training of security practitioners (22 states require basic training for licensed security personnel)[75]

Another long-standing issue—prompted by many shootings involving private security personnel—concerns whether or not private police should be armed. These concerns are prompted by training requirements that should be met (in 15 states, security officers can carry weapons on the job without any firearms training at all) as well as the balance that must be struck between providing a safe environment for the officers and their clients, while ensuring that their workspace does not devolve into one in which, as two authors put it, the officers "themselves pose the greatest threat."[76]

/// IN A NUTSHELL

- All four of the primary criminal justice officials of early England—sheriff, constable, coroner, and justice of the peace—either still exist or existed until recently in the United States.
- In 1829, Sir Robert Peel established professional policing as it is known today, under the Metropolitan Police Act of 1829. Later, in 1844, the New York state legislature passed a law establishing a full-time preventive police force for New York City. However, this new body took a very different form than in Europe. The U.S. version was placed under the control of the city government and city politicians, thus launching the first era of policing: the political era, from the 1840s to the 1930s.
- The reform (also known as the professional) era of policing, from the 1930s to the 1980s, sought to remove police from political control. August Vollmer was a major contributor to this era, being the architect of many developments in policing that paved the way for today's practices. Civil service systems were created, and soon the crime-fighter image was projected; much police work was driven by *numbers*—arrests, calls for service, response time to calls for service, number of tickets written, and so on. The police were the "thin blue line," but crimes continued to increase, and the police were becoming increasingly removed from their communities.
- The community era of policing, 1980s to the present, was born because of the problems that overwhelmed the reform era. The police were trained to work with the community to solve problems by looking at their underlying causes and developing tailored responses to them.
- At the federal level, agencies of the Department of Homeland Security were formed in 2003 to combat terrorism; agencies of the Department of Justice and other federal agencies support that and other efforts as well.
- State law enforcement agencies are of two primary types: general law enforcement agencies engaged in patrol and related functions, and state bureaus of investigation.
- Today there are approximately 18,000 police agencies, each with an organizational structure divided into a number of operations and support functions.
- Although there are some advantages to a nation's having a single, national police force, municipal and county policing in the United States has always been primarily decentralized, fragmented, and localized.
- INTERPOL is the oldest, the best-known, and probably the only truly international crime-fighting organization for crimes committed on an international scale; its agents do not patrol the globe, nor do they make arrests or engage in shootouts. They are basically intelligence gatherers who have helped many nations work together in attacking international crime since 1923.
- There are a number of common elements—often termed KSAs (for knowledge, skills, and abilities)—of employment in many federal, state, and local law enforcement positions; many such positions (particularly at the federal level) now involve the enforcement of immigration laws.

- The field of private policing or security is much larger than public policing; it is divided into in-house security services, called *proprietary services*, and *contract services*, where outside firms or individuals are hired by the individual or company to provide security services for a fee. Private security officers are not bound by court decisions that govern the public police, and generally they possess the same authority to make an arrest as a private citizen.

/// KEY TERMS & CONCEPTS

Review key terms with eFlashcards at **edge.sagepub.com/peakbrief**.

Chain of command 92
Community era 81
Constable 77
Contract services 96
Coroner 77
County sheriff's office 91
Federal law enforcement agencies 82
INTERPOL 85
Justice of the peace (JP) 77
Municipal police department 91
Organization 76
Organizational structure or chart 91
Political era 79
Private police/security officers 96
Proprietary services 96
Reform era 79
Sheriff 76
Special-purpose state agencies 90
State bureau of investigation 90
State police 89

/// REVIEW QUESTIONS

Test your understanding of chapter content. Take the practice quiz at **edge.sagepub.com/peakbrief**.

1. What were the four major police-related offices and their functions during the early English and colonial periods?
2. What are the three eras of local policing, what were August Vollmer's contributions to policing's reform era, and what primary problems of the two initial eras led to the development of the current community era?
3. What are the major agencies within the Department of Homeland Security and the Department of Justice, and what are their primary functions?
4. What general qualifications, job requirements, and benefits are involved in careers in federal law enforcement agencies?
5. How would you describe the primary differences between federal and state law enforcement agencies?
6. What functions do the Central Intelligence Agency and the Internal Revenue Service perform?
7. How does INTERPOL function, and what are its primary contributions to crime fighting?
8. What are some advantages and disadvantages of a nation's having a single, national, centralized police force?
9. What are some of the differences between municipal police agencies and county sheriff's departments?
10. Using a simple organizational structure that you have drawn, how would you describe the major functions of a local (i.e., municipal or county) police agency?
11. Why were the private police organizations developed, and what are their contemporary purposes and issues?

/// LEARN BY DOING

1. Assume that you are part of a group that is studying the feasibility of a single, national police force in the United States, such as those found in many countries around the world. All state, county, and municipal police organizations would be abolished and absorbed into this single centralized agency, with one governing board, one set of laws to enforce and agency policies to uphold, and standardized training and pay/benefits. Develop arguments both for and against this proposal, perhaps including the history of policing, all possible positive and negative consequences that might occur from this single entity, its political pitfalls and favor with the general public, and whether or not you would support this proposal.

2. Do an Internet search to determine the extent to which the following four police-related offices are still authorized and used in the United States and in your state: sheriff, constable, justice of the peace, and coroner.

3. You have been assigned to describe your state's special-purpose law enforcement agencies for a class presentation. Prepare a lecture outline covering those agencies and the services they provide.

/// STUDY SITE

Get the tools you need to sharpen your study skills. SAGE edge offers a robust online environment featuring an impressive array of free tools and resources.

Access practice quizzes, eFlashcards, video, and multimedia at **edge.sagepub.com/peakbrief**

CHAPTER 5

©AP Photo/ASSOCIATED PRESS, J. Scott Applewhite

POLICE TRAINING, PATROLLING, INVESTIGATING

Forming, Reforming, and Solving

Down these mean streets a man must go who is not himself mean, who is neither tarnished nor afraid. He is the hero, he is everything. He must be . . . a man of honor.

—Raymond Chandler, *The Simple Art of Murder* (1950)

Murder, though it have no tongue, will speak.

—Shakespeare, *Hamlet*, Act II, Scene 2

LEARNING OBJECTIVES

As a result of reading this chapter, you will be able to

1. Explain the kinds of topics that are taught in the recruit academy and overall methods for preparing recruits for a career in policing
2. Explain the roles, styles, four basic tasks, and dangers of policing
3. Define police discretion, how and why it functions, and some of its advantages and disadvantages
4. Explain the current era of policing, the community era, and the prevailing philosophy and strategies of community policing and problem solving
5. Review the use of force by the police, the current state of police-community relations regarding such force, and the distinction between police being viewed as guardians versus soldiers
6. Define forensics and criminalistics, and describe the qualities and methods of investigative personnel

ASSESS YOUR AWARENESS

Test your current knowledge of police patrol and investigations by responding to the following eight true-false items; check your answers after reading this chapter.

1. Normally there are only three types of tests one has to pass in order to become a police officer: a background check, a criminal record check, and a polygraph exam.
2. The four basic tasks of policing that involve the public include patrolling, tracking, arresting, and appearing in court.
3. Studies have shown that doubling a city area's level of police patrol, or greatly decreasing such levels, has little significant deterrent effect on that area's crime rates.
4. Police officers, being governed by laws and procedure manuals, actually have very little discretion in how they perform their jobs.
5. Policing is now in what is termed the community era, where great emphasis is placed on the public's assistance in preventing and solving crimes.
6. The use of force by the police is now highly controversial, with the public demanding that police act more like guardians and less like soldiers.
7. DNA is the most sophisticated and reliable type of physical evidence.
8. Police work as it concerns juvenile offenders and varies greatly from that which concerns adults.

<< Answers can be found on page 293.

The Golden State Killer terrorized all of California in the 1970s and 1980s—a period of time when there were no extensive DNA, mapping, automated fingerprinting, or other databases and forensic technologies and investigative tools that are taken for granted today. The murderer was clever and sadistic, staying one step ahead of police while killing at least 12 people, raping more than 50 (many in very cruel fashion), and committing some 100 burglaries over a 10-county area between 1974 and 1986. But one persistent investigator and DNA expert, Paul Holes, created a road map to the killer through his genetics.

Holes used DNA recovered from a 1980 rape/murder crime scene to begin the investigation: an evidence kit from a rape and murder of a couple that had been placed in a freezer in 1980. A lab converted the sample into a format that could be read by a website that allows users to upload their genetic information for family history searches and analyzes hundreds of thousands of DNA data points to determine relatedness. Holes found 10 to 20 distant relatives of the killer; he knew that if he traced back the lineages of distant cousins far enough, he could find a common ancestor they shared with the killer.

Holes and his team used census data, old newspaper clippings and a gravesite locator for finding the deceased relatives. The family trees were created using a tool on Ancestry.com. After four months of work examining family trees, they had pieced together about 25 distinct family trees from the great-great-great grandparents; the trees composed roughly 1,000 family members and would include the suspect. The team began scouring the trees for men about the killer's age who had connections to Sacramento and other locations of the crimes. They found two. One was eventually eliminated by a DNA test of a relative; the other was Joseph James DeAngelo, a 72-year-old retired police officer who was quietly living near Sacramento. Deputies surveilled DeAngelo, and in April 2018 they pounced on an item discarded by DeAngelo that contained his DNA, producing a match. More than four decades after he commenced his horrible crime spree, DeAngelo was arrested in Citrus Heights on April 24. He now faces eight counts of murder.[1] (In the aftermath of the DeAngelo case, several genealogical database firms changed their terms of service, allowing users to deny law enforcement from accessing their genetic information.)

As you read this chapter, consider the essential traits that one must possess, the knowledge that must be obtained, and the skills one must have in order to be a successful police officer as well as criminal investigator.

INTRODUCTION

How do I become a police officer? What would I do if hired as one? Could I be a detective? A criminal profiler? How do I qualify to work in forensics? Do police have to arrest everyone they see breaking the law?

These are all legitimate, often-heard questions as posed by university students. Unfortunately, owing in large measure to Hollywood's portrayals of police work and investigations, there are many misperceptions about the field. First, the odds of one's becoming a criminal profiler are virtually nil—as are the odds of some federal agent academy trainee being brought out to help investigate a serial killer case (as was the plot in the movie *The Silence of the Lambs*). Second, one who wishes to work in a forensics lab must have a background in chemistry, biology, or a related natural science field (e.g., biochemistry or microbiology). Finally, to become a detective, one must typically begin as a regular officer; then, with years of training, experience, and often testing or at least an oral examination, one might be deemed worthy of being an investigator. This chapter hopes to remedy those misperceptions, at least in part, by looking at some of the primary roles and functions of the individuals who work in police organizations.

We begin with a review of the initial training received by police recruits at the academy; included are considerations of the roles, styles, and tasks involved in policing, as well as discretionary use of authority by police in enforcing the law, and the community era of policing (and its new methods of addressing crime for officers). Next we examine the current chasm between the police and the public regarding their use of force, particularly the lethal type; the public now demands greater accountability with force, and we review how many agencies are turning to constitutional policing and procedural justice to bridge this police-community gap. Then we consider the work of forensics laboratories and the attributes and methods of detectives; included is a review of how DNA analyses are performed. The chapter concludes with a review of the different kinds of knowledge and abilities one must possess in order to work with juvenile offenders and those in need of services.

FROM CITIZEN TO PATROL OFFICER

The idea of a police subculture was first proposed by William Westley in his 1950 study of the Gary, Indiana, Police Department, where he found, among many other things, a high degree of secrecy and violence.[2] The police develop traditions, skills, and attitudes that are unique to their occupation because of their duties and responsibilities. In this section, we consider how citizens are brought into, and socialized within, the police world.

Recruiting the Best

Recruiting an adequate pool of applicants is an extremely important facet of the police hiring process. August Vollmer, the renowned Berkeley, California, police innovator and administrator discussed in Chapter 4, said that law enforcement candidates should

> have the wisdom of Solomon, the courage of David, the patience of Job and leadership of Moses, the kindness of the Good Samaritan, the diplomacy of Lincoln, the tolerance of the Carpenter of Nazareth, and, finally, an intimate knowledge of every branch of the natural, biological and social sciences.[3]

Police officers are solitary workers, spending most of their time on the job unsupervised. At all times they must be able to make sound decisions and adjust quickly to changing situations during periods that are unpredictable and unstable, chaotic, or high stress—all the while acting ethically and in keeping with the U.S. Constitution, their state statutes, and their agency's policy and procedures manual. For these reasons, police agencies must attempt to attract the best individuals possible.

Figure 5.1 shows the general kinds of screening and testing methods used with new police recruits in the United States. Even after a person meets the minimum qualifications for

Figure 5.1 /// Local Police Officers' Selected Screening Methods in the Hiring Process

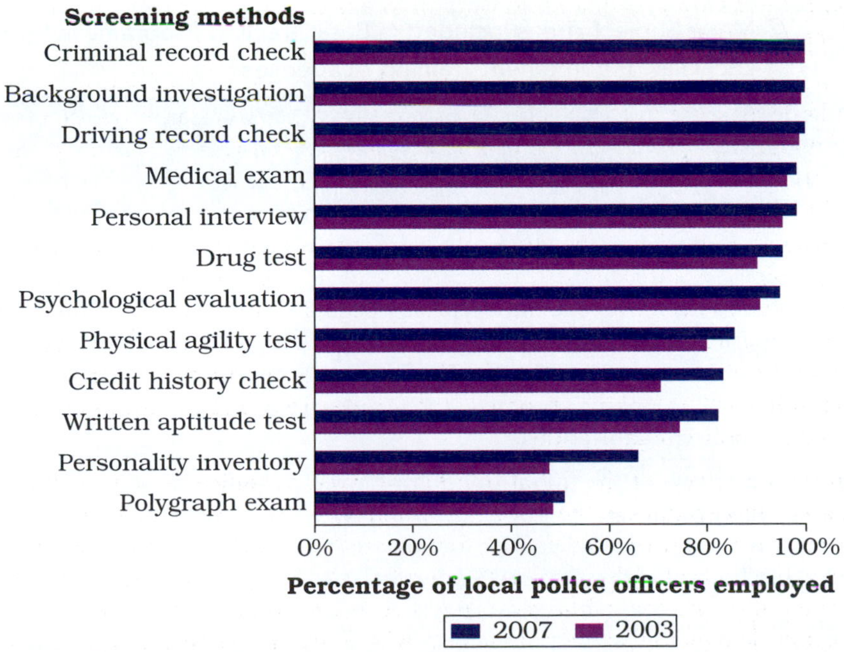

Source: Brian A. Reaves, *Hiring and Retention of State and Local Law Enforcement Officers* (Washington, D.C.: U.S. Department of Justice, Bureau of Justice Statistics, October 2012), p. 14, https://www.bjs.gov/content/pub/pdf/hrslleo08st.pdf.

Note: Most current data available.

Figure 5.2 /// Major Elements of the Police Hiring Process

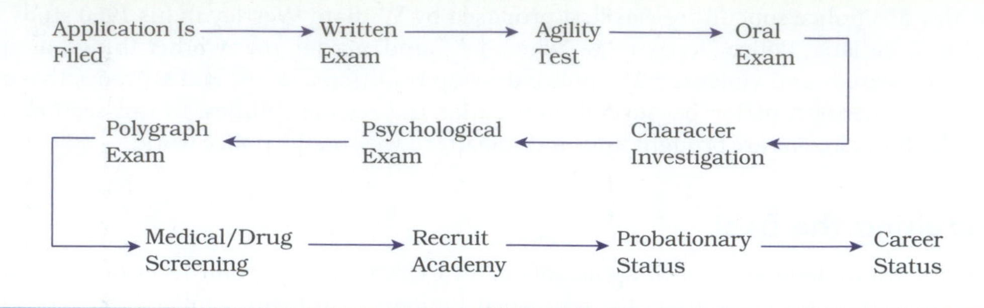

being a police officer (age, education, no disqualifying criminal record), and is recruited into the testing process, much work and testing remain before he or she is ready to be sent to the streets as a police officer. This so-called hurdle process is shown in Figure 5.2. Not all types of tests shown in the figure are employed by all of the 18,000 federal, state, county, and municipal law enforcement agencies in America, nor are these tests necessarily given in the sequence shown. Furthermore, this gamut of testing can easily require several months to complete, depending on the number and types of tests used and the ease of scheduling and performing them.

Academy training: where police and corrections personnel are trained in the basic functions, laws, and skills required for their positions

Sixth sense: in policing, the notion that an officer can "sense" or feel when something is not right, as in the way a person acts, talks, and so on.

Recruit Training

Once hired into an agency, police recruits are taught a variety of subjects in **academy training** (see Table 5.1). They are taught to nurture a **sixth sense**: suspicion. A suspicious nature is as important to the street officer as a fine touch is to a surgeon. The officer should be able to visually recognize when something is wrong or out of the ordinary.

Table 5.1 /// Major Subject Areas Included in Basic Training Programs in State and Local Law Enforcement Training Academies

Training Area	Percentage of Academies With Training	Average Number of Hours of Instruction Required per Recruit*
Operations		
Report writing	99%	25 hrs.
Patrol procedures	98	52
Investigations	98	42
Traffic accident investigations	98	23
Emergency vehicle operations	97	38
Basic first aid/CPR	97	24
Computers/information systems	61	9
Weapons/defensive tactics/use of force		
Defensive tactics	99%	60 hrs.

Training Area	Percentage of Academies With Training	Average Number of Hours of Instruction Required per Recruit*
Firearms skills	98	71
Use of force	98	21
Nonlethal weapons	88	16
Self-improvement		
Ethics and integrity	98%	8 hrs.
Health and fitness	96	49
Communications	91	15
Professionalism	85	11
Stress prevention/management	81	6
Legal education		
Criminal/constitutional law	98%	53 hrs.
Traffic law	97	23
Juvenile justice law/procedures	97	10

*Excludes academies that did not provide this type of instruction.

Source: Brian A. Reaves, *State and Local Law Enforcement Training Academies, 2013* (Washington, D.C.: U.S. Department of Justice, Bureau of Justice Statistics, February 2016), p. 5, https://www.bjs.gov/content/pub/pdf/slleta13.pdf.

Table 5.1 displays the predominant subjects included in academy training of new officers and the average length of time devoted to each. The table shows that large blocks of time are devoted to defensive tactics/weapons skills, criminal/constitutional law, patrol procedures, and health and fitness.

Field training officer (FTO): one who is to oversee and evaluate the new police officer's performance as he or she transitions from the training academy to patrolling the streets.

Field Training Officer

Once the recruits leave the academy, they are not merely thrown to the streets to fend for themselves in terms of upholding the law and maintaining order. Another important part of this acquisition process involves being assigned to a veteran officer for initial field instruction and observation. This veteran is sometimes called a **field training officer (FTO)**.[4] This training program provides recruits with an opportunity to make the transition from the academy to the streets under the protective arm of a veteran officer. Recruits are on probationary status, typically ranging from six months to one year; their employment may be terminated immediately if their overall performance is unsatisfactory during that period.

Helen H. Richardson/Denver Post/Getty Images

Police recruits undergo training in a variety of subjects during their many weeks in the academy.

Most FTO programs consist of three identifiable phases:

1. Introductory phase (the recruit learns agency policies and local laws)
2. Training and evaluation phase (the recruit is introduced to more complicated tasks that patrol officers confront)
3. Final phase (the FTO acts strictly as an observer and evaluator while the recruit performs all the functions of a patrol officer)[5]

The length of time rookies are assigned to FTOs will vary. A formal FTO program might require close supervision for a range of 1–12 weeks.

Most police officers also receive in-service training throughout their careers, because their states require a minimum number of hours of such training. News items, court decisions, and other relevant information can also be covered at roll call before the beginning of each shift. Short courses ranging from a few hours to several weeks are available for in-service officers through several means such as videos and nationally televised training programs.

HAVING THE "RIGHT STUFF:" DEFINING THE ROLE

Policing role: the function of the police in contemporary society.

Policing styles: James Q. Wilson argued that there are three styles of policing: watchman, legalistic, and service.

In addition to possessing the "bare essentials" mentioned earlier—a high degree of ethics/integrity (determined via polygraph, record check, and other means), knowledge of applicable laws and procedures (obtained at the basic academy), communication skills, physical fitness, and so on—who are the police and what are they supposed to do? Although a seemingly straightforward question, the **policing role** is far more complex than merely saying they "enforce the law" or "serve and protect."

One of the greatest obstacles to understanding the American police is the crime-fighter image. Because of film and media portrayals, many people believe that the role of the police is confined to the apprehension of criminals.[6] However, only about 20 percent of the police officer's typical day is devoted to fighting crime per se.[7] And, as Jerome Skolnick and David Bayley point out, the crimes that terrify Americans the most—robbery, rape, burglary, and homicide—are rarely encountered by police on patrol. In their words:

> Only "Dirty Harry" has his lunch disturbed by a bank robbery in progress. Patrol officers individually make few important arrests. The "good collar" is a rare event. Cops spend most of their time passively patrolling and providing emergency services.[8]

©AP Photo/Matt Rourke

To assist in their transition from the training academy to the street, new officers are typically assigned to work with a veteran field training officer for a period of time before being released to act on their own.

Also, many individuals enter police work expecting it to be exciting and rewarding, as depicted on television and in the movies. Later they discover that much of their time is spent with boring, mundane, and trivial tasks—and that paperwork is seldom stimulating.

Three Distinctive Styles

James Q. Wilson also attempted to clarify what it is that the police are supposed to do; Wilson maintained that there are three distinctive **policing styles**:[9]

- The *watchman* style involves the officer as a "neighbor." Here, officers act as if order maintenance (rather than law enforcement) is their primary function. The emphasis is on using the law as a means of maintaining order rather than regulating conduct through arrests. Police ignore many common minor violations, such as traffic and juvenile offenses. These violations and so-called victimless crimes, such as gambling and prostitution, are tolerated; they will often be handled informally. Thus the individual officer has wide latitude concerning whether to enforce the letter or the spirit of the law; the emphasis is on using the law to give people what they "deserve."

Mark Wilson/Getty Images News/Getty Images

As opposed to Hollywood's portrayals, very little police work is action packed. Here, Washington, D.C., police cite people who are talking on cell phones while driving.

- The *legalistic* style casts the officer as a "soldier." This style takes a much harsher view of law violations. Police officers issue large numbers of traffic citations, detain a high volume of juvenile offenders, and act vigorously against illicit activities. Large numbers of other kinds of arrests occur as well. Chief administrators want high arrest and ticketing rates not only because violators should be punished but also because it reduces the opportunity for their officers to engage in corrupt behavior. This style of policing assumes that the purpose of the law is to punish.

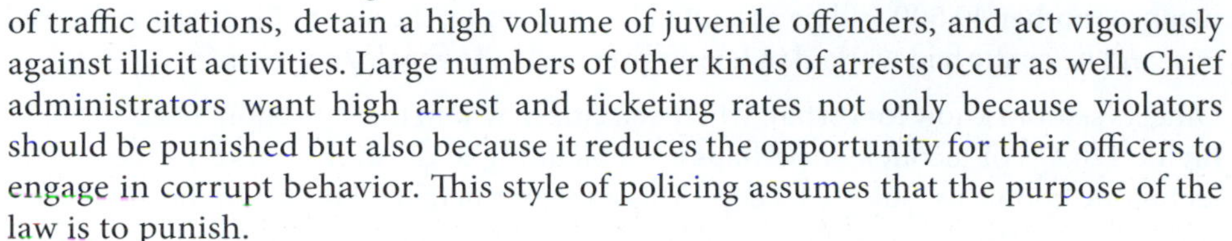

- The *service* style views the officer as a "teacher." This style falls in between the watchman and legalistic styles. The police take seriously all requests for either law enforcement or order maintenance (unlike in the watchman-style department) but are less likely to respond by making an arrest or otherwise imposing formal sanctions. Police officers see their primary responsibility as protecting public order against the minor and occasional threats posed by unruly teenagers and "outsiders" (tramps, derelicts, out-of-town visitors). The citizenry expects its service-style officers to display the same qualities as its department store salespeople: They should be courteous, neat, and deferential. The police will frequently use informal sanctions instead of making arrests.

Four Basic Tasks

Patrol officers may be said to perform four basic **tasks of policing**:

Tasks of policing (four basic): enforce the law, perform welfare tasks, prevent crime, and protect the innocent.

1. *Enforce the laws:* Although this is a primary function of the police, as we saw earlier, they actually devote a very small portion of their time to "chasing bad guys."
2. *Perform welfare tasks:* Throughout history, the police have probably done much more of this type of work than the public (or the police themselves) realize; following are some of them:
   - "Check the welfare of" kinds of calls, where someone has not been seen or heard from for some time and may be deceased, ill, missing, or in distress
   - "Be on the lookout" (BOLO) calls, where someone has wandered away from a nursing home or an assisted living home, is a juvenile runaway, and so forth
   - Delivering death notifications/messages
   - Delivering blood to hospitals (particularly in more rural areas where blood banks are not available)
   - Assisting firefighters and animal control units
   - Reporting burned-out street lights or damaged traffic signs

- Performing all manner of errands simply because they are available—locking and unlocking municipal parking lots, collecting receipts from municipal entities such as golf courses, delivering agendas to city/county commissioners, and so forth

3. *Prevent crime:* This function of police involves engaging in random patrol and providing the public with crime prevention information.
4. *Protect the innocent:* By investigating crimes, police are systematically removing innocent people from consideration as crime suspects.

Perils of Patrol

Although workers die at much higher rates in occupations other than policing—such as commercial fishing, logging, piloting airplanes, roofing, and recycling materials[10]—police officers' lives are still rife with occupational hazards. Each year, an average of about 150 police officers die while on duty, and about 25 to 50 of them will be killed through felonious means; another 49,500 will be assaulted.[11] On average, the slain officer was in his or her late 30s, and he or she had worked in law enforcement about a dozen years.[12]

Officers seldom know for certain whether a citizen whom they are about to confront is armed, high on drugs or alcohol, or perhaps even planning to die at the hands of the police using a technique known as "suicide by cop." This danger is heightened during the graveyard (night) shift, when patrol officers encounter burglars looking to invade homes and businesses, people who are intoxicated from a night of partying, and so on, all under cover of darkness.

Kansas City Preventive Patrol Experiment: in the early 1970s, a study of the effects of different types of patrolling on crime—patrolling as usual in one area, saturated patrol in another, and very limited patrol in a third area; the results showed no significant differences.

A Study of Patrol Effectiveness

The best-known study of patrol effectiveness, the **Kansas City Preventive Patrol Experiment**, was conducted in Kansas City, Missouri, in 1973. Researchers—wanting to know if random patrol had any significant effect on crime, police response rates to crime, or citizen fear of

Figure 5.3 /// Schematic Representation of the Kansas City Preventive Patrol Experiment

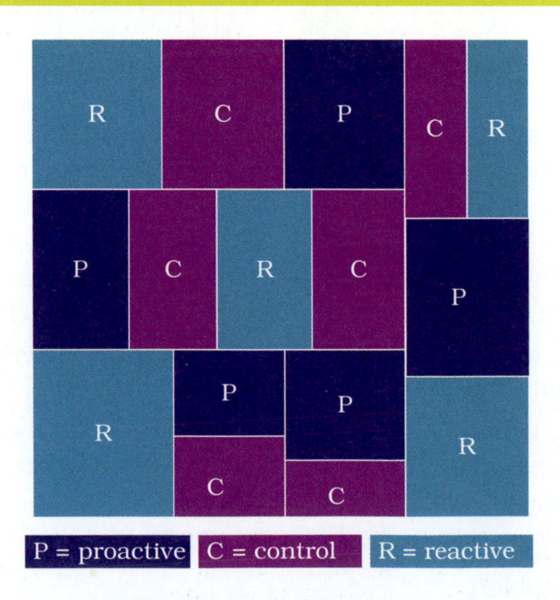

Source: George L. Kelling, Tony Pate, Duane Dieckman, and Charles E. Brown, *The Kansas City Preventive Patrol Experiment: A Summary Report* (Washington, D.C.: Police Foundation, 1974). Reprinted with permission of the Police Foundation.

crime—divided the city into 15 beats, which were then categorized into five groups of three matched beats each. Each group consisted of neighborhoods that were similar in terms of population, crime characteristics, and calls for police services. In one beat area, there was no preventive patrol (police responded only to calls for service); another beat area had increased patrol activity (two or three times the usual amount of patrolling); and in the third beat there was the usual level of patrol service. Citizens were interviewed and crime rates were measured during the year the experiment was conducted.

The study found that the deterrent effect of policing was not reduced by the elimination of routine patrolling; nor were citizens' fear of crime, their attitudes toward the police, or the ability of the police to respond to calls reduced. The Kansas City Preventive Patrol Experiment (depicted in Figure 5.3) indicated that the traditional assumption of "Give me more cars and more money, and we'll get there faster and fight crime" was probably not a viable argument.[13]

USE OF DISCRETION

Chapter 1 discussed how discretionary use of authority permeates the entire U.S. criminal justice system; here we focus on how discretion is employed in policing. The power to use discretion in performing one's role is at the very core of policing. However, as is discussed in this section, this power can be controversial—and used for both good and bad.

The Myth of Full Enforcement

The law is written in black and white; however, the manner in which most laws are enforced by the police can be said to be colored gray. Noted police scholar Herman Goldstein wrote in 1963 that the police should be "Enforcing the law without fear or favor."[14] However, the truth is anything but (except for laws or policies related to domestic violence and driving under the influence, for example, where the police are given no discretion but to arrest).

Consider this scenario: The municipal police chief or county sheriff is asked during a civic club luncheon speech which laws are and are not enforced by his or her agency. The official response will inevitably be that *all* of the laws are enforced equally, all of the time. Yet the chief or sheriff knows that full enforcement of the laws is a myth—that there are neither the resources nor the desire to enforce them all, nor are all laws enforced impartially. There are legal concerns as well. For example, releasing some offenders (to get information about other crimes, because of a good excuse, etc.) cannot be the official policy of the agency (and letting an offender go is a form of discretion as well); however, the chief or sheriff cannot broadcast that fact to the public. Indeed, it has been stated that "the single most astonishing fact of police behavior is the extent to which police do *not* enforce the law when they have every legal right to do so."[15]

Attempts to Define Discretion

The way police make arrest decisions is largely unknown (see possible determining factors, described in the next section). What *is* known, however, is that when police observe something of a suspicious or an illegal nature, two important decisions must be made: (1) whether to intervene in the situation and (2) how to intervene. The kinds, number, and possible combinations of interventions are virtually limitless. What kinds of decisions are available for an officer who makes a routine traffic stop? David Bayley and Egon Bittner observed long ago that officers have as many as 10 actions to select from at the initial stop (e.g., order the driver out of the car), 7 strategies appropriate during the stop (such as a roadside sobriety test), and 11 exit strategies (including releasing the driver with a warning), representing a total of *770* different combinations of actions that might be taken![16]

Criminal law has two sides—the formality and the reality. The formality is found in the statute books and opinions of appellate courts. The reality is found in the practices of enforcement officers. In some circumstances, the choice of action to be taken is relatively easy, such as arresting a bank robbery suspect. In other situations, such as quelling a dispute between neighbors or determining how much party noise is too much, the choice is more difficult.[17] Our system tends to treat people as individuals. One person who commits a robbery is not the same as another person who commits a robbery. Our system also takes into account why and how a person committed a crime (his or her intent, or *mens rea*, discussed in Chapter 2). The most important decisions take place on the streets, day or night, generally without the opportunity for the officer to consult with others or to carefully consider all the facts.

Determinants of Discretionary Actions

The power of discretionary policing can be awe-inspiring. Kenneth Culp Davis, an authority on police discretion, writes, "The police are among the most important policy makers of our entire society. And they make far more discretionary determinations in individual cases than does any other class of administrators; I know of no close second."[18]

What determines whether the officer will take a stern approach (enforcing the letter of the law with an arrest) or will be lenient (issuing a verbal warning or some other outcome short of arrest)? Several variables enter into the officer's decision:

1. The *law* is indeed a factor. For example, many state statutes and local ordinances now mandate that the police arrest for certain suspected offenses, such as driving under the influence or domestic violence.
2. The *officer's attitude* can also be a factor. First, some officers are more willing to empathize with offenders who feel they deserve a break than others. Furthermore, police, being human, can bring to work either a happy or an unhappy disposition. Personal viewpoints can also play a role, such as when, for example, the officer is fed up with juvenile crimes that have been occurring of late and thus will not give any leniency to youths that he or she encounters. Also, as Carl Klockars and Stephen Mastrofski observed, although violators frequently offer what they feel are very good reasons for the officer to overlook their offense, "every police officer knows that, if doing so will allow them to escape punishment, most people are prepared to lie through their teeth."[19]
3. Another major consideration in the officer's choice among options is the *citizen's attitude*. If the offender is rude and condescending, denies having done anything wrong, or uses some of the standard clichés that are almost guaranteed to rankle the officer—such as "You don't know who I am" (someone who is obviously very important in the community), "I'll have your job," "I know the chief of police," or "I'm a taxpayer, and I pay your salary"—the probable outcome is obvious. By contrast, the person who is honest with the officer, avoids attempts at intimidation and sarcasm, and does not try to "beat the rap" may fare better.

In addition to these considerations, other factors that an officer may take into account when deciding whether or not to arrest might include injury to and preference of the victim; prior criminal record of the offender; amount and strength of evidence; peer and agency pressure regarding certain kinds of crimes; media coverage of this and other related crimes; and availability and credibility of witnesses. The officer's specific assignment may also come into play. For example, homicide investigators would care little about a driver's tendency to disobey traffic laws, whereas one who is an assigned and dedicated traffic officer will likely treat such offenses much more seriously.

Pros and Cons of Discretion

Having discretionary authority carries several advantages for the police officer: First, because the law cannot (and should not) cover every sort of situation the officer encounters,

You Be the . . .

Police Officer

Assume a police officer pulls over a vehicle for swerving across the center line. The driver, a 17-year-old college student who is an elementary education major, admits she's been drinking at a party. The breathalyzer test reveals a .07 blood alcohol concentration. In this state .07 can result in a charge of either "driving after having consumed alcohol" or a more serious "driving while ability is impaired." There are two others in the car: One young woman has a badly swollen jaw after having fallen at the party; the driver is attempting to get her to the urgent care facility. The injured passenger, an older sister of the driver, is visibly pregnant.

1. What discretionary issues are presented?
2. Given all of the facts at hand, how do you believe the officer should deal with the driver?

discretion allows the officer to have the flexibility to treat different situations in accordance with humanitarian and practical goals. For example, assume an officer pulls over a speeding motorist, only to learn that the car is en route to the hospital with a woman who is about to deliver a baby. While the agitated driver is endangering everyone in the vehicle as well as other motorists on the roadway, discretion allows the officer to be compassionate and empathetic, giving the car a safe escort to the hospital rather than issuing a citation for speeding. In short, discretionary use of authority allows the police to employ a philosophy of "justice tempered with mercy."

One disadvantage of discretionary authority is that those officers who are the least trained and experienced have the greatest amount of discretion to exercise. In other words, as the rank of the officer *increases*, the amount of discretion that he or she can employ typically *decreases*. The patrol officer or deputy, being loosely supervised on the streets, makes many discretionary decisions about whether or not to arrest, search, frisk, and so forth. Conversely, the chief of police or sheriff will be highly constrained by department policies and procedures, union agreements, affirmative action laws, and/or governing board guidelines and policies. Another disadvantage is that allowing police to exercise such discretion belies their need to appear impartial—treating people differently for committing essentially the same offense. Critics of discretion also argue that such wide latitude in decision making may serve as a breeding ground for police corruption; for example, an officer may be offered a bribe to overlook an offense.

See the case study in the accompanying "You Be the . . . Police Officer" box and respond to the questions posed.

COMMUNITY POLICING AND PROBLEM SOLVING

Community policing and problem solving: a proactive management philosophy that involves police-community collaboration and a four-step process (scanning, analyzing, response, and assessment) to focus police activities and thus enable officers to respond more effectively to crime and disorder with arrests or other appropriate actions.

Chapter 4 discussed the three eras of policing, including today's "community era." The seeds of **community policing and problem solving** were sown in London in 1829 when, as stated in Chapter 4, Sir Robert Peel offered that "the police are the public and . . . the public are the police," and that by establishing patrol beats, officers could get to know citizens and thus be better able to gather information about neighborhood crime and disorder. As you saw in Chapter 4, however, in the United States that close police-public association over time often led to powerful political influences and corruption in terms of who was hired, who was promoted, and who could bring elected officials the most votes. This led to the onset of the reform era in the 1930s. Reforms included the removal of police from the influence of the community and politics through the creation of civil service

Robert Alexander/Archive Photos/Getty Images

As part of their community-policing and problem-solving efforts, many agencies use bicycle patrols to focus on crime prevention and greater interaction with the community.

systems. The community era of policing recognized that the public has a vested interest in addressing—as well as vital information concerning—neighborhood crime and disorder, and thus a return to Peel's principles was needed and the two entities should work hand in glove to resolve problems.

Problem-oriented policing, which began to develop in the mid-1980s, was grounded in principles different from, but complementary to, those of community-oriented policing. Problem-oriented policing is a strategy that puts the community policing philosophy into practice. It advocates that police examine the underlying causes of recurring incidents of crime and disorder. The problem-solving process helps officers identify problems, analyze them completely, develop response strategies, and assess the results. Police must be equipped to define more clearly and to understand more fully the problems they are expected to handle. They must recognize the relationships between and among incidents—for example, incidents involving the same behavior, the same address, or the same people. The police must therefore develop a commitment to analyzing problems—gathering information from police files, the minds of experienced officers, other agencies of government, and private sources as well. It can also require conducting house-to-house surveys and talking with victims, complainants, and offenders. It includes an uninhibited search for the most effective response to each problem; in sum, police must try to design a customized response that holds the greatest potential for dealing effectively with a specific problem in a specific place under specific conditions.

Thousands of police agencies have thus broken away from their reactive, incident-driven methods that characterized the reform era—where police would race from call to call, take an offense report, and leave the scene without seeking any resolution to problems or achieving any long-term benefits.

Finally, three elements must be present in order for a crime to occur: an offender, a victim, and a location, as shown in Figure 5.4. The problem analysis triangle helps officers visualize the problem and understand the relationships among these three elements. Additionally, it

Figure 5.4 /// Problem Analysis Triangle

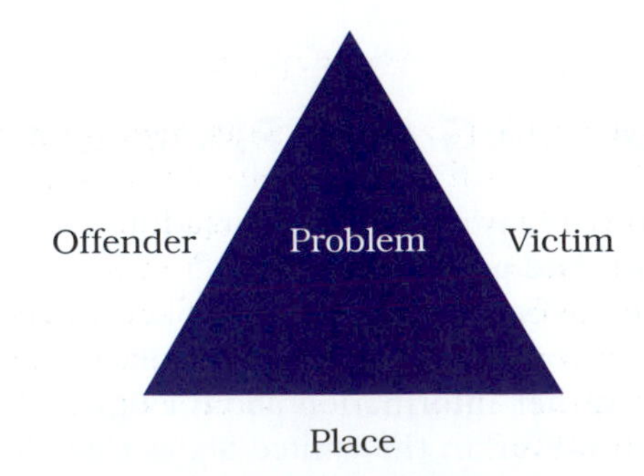

Source: Theory for Practice in Situational Crime Prevention, edited by Martha J. Smith and Derek B. Cornish. Copyright © 2003 by Lynne Rienner Publishers, Inc. Used with permission of the publisher.

helps officers analyze problems, it suggests where more information is needed, and it assists with crime control and prevention. Simply put, if there is a victim and he or she is in a place where crimes occur but there is no offender, no crime occurs. If there is an offender and he or she is in a place where crimes occur but there is nothing or no one to be victimized, then no crime will occur. If an offender and a victim are not in the same place, there will be no crime. Police must learn as much as possible about the victims, offenders, and locations where problems exist in order to understand what is prompting the problem and what can be done about it.

USE OF FORCE: A SACRED TRUST

The police must use force at times to protect themselves or others; such force can include verbal directives, empty-hand control (grabs, holds), less lethal means (e.g., chemical sprays, baton, electrical device), and lethal means. The issue that is often raised in a given situation is whether or not the type of force employed was reasonable under the circumstances—considering the severity of the crime, whether the suspect posed a threat, and whether the suspect was resisting or attempting to flee.

Given the unusually high number of police shootings—often involving unarmed minorities—that came to light in the United States beginning in August 2014 with the shooting death of Michael Brown in Ferguson, Missouri, subsequent police shootings of minorities in other cities have fostered thousands of demonstrations across the country. The "Black Lives Matters" movement initiated national scrutiny of police shootings.

A Threshold Question: Police as "Guardians" or "Soldiers"?

The contemporary chasm between police and minorities, and many people in society-at-large, often revolves around how the police are now too often being seen as "soldiers" or "warriors." The previously mentioned killings of young African American men and others have caused many people to ask whether or not there exists within the police culture a "warrior mindset." As a former police chief put it:

> Why are we training police officers like soldiers? Although police officers wear uniforms and carry weapons, the similarity ends there. The missions and rules of engagement are completely different. The soldier's mission is that of a warrior: to conquer. The rules of engagement are decided before the battle. The police officer's mission is that of a guardian: to protect. Soldiers come into communities as an outside, occupying force. Guardians are members of the community, protecting from within.[20]

Constitutional Policing and Procedural Justice

Several of the aforementioned race-related incidents beginning in 2014 have led to a careful review of police practices and calls for reform. Many police chief executives are now becoming much more involved in what is termed constitutional policing—a cornerstone of community policing and problem-solving efforts. When a police agency develops policies and practices that advance the constitutional goals of protecting citizens' rights and provides equal protection under the law, then, as New Haven, Connecticut, police chief Dean Esserman put it, "The Constitution is our boss. We are not warriors, we are guardians. The [police] oath is to the Constitution."[21]

A related concept is procedural justice, which is the guiding principle for building trust and confidence in the community. Procedural justice involves the principle of fairness to all, citizens having a voice and the ability to express their concerns, transparency and

You Be the . . .

Police Officer

As you read the following scenario, consider the "guardian versus soldier" distinction as well as the concept of constitutional policing:

> A police officer gave a woman a ticket for making an illegal turn. When the woman protested that there was no sign prohibiting the turn, the officer pointed to one that was bent out of shape, leaning over, and barely visible from the road. Furious and feeling the officer hadn't listened to her, the woman decided to appeal the ticket by going to court. On the day of her hearing, she was anxious to tell her side of the story. However, when she began to speak, the judge (being familiar with the location and the stop sign issue) interrupted her and summarily ruled in her favor, dismissing the case.

How do you believe the woman felt? Vindicated? Victorious? Satisfied? How would the officer feel? Was justice served in this case?

openness of process (meaning that the processes by which police decisions are made do not rely on secrecy or deception), and impartiality and unbiased decision making (decisions are made based on relevant evidence or data rather than on personal opinion, speculation, or guesswork).[22]

FROM PATROL OFFICER TO DETECTIVE: THE WORK OF FORENSICS AND INVESTIGATORS

Policies concerning time on the job, qualifications (knowledge, skills, abilities), and testing/interviewing will vary from agency to agency, but at some point the police officer may wish to be laterally assigned to a detective assignment (or to SWAT, K-9, gangs, drugs, motorcycle/bicycle patrol, or other specialty assignments). Aside from the status that a detective assignment might bring, it also often carries a raise in salary, a civilian clothing allowance, on-call pay, day-shift work, the ability to generally manage one's own time and activities (as opposed to being closely supervised), and other perquisites. But not everyone is destined to be good in this role because, as will be seen in the next section, the successful investigator must possess several personal attributes.[23]

The challenges involved with investigating crimes may well be characterized by a quote from Ludwig Wittgenstein: "How hard I find it to see what is right in front of my eyes!"[24] The art of sleuthing has long fascinated the American public, and news reports on the expanding uses of DNA and television series such as *CSI: Crime Scene Investigation* have done much to capture the public's fascination with criminal investigation and forensic science in the 21st century. This interest in "sleuthing" is not a recent phenomenon. For decades, Americans have feasted on the exploits of dozens of fictional masterminds, like Sherlock Holmes, Agatha Christie's Hercule Poirot and Miss Marple, Clint Eastwood's portrayal of Detective "Dirty Harry" Callahan, and Peter Falk's Columbo, to name a few.

Next we briefly review the important work that is accomplished by law enforcement personnel in both the forensic laboratories as well as in the field in the role of detective/criminal investigator; included is a look at the use of DNA.

Forensic Science and Criminalistics: Defining the Terms

The terms *forensic science* and *criminalistics* are often used interchangeably. **Forensic science** is the broader term; it is that part of science used to answer legal questions. It is the examination, evaluation, and explanation of physical evidence in law. Forensic science encompasses pathology, toxicology, physical anthropology, odontology (development of dental structure and dental diseases), psychiatry, questioned documents, ballistics, tool work comparison, and serology (the reactions and properties of serums), among other fields.[25]

Forensic science: the study of causes of crimes, deaths, and crime scenes.

Criminalistics is one branch of forensic science; it deals with the study of physical evidence related to crime. From such a study, a crime may be reconstructed. Criminalistics is interdisciplinary, drawing on mathematics, physics, chemistry, biology, anthropology, and many other scientific fields.[26] Basically, the analysis of physical evidence is concerned with identifying traces of evidence, reconstructing criminal acts, and establishing a common origin of samples of evidence. The types of information that physical evidence can provide are as follows:[27]

Criminalistics: the interdisciplinary study of physical evidence related to crime; drawing on mathematics, physics, chemistry, biology, anthropology, and many other scientific fields.

- Information on the *corpus delicti* (or "body of the crime") is physical evidence showing that a crime was committed, such as tool marks, a broken door or window, a ransacked home, and missing valuables in a burglary, or a victim's blood, a weapon, and torn clothing in an assault.
- Information on the *modus operandi* (or method of operation) is physical evidence showing means used by the criminal to gain entry, tools that were used, types of items taken, and other signs—items left at the scene, an accelerant used at an arson scene, the way crimes are committed, and so forth.
- *Linking a suspect with a victim* is one of the most important linkages, particularly with violent crimes. It includes hair, blood, clothing fibers, and cosmetics that may be transferred from victim to perpetrator. Items found in a suspect's possession can also be linked to a victim.

Crime scene: any location where a crime occurred and that may contain forensic evidence relating to and supporting a criminal investigation.

- *Linking a person to a crime scene* is also a common and significant linkage. It includes fingerprints, glove prints, blood, semen, hairs, fibers, soil, bullets, cartridge cases, tool marks, footprints or shoeprints, tire tracks, and objects that belonged to the criminal. Stolen property is the most obvious example.
- In terms of *disproving or supporting a witness's testimony*, evidence can indicate whether or not a person's version of events is true. An example is a driver whose car matches the description of a hit-and-run vehicle. If blood is found on the underside of the car and the driver claims that he hit a dog, tests on the blood can determine whether the blood is from an animal or from a human.
- One of the best forms of evidence for *identification of a suspect* is DNA evidence, which proves "individualization." It can prove, without a doubt, that the person whose DNA was found was at the **crime scene**.

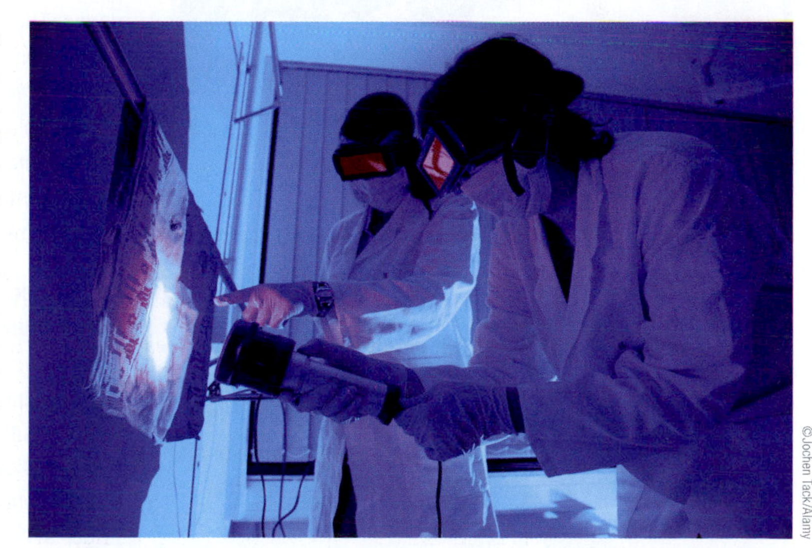

Forensic scientists examine all kinds of articles and substances in their search for physical evidence that will link persons to their crimes.

Attributes of Detectives

Detective/criminal investigator: a police officer who is assigned to investigate reported crimes, to include gathering evidence, completing case reports, testifying in court, and so on.

To begin to understand the role of **detectives/criminal investigators**, one must first understand that these individuals typically begin their careers as police officers before being promoted to detective. In other words, unless hired by a state or federal agency, such as a state bureau of investigation or the Federal Bureau of Investigation and specifically to be trained as an investigator, one must go through the traditional hurdle process outlined previously and be hired as a patrol officer. Then, having distinguishing himself or herself in that role and (as is commonly done) having worked for a specified number of years and obtained enough experience, one may then test and interview for "the gold badge." To be effective, detectives must also be trained in general areas such as the laws of arrest, search, and seizure; investigative principles and practices; judicial proceedings; and oral and written communications. More specialized training will often be required if individuals are

Practitioner's Perspective

Crime Scene Investigator

Name: Angela Benford
Position: Crime Scene Investigator
Location: Aurora, Colorado

What is your career story? I started as an Explorer with the Arapahoe County Sheriff's Department, which is a program for the Boy Scouts. I started there at 15. There, I got to work with police officers and the Crime Prevention Unit and the crime lab, and I was exposed to all of the different criminal justice programs. When I started that, I was going to be a police officer, so I went to college and majored in criminal justice. But then, I took the science class that they have designed for the criminal justice students, Intro to Criminalistics, and I got hooked. We had a person come in and talk to us about blood spatter interpretation, and we did shoe print castings. And I decided that's what I wanted to do. So I added another major and took a bunch of chemistry and double majored in criminal justice and chemistry.

What are some challenges and misconceptions you face in this position? Some of the biggest myths are perpetuated by the media, with TV shows like *CSI* and *NCIS*, and others on reality TV. In these shows, if you're a crime scene investigator, you show up on the scene, you collect all the evidence, you process all the evidence, and you arrest the suspect. But in reality, there are multiple jobs for each of those actions. The crime scene investigators are the ones that go out to the scenes and document the evidence. Then we bring it back to a lab specialist, and depending on what type of evidence it is, it could go to a latent print examiner or a question document examiner, so there are multiple jobs in that. And then it's a detective's job to go out and interview suspects and arrest them. In some police departments the roles are more merged, but the larger the police department, the more individuated the roles are.

What directions do you envision your department going in the future? What I see changing in the future of crime scene investigation is that technology is going to keep growing and growing. Already we have a 3-D scanner that we take out to a scene, and it documents the measurements and takes photographs, and we can then create a model of the scene back at the lab. These technologies are going to get even better. To the point where you might be a jury member in court and they bring in a computer, and you can walk through the crime scene as they're talking about it.

To learn more about Angela Benford's experiences as a Crime Scene Investigator, watch the Practitioner's Perspective video in SAGE Vantage.

You Be the . . .

Detective

As the coach opened the door to the locker room, the only light that shone was from the players' large shower area. Upon flipping the light switch, he saw the body of his once-"ace" pitcher, Hines, lying on the floor in the shower. In his pale left hand he held a gun. There was a bullet wound in his left temple. Under his tanned right hand was a note saying, "My pitching days are gone, my debts and humiliation more than I can bear. Sorry." His nearby locker contained a half-empty bottle of beer, his uniform, an uneaten stadium hot dog, a picture of his two children, and his Acme-brand ball glove with "RH" stamped in the webbing. Wet footprints were observed walking in and out of the shower. Upon surveying the scene, the responding detective said, "I do not believe this was a suicide."

1. Was this a suicide or a murder?
2. What fact(s) led you to this conclusion?

specializing in areas such as sex crimes, family crimes, homicide investigation, and gang and drug enforcement.

In sum, to be successful, the investigator must possess four personal attributes to enhance the detection of crime: an unusual capability for observation and recall; extensive knowledge of the law, rules of evidence, scientific aids, and laboratory services; power of imagination; and a working knowledge of social psychology.[28]

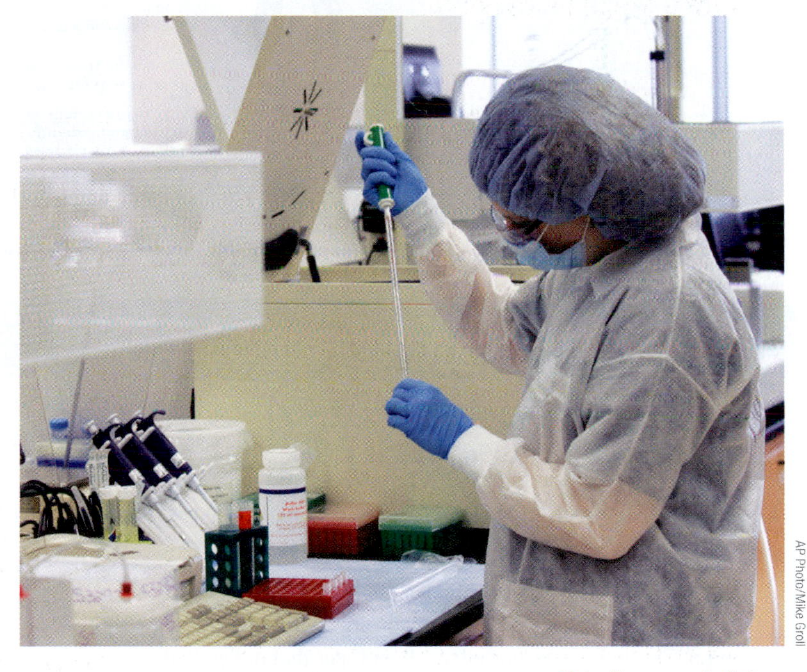

AP Photo/Mike Groll

Today the power of what can be done with DNA, as well as its variety of uses, is incredible.

DNA: deoxyribonucleic acid, which is found in all cells; used in forensics to match evidence (hair, semen) left at a crime scene with a particular perpetrator.

Using DNA Analysis

Today **DNA** is the most sophisticated and reliable type of physical evidence (see Figure 5.5). Police are now able to submit to laboratories evidence that until recently was not even possible to examine. For example, "touch" DNA evidence can now be examined and requires very small amounts of skin cells left on an object after it has been touched or handled; it can be used to determine whether a defendant merely touched a weapon, or whose hand threw drugs to the floor of a room.[29]

A testimonial to DNA's promise in investigations is offered by a former supervising criminalist of the Los Angeles County Sheriff's Department:

> The power of what we can look for and analyze now is incredible. It's like magic. Every day we discover evidence where we never thought it would be. You almost can't do anything without leaving some DNA around. DNA takes longer than fingerprints to analyze, but you get a really big bang for your buck.[30]

Furthermore, DNA has allowed investigative personnel to exonerate people who were convicted in the past for crimes they did not commit. Indeed, the National Registry of Exonerations at the University of Michigan Law School lists more than 2,400 exonerations since 1989 with the help of DNA evidence.[31]

Figure 5.5 /// "What Is DNA?"

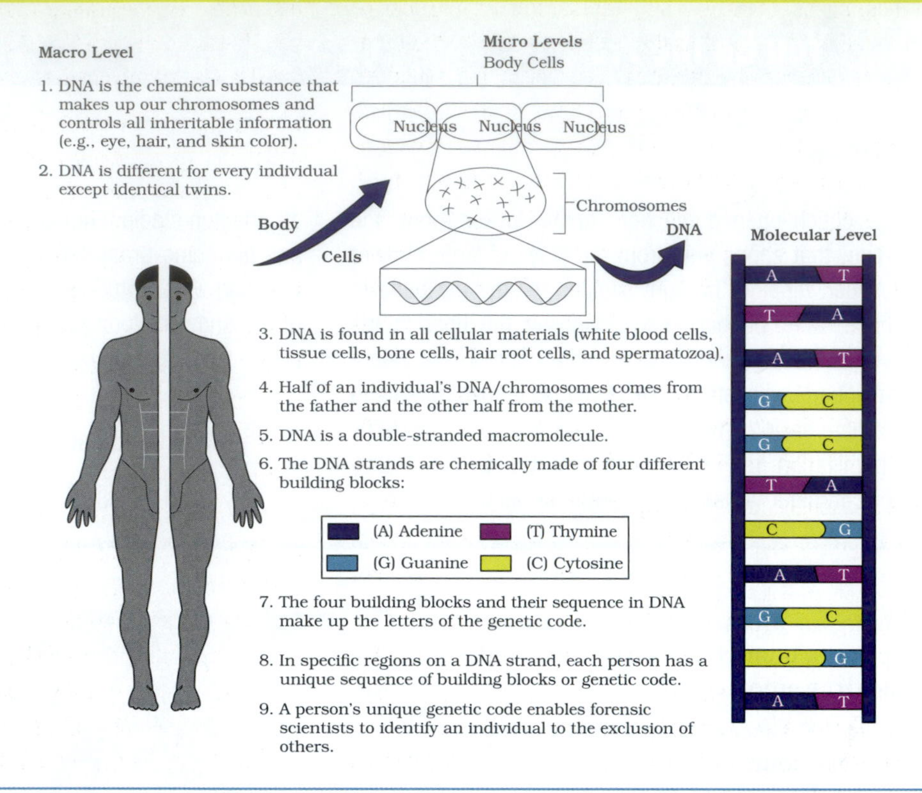

Macro Level
1. DNA is the chemical substance that makes up our chromosomes and controls all inheritable information (e.g., eye, hair, and skin color).
2. DNA is different for every individual except identical twins.
Micro Levels
Body Cells
Nucleus
Nucleus
Nucleus
Chromosomes
Body
Cells
DNA
Molecular Level
3. DNA is found in all cellular materials (white blood cells, tissue cells, bone cells, hair root cells, and spermatozoa).
4. Half of an individual's DNA/chromosomes comes from the father and the other half from the mother.
5. DNA is a double-stranded macromolecule.
6. The DNA strands are chemically made of four different building blocks:
(A) Adenine
(T) Thymine
(G) Guanine
(C) Cytosine
7. The four building blocks and their sequence in DNA make up the letters of the genetic code.
8. In specific regions on a DNA strand, each person has a unique sequence of building blocks or genetic code.
9. A person's unique genetic code enables forensic scientists to identify an individual to the exclusion of others.

Juvenile Justice Journal

Differential Treatment by the Police

Assume it is about 10:00 on a warm summer's night. A municipal police officer is dispatched to a residence to take a theft report. Upon arrival, she is informed by the residents that a very expensive bicycle has been stolen from their front porch. The victims further inform the officer that earlier that afternoon they observed a juvenile—whom they know by name, because he lives a few blocks up the street—walking on the sidewalk across the street and looking furtively at the bicycle. The officer recognizes the youth's name by reputation (i.e., prior involvement with police) and is aware of his home address. She drives her patrol car by the youth's home and, through the open front door, observes that the living room is dark but the television is turned on. She goes to the front door, and the juvenile is alone watching television; he comes to the door and tells the officer that his parents are sleeping. The officer knows that if she wakes the parents and (in their presence) asks the boy if he knows anything about the stolen bicycle, he will deny any such knowledge.

1. How should the officer proceed?
2. Is there an option (i.e., an extralegal means) available for handling the matter that will have a positive outcome for the juvenile, the victims, and the police officer?

Note: This relatively simple case study represents police work as it often occurs on the streets, where there is a solitary officer with little or no opportunity to immediately seek a search warrant. This case incorporates several topics discussed in other chapters: the Fourth Amendment (probable cause, arrest, search and seizure); the exercise of discretion by police (i.e., knowledge of someone's past criminal activity may influence the officer's decision-making process); police policy and procedures (regarding treatment of juveniles); and police ethics. Also implicated are James Q. Wilson's "watchman" style of policing (discussed earlier in this chapter) and Herbert Packer's crime control/due process dichotomy (discussed in Chapter 1).

In most cases, the officer is given several cues for deciding whether to take official or unofficial action: the complainant's wishes; the nature of the violation; the offender's race, attitude, and gender; knowledge of the offender's prior police contacts and crimes; laws, statutes, and ordinances; agency policies and procedures; and so on. In some instances, there might be a combination of cues: If the offense is minor and the complainant does not desire to pursue the matter, the officer may prefer to handle the case unofficially (and, as well, avoid the paperwork, trips to court, and so on). Furthermore, if certain conditions are present (a minor offense, a youth who is remorseful and from a good family, and so on), the officer might try to convince the complainant not to take official action.

Male and female juveniles may be treated differently as well, and an officer might take a hard-nosed view toward certain offenses. Similarly, a juvenile suspect who displays a very sarcastic or machismo attitude will likely not be given any benefit.

Concerning outcomes, the officer may simply release the juvenile in question, release the juvenile but submit a report describing the encounter to the juvenile probation office or the police department, verbally reprimand the juvenile and release him or her, take the juvenile into custody to attempt an attitude adjustment, or make a formal arrest of the juvenile (note that only the last two alternatives involve official action).[32]

Of course, with respect to abused or neglected children, police are typically much more concerned with the safety and well-being of the child or youth but also may well find that options are more restricted by laws and agency policies. In some cases, training and expertise will be required to recognize and report suspected incidents of child abuse or neglect, while special-assignment detectives (e.g., family crimes unit) or the state or county department of child services may have to be called in for abused or neglected children.

If the officer opts to handle a juvenile situation under the official umbrella of the law, he or she must know the relevant law (as set forth in Chapter 6) but also must decide whether or not to notify a juvenile's parents about the incident, release the juvenile to his or her parents, or detain the youth (detention in a lockup is rarely used and can even be illegal, however). Other considerations include whether or not to obtain fingerprints and photographs, and whether transportation for the youth is necessary.

Also, DNA is used to solve "cold" cases. These are cases selected for investigation that are usually at least a year old and cannot be addressed by the original investigative personnel because of workload, time constraints, or the lack of viable leads. Cases are prioritized on the basis of the likelihood of an eventual solution, with the highest priority cases being those in which there is an identified homicide victim, suspects were previously named or identified through forensic methods, an arrest warrant was previously issued, significant physical evidence can be reprocessed, newly documented leads have arisen, and critical witnesses are available and willing to cooperate.

/// IN A NUTSHELL

- The idea of a police subculture was first proposed in 1950 by William Westley, who found a high degree of secrecy and violence.
- Recruit academy training covers a variety of subjects; neophyte officers learn how to use lethal and less lethal weapons, and how to deal with criminal suspects, offenders, victims, and witnesses.
- After leaving the academy, new officers are assigned to a veteran officer—a field training officer—for initial field instruction and observation; this phase of training helps recruits make the transition from the academy to the streets under the protective arm of a veteran officer, while on probationary status.
- Police have four basic tasks: enforcing the laws, performing welfare tasks, preventing crimes, and protecting the innocent.
- James Q. Wilson maintained that there are three distinctive policing styles: the watchman style, the legalistic style, and the service style.
- Although workers in several other occupations die at much higher rates than those in policing, this occupation still has many occupational hazards.
- Full enforcement of the laws by police is a myth; they typically have considerable discretion in whether or not to arrest someone. Determining factors include the law (e.g., some ordinances mandate arrest for certain offenses, such as domestic violence), the officer's attitude (concerning the law that is violated, as well as toward the offender), and the citizen's attitude toward the officer.

- Today, under the community-policing philosophy, officers are trained to examine the underlying causes of problems in the neighborhoods on their beats and to involve citizens in the long-term resolution of neighborhood problems.
- The fields of forensic science and criminalistics are the most rapidly developing areas in policing—and probably in all of criminal justice.
- DNA is the most sophisticated and reliable type of physical evidence, used to obtain convictions in contemporary as well as "cold" cases and to exonerate many arrestees.
- The police officer's knowledge (i.e., of the federal, state, and local laws, statutes, ordinances) as well as agency policies concerning juvenile offenders and those in need of services is vastly different from the kinds of knowledge and practices used with adults; still, there is much discretion at the officer's disposal.

/// KEY TERMS & CONCEPTS

Review key terms with eFlashcards at **edge.sagepub.com/peakbrief**.

Academy training 104

Community policing and problem solving 111

Crime scene 115

Criminalistics 115

Detective/criminal investigator 116

DNA 117

Field training officer (FTO) 105

Forensic science 115

Kansas City Preventive Patrol Experiment 108

Policing role 106

Policing styles 106

Sixth sense 104

Tasks of policing (four basic) 107

/// REVIEW QUESTIONS

Test your understanding of chapter content. Take the practice quiz at **edge.sagepub.com/peakbrief**.

1. What ideal traits are sought among those persons wishing to enter policing?
2. What are some topics that police trainees study while attending the academy?
3. What are the methods and purposes of the FTO concept?
4. What are the prevailing styles of policing? The four basic tasks of the police?
5. How do you define police discretion, and what are some examples of its use? What are some pros and cons of police discretion?
6. What are the attributes possessed and methods used that tend to be present in successful investigative personnel?
7. What is DNA, and how is it used in policing?
8. What kinds of special knowledge, skills, and abilities must police officers generally possess in dealing with juvenile offenders?

/// LEARN BY DOING

1. You are a patrol sergeant lecturing to your agency's Citizens' Police Academy about the patrol function. Someone asks, "Sergeant, do your officers enforce all of the laws all of the time? If not, which laws are always enforced, and which ones are not?" How do you respond to her? How would you fully explain the use of police discretion to the group?
2. You have been assigned to develop an examination for persons seeking a detective assignment in your agency. What will be the content of your exam? (For guidance, you may wish to begin by reviewing an excellent study guide for the investigator's exam used by the Los Angeles County District Attorney's Office, Examination Unit, at http://da.co.la.ca.us/sites/default/files/invstudyguide.pdf.)
3. For a practical view of "what works" in terms of addressing all kinds of criminal matters, visit the website of the Center for Problem-Oriented Policing at http://www.popcenter.org/problems/. The center has published more than 70 POP guides, which summarize how a variety of problems of crime and disorder can be examined and addressed.
4. If you possess an interest in working with juveniles, contact a police officer/detective specializing in family crimes, a juvenile probation officer, social services office, or so on, in order to learn more about their methods and challenges.

/// STUDY SITE

SAGE edge™

Get the tools you need to sharpen your study skills. SAGE edge offers a robust online environment featuring an impressive array of free tools and resources.

Access practice quizzes, eFlashcards, video, and multimedia at **edge.sagepub.com/peakbrief**

©REUTERS/Jim Young

CHAPTER 6

EXPOUNDING THE CONSTITUTION

Laws of Arrest, Search, and Seizure

We must never forget that it is a constitution we are expounding.

—John Marshall, *McCulloch v. Maryland* (1819)[1]

LEARNING OBJECTIVES

As a result of reading this chapter, you will be able to

1. Define and provide examples of probable cause
2. Explain the rationale for, ramifications of, and exceptions to the exclusionary rule
3. Distinguish between arrests and searches and seizures with and without a warrant
4. Review what is meant by "stop and frisk"
5. Explain generally to what extent the police may use electronic devices in order to obtain information from members of the public
6. Discuss some significant ways in which the *Miranda* decision has been eroded
7. Describe what rights are held by criminal defendants regarding the right to counsel, for both felony and misdemeanor arrests

ASSESS YOUR AWARENESS

Test your knowledge of your constitutional rights regarding arrest, search, and seizure by responding to the following eight true-false items; check your answers after reading this chapter.

1. The standard for a legal arrest, search, and seizure is probable cause.
2. The exclusionary rule requires that evidence obtained by police in violation of the Fourth Amendment must *not* be used against the defendant in a criminal trial.
3. There are essentially two ways for the police to make an arrest: with a warrant and without a warrant.
4. The courts grant police greater latitude in searching automobiles, which can be used to help make a getaway and to hide evidence.
5. The *Miranda* warning remains intact from its original version; no deviations in its explanation or practice by police are permitted.
6. Generally, police do not have to "search" for things they see in plain view.
7. The Sixth Amendment guarantees the right to a speedy and public trial, as well as the assistance of counsel to those accused of crimes by our justice system.
8. The burden of proof standard is the same for both juvenile and adult criminal trials.

<< Answers can be found on page 293.

A group of armed robbers targeted Radio Shack and T-Mobile stores in Michigan and Ohio between late 2010 and early 2011. The offenders forced employees to the back of the stores and filled their bags with new smartphones. A few of the robbers were eventually arrested; one confessed to the crimes and turned his cell phone over to police. Investigators reviewed the calls made from his phone around the time of the robberies and identified the cell phone numbers of other individuals they believed participated in the crimes. Although the evidence was not strong enough to obtain a warrant, police gained access to the cell phone records through the Stored Communications Act. Through these records, they determined that Timothy Carpenter was within a two-mile radius of the stores at the time of the crimes. Carpenter was arrested and charged with armed robbery.

Just a few years prior, in August 2009, a San Diego police officer pulled over David Riley for expired registration tags. The officer then found that Riley's driver's license was suspended, and, following department policy for such cases, he searched the car before having it towed and impounded. He found two illegal handguns under the hood and arrested Riley. As is standard practice during a lawful arrest, the officer searched Riley and found his cell phone in a pocket. The phone contained evidence—pictures, cell phone contacts, texts messages, and video clips—indicating Riley's membership in a local gang. The cell phone also included a picture of a different vehicle that Riley owned and that had been involved in a gang shooting a few days earlier. Ballistics tests would later show that the handguns found in Riley's

©Tim Gainey/Alamy

Cell phones have changed everything, including the law that governs police officers when they arrest someone.

car were linked to the earlier shooting. Based in part on the evidence recovered from his cell phone, Riley was charged in connection with a gang shooting.

Like most people around the world today, Timothy Carpenter and David Riley used their cell phones in such a way that the phones contained many private details about their lives—pictures, contacts, communications with other people, videos, and even records of their physical locations. Some of those details were evidence of criminal activity, but surely other "evidence" on these phones was simply the typical private, personal information we all carry with us wherever we go.

The Fourth Amendment to the U.S. Constitution protects us from unreasonable police searches and seizures, and the long line of U.S. Supreme Court cases interpreting the Fourth Amendment repeatedly recognize that we have reasonable expectations of privacy in certain circumstances. Sometimes, that expectation of privacy requires police to get a warrant before they can search people and places.

Did Timothy Carpenter have an expectation of privacy concerning the places in which his phone was used, even if the phone produced a record of his movements? Did David Riley have an expectation of privacy in the contents of his cell phone? Did Carpenter's and Riley's crimes change those expectations? Did the police have a legitimate, overriding interest in seizing evidence from these phones? Should the government be able to use personal cell phone evidence against suspects in criminal trials?

Think about these questions and about the contents on your own cell phone as you read further and learn about the critically important area of constitutional law in the criminal justice system.

INTRODUCTION

What constitutional rights do Americans possess regarding arrest, search, and seizure? How are the police constrained by those rights? If the police have a search warrant for a friend you are accompanying, may they search you as well? If the police knock on your apartment door, must you allow them to enter if you do not see a search warrant in their possession? And if they arrest you, should they be able to search your person, your car, your apartment—your cell phone? These are challenging questions if someone is not familiar with the Bill of Rights. After you read this chapter, however, the answers should be clear.

Most college and university students probably have few opportunities to witness police actions that concern the U.S. Constitution. Although they may become directly and innocently involved with arrest, search, and seizure by being present at parties, in friends' vehicles, or at their places of employment, most Americans' knowledge of the Bill of Rights probably comes from movies and television crime shows—which are questionable at best in terms of how they portray police conduct. This chapter will clarify any such confusion.

We focus in this chapter on three of the ten amendments that constitute the Bill of Rights: the Fourth Amendment (probable cause, the exclusionary rule, arrest, search and seizure, electronic surveillance, and lineups), the Fifth Amendment (confessions and interrogations), and the Sixth Amendment (right to counsel and interrogation). Chapter 7 covers the Eighth Amendment, prohibiting cruel and unusual punishment; furthermore, Chapter 1 covers the concept of due process, which is guaranteed by the Fourteenth Amendment. As such, these two amendments are not examined in detail in this chapter. Also note that the law as it pertains to juvenile offenders is quite different from that for adults. Significant legislation and court cases that have shaped the U.S. juvenile justice system are covered in the "Juvenile Justice Journal" box at the end of this chapter.

Some Caveats

Students of criminal justice might do well to remember the classic words of John Adams, who said in 1774 that a republic is "a government of laws, and not of men."[2] The Bill of Rights was enacted largely to protect all citizens from excessive governmental power.

As was seen in previous chapters of this book, criminal justice practitioners have far-reaching powers. Furthermore, agencies of criminal justice have the added benefit of using expert witnesses, forensic crime laboratories, undercover agents, informants, and so forth. Therefore, the Bill of Rights serves as an important means of "balancing the scales"—controlling the police and others so that they conduct themselves in a manner suited to a democratic society. Criminal justice professionals must conform their behavior to the rule of law as set forth not only in the U.S. Constitution but also in the state constitutions, statutes enacted by state legislatures, municipal ordinances, and the precedent of prior interpretations by the courts.

Remember also that the law is *dynamic*—that is, constantly changing—by virtue of acts by federal and state courts as well as their legislative bodies. Therefore, criminal justice practitioners must stay abreast of such changes and have formal mechanisms for imparting these legal changes to their employees. In the best case, agencies will have an in-house assistant district attorney—or, at the least, one who is on call—to render advice concerning legal matters.

A final note: Anyone who believes our legal system to be unduly harsh and restrictive might wish to research police tactics used in Saudi Arabia and China. Remember that governments—through their criminal justice systems (and, too often, their military forces)—may employ many methods to maintain order and attempt to maintain "justice." In Saudi Arabia or China, however, the methods used are far different from those in the United States. But would "justice" really result? And would many Americans want to *live* in one of those venues?

THE FOURTH AMENDMENT

> The right of the people to be secure in their persons, papers, and effects, against unreasonable searches and seizures, shall not be violated, and no Warrants shall issue, but upon probable cause, supported by Oath or affirmation, and particularly describing the place to be searched, and the persons or things to be seized.

The **Fourth Amendment** is intended to limit overzealous behavior by the police. Its primary protection is the requirement that a neutral, detached magistrate, rather than a police officer, issue **warrants for arrest and search**. The all-important principle of separation of powers in our American government demands that the legal right to search be decided by someone other than an agent for the police, because the police are not expected to be neutral or objective with respect to police matters. Instead, a neutral judicial officer oversees this important police function—the judicial branch oversees the executive branch to ensure law enforcement actions are constitutional.[3]

Probable Cause

The standard for a legal arrest, as well as for **search and seizure**, is **probable cause**. This important concept is elusive at best; it is often quite difficult for professors to explain and even more difficult for students to understand. One way to define probable cause is to say that for an officer to make an arrest or conduct a search of someone's person or effects, he or she must have a reasonable basis to believe that a crime has been or is about to be committed by that individual (which, in the case of a search, can include the individual's mere possession of some form of illegal contraband).

Of course, the facts and probable cause indicators of each case are different. The court will examine the type and amount of probable cause that the police officer had, but the officer cannot add to the probable cause used to make an arrest *after* making the arrest. The judge will then determine whether or not sufficient probable cause existed to arrest the individual

Fourth Amendment: in the Bill of Rights, it contains the protection against unreasonable searches and seizures and protects people's homes, property, and effects.

Warrant, arrest: a document issued by a judge directing police to immediately arrest a person accused of a crime.

Warrant, search: a document issued by a judge, based on probable cause, directing police to immediately search a person, a premises, an automobile, or a building for the purpose of finding illegal contraband felt to be located therein and as stated in the warrant.

Search and seizure: in the Fourth Amendment, the term refers to an officer's searching for and taking away evidence of a crime.

Probable cause: a reasonable basis to believe that a crime has been, or is about to be, committed by a particular person.

AP Photo/Greg Barnette/Record Searchlight

The exclusionary rule requires that evidence obtained improperly by the police must not be used in a criminal trial. These two officers wait outside a jewelry store for a search warrant to be signed by a judge so that they may then search the property for stolen goods.

Exclusionary rule: the rule (see *Mapp v. Ohio*, 1961) providing that evidence obtained improperly cannot be used against the accused at trial.

based on the officer's knowledge of the facts *at the time of* the arrest.

The U.S. Supreme Court has upheld convictions when probable cause was provided by a reliable informant,[4] when it came in an anonymous letter,[5] and when a suspect fit a Drug Enforcement Administration profile of a drug courier.[6]

The Exclusionary Rule

The Fourth Amendment protects people's right to be secure against unreasonable searches and seizures in their houses, papers, and effects. However, simple as that may sound, the manner in which the amendment is applied on the street is more complicated. First of all, not *all* searches are prohibited—only those that are unreasonable. Another issue has to do with how to handle evidence obtained illegally. Should murderers be released, Justice Benjamin Cardozo asked, simply because "the constable blundered"?[7] The Fourth Amendment says nothing about how it is to be enforced; this is a problem that has stirred a good amount of debate for a number of years. Most of this debate has focused on the constitutional necessity for the **exclusionary rule**, which basically requires that all evidence obtained in violation of the Fourth Amendment must be excluded from government use in a criminal trial.

The exclusionary rule first appeared in the federal criminal justice system when the U.S. Supreme Court ruled in *Weeks v. United States* (1914) that all illegally obtained evidence was barred from use in federal prosecutions.[8] Then, in *Mapp v. Ohio* (1961), the Court applied the doctrine to the states' courts.[9] In May 1957, three Cleveland police officers went to the home of Dollree Mapp to investigate intelligence that suggested the owner was hiding a recent bombing suspect and engaged in an illegal gambling operation. Mapp, after telephoning her lawyer, refused the police entry without a search warrant. Mapp refused entry again three hours later, but police forcibly entered the home. Mapp demanded to see a search warrant; an officer waved a piece of paper at her, which she grabbed and placed down her shirt. The officers struggled with Mapp to retrieve the piece of paper, at which time Mapp's attorney arrived at the scene. The attorney was not allowed to enter the house or to see his client. One officer searched her upstairs bedroom found a brown paper bag containing books that he deemed to be obscene. A jury convicted Mapp of possession of obscene, lewd, or lascivious materials, and she was sentenced to an indefinite term in prison. In June 1961, the U.S. Supreme Court overturned the conviction, holding that the Fourth Amendment's prohibition against unreasonable search and seizure had been violated and that as

> the right to be secure against rude invasions of privacy by state officer is . . . constitutional in origin, we can no longer permit that right to remain an empty promise. We can no longer permit it to be revocable at the whim of any police officer who, in the name of law enforcement itself, chooses to suspend its enjoyment.[10]

Although the exclusionary rule continues to safeguard citizens against unreasonable search and seizure, a 2016 U.S. Supreme Court decision limited the scope of this protection, particularly for those who have outstanding arrest warrants. Acting on an anonymous tip, Utah detective Douglas Fackrell observed Edward Joseph Strieff Jr. leaving a residence where drug sales were suspected. Without sufficient evidence, Detective Fackrell illegally stopped Strieff for questioning. During the stop, Fackrell discovered that Strieff had an outstanding warrant and arrested him. Fackrell conducted a lawful search following the arrest and found that Strieff was carrying a drug pipe and methamphetamine. In a 5–3 decision,[11] the U.S. Supreme

Court ruled that, even though the initial investigatory stop was unconstitutional, exclusion of the evidence obtained during the search incident to arrest was not justified and the items found during the search could be lawfully considered as evidence for trial since the defendant had an outstanding warrant, which led directly to the subsequent arrest and search.

The Timothy Carpenter and David Riley Cases: Now that you understand the Fourth Amendment, expectations of privacy, and the landmark *Mapp* case, reconsider the two cell phone cases that opened this chapter and how police obtained "evidence" from these phones. Was the police action an invasion of privacy in either case? Did officers have time or a legal obligation to obtain warrants first? The defendants' attorneys argued that the cell phone evidence was obtained illegally and that, under the exclusionary rule established in the *Mapp* case, the government should not have been able to use it in court. How do you think the U.S. Supreme Court ruled in these cases? We revisit both cases later in this chapter.

Arrests With and Without a Warrant

It is always best for a police officer to arrest someone with a warrant, because that means a neutral magistrate—rather than a police officer—has examined the facts and determined that the individual should be arrested in order to make an accounting of the charge(s). An arrest made with a valid warrant will be presumed constitutional or legal, so an officer will always opt to act with a warrant when possible.

To obtain an arrest warrant, the officer or a citizen swears in an **affidavit** (as an "affiant") that he or she possesses certain knowledge that a particular person has committed an offense. This person might be, as an example, a private citizen who informs police or the district attorney that he or she attended a party at a residence where drugs or stolen articles were present. Or, as is often the case, a detective gathers physical evidence or interviews witnesses or victims and determines that probable cause exists to believe that a particular person committed a specific crime. In either case, the affidavit is presented to a judge, and if the judge finds that probable cause exists, he or she will issue the arrest warrant. Officers will execute the warrant, taking the suspect into custody to answer the charges.

Affidavit: any written document in which the signer swears under oath that the statements in the document are true.

Furthermore, the Supreme Court has held that police may not make "a warrantless and nonconsensual entry into a suspect's home in order to make a routine felony arrest" except where there are **exigent circumstances**.[12] Exceptions made based on exigent circumstances apply only when the officer possesses probable cause and immediate action is required—to prevent danger to life or serious damage to property, the escape of a suspect, the destruction of evidence, or some other such action.[13] Patrol officers, unlike detectives, rarely have the time or opportunity to perform an arrest with a warrant in hand because the suspect is generally trying to escape, dispose of evidence, and so on. Although the situation described in the accompanying "You Be the . . . Police Officer" box involves consent to search, it serves as a good example.

Exigent circumstance: an instance in which quick, emergency action is required to save lives, protect against serious property damage, or prevent suspect escape or evidence destruction; in such cases, officers can enter a structure without a search warrant .

Police may not randomly stop a vehicle to spot-check the driver's license and registration; to do so they must at a minimum possess articulable, reasonable suspicion (e.g., that a motorist is unlicensed, the automobile is not registered, or the vehicle or an occupant is subject to arrest for violation of law).[14] However, the Supreme Court ruled in 1990 that the stopping of all vehicles passing through sobriety checkpoints—a form of seizure—did not violate the Constitution, although singling out individual vehicles for random stops without probable cause is not authorized.[15] Several days later, it ruled that police were not required to give drunk-driving suspects a *Miranda* warning and could videotape their responses.[16] The Supreme Court held also that police may arrest *everyone* in a vehicle in which drugs are found,[17] and that police may set up roadblocks to collect information from motorists about crime. Short stops, "a very few minutes at most," are not too intrusive, considering the value in crime solving, the Court noted.[18]

You Be the . . .

Police Officer

One spring afternoon a police officer was dispatched to the residence of several university students, who were hosting a keg party for about 20 guests. They reported that three unknown men (two white, one African American) entered the home, had a few beers, and quickly left. Soon thereafter, a party guest also left and discovered that the stereo had been taken from his car, parked in the back yard. A description of the men and their vehicle was given to the officer, who soon (within 20 minutes and a mile from the crime scene) observed a vehicle and three men, all of whom matched the description that was given. The officer stopped the vehicle, informed the dispatcher of the stop and location, and approached the vehicle.

1. Are exigent circumstances and probable cause sufficient for the officer to search the vehicle without a warrant or the men's consent? If you believe not, how do you propose the three men be handled while a warrant is sought from a judge?
2. Assume the officer asks the vehicle's driver for permission to search the vehicle, and consent is given; what is the officer's next move? Instead, assume the driver of the vehicle refuses to give the officer permission to search the vehicle; what action(s) do you believe the officer can then legally take?
3. Assume the officer, with or without a warrant, locates the stolen stereo equipment inside the vehicle under the driver's seat. Can the officer then arrest only the vehicle's driver, or should all other occupants be arrested as well?
4. Following the arrest of one or more of the occupants, can they then be searched (if necessary, see the "Searches Incident to Lawful Arrest" section, later in this chapter)?
5. What other issues arise for the officer (e.g., his or her safety) once an arrest is made of one or more of the occupants?

©iStockphoto.com/kali9

Police officers may perform warrantless arrests and searches under certain circumstances if they have probable cause to do so.

Search and Seizure in General

An old saying holds that "a person's home is his castle," and the police are held to a high Fourth Amendment standard when wanting to enter and search someone's domicile. Regarding what have become known as "knock and announce" cases, the U.S. Supreme Court has determined that the Fourth Amendment requires the police to first knock and announce their presence before entering a person's home for the purpose of executing a warrant; the Court allows exceptions, however, if knocking and announcing would be likely to endanger the officers or lead to the destruction of evidence. In mid-2011, the Supreme Court expanded the ability of police to enter a home without a warrant—but under exigent circumstances. In *Kentucky v. King,* police in Lexington were pursuing a drug suspect and banged on the door of an apartment where they thought they smelled marijuana.[19] After identifying themselves, the officers heard movement inside the apartment and, suspecting that evidence was being destroyed, kicked in the door and found King smoking marijuana.

King, convicted of multiple drug crimes, appealed, and Kentucky's highest court ruled there were no "emergency circumstances" present and thus the drugs were inadmissible because police should have sought a search warrant. The Supreme Court disagreed, saying that the police acted reasonably: When police knock on a door and there is no response, and then hear movement inside that suggests evidence is being destroyed, they are justified in breaking in.

In other Fourth Amendment cases, the Court has upheld a warrantless search and seizure of garbage in bags outside the defendant's home,[20] as well as approaching seated bus passengers and asking permission to search their luggage for drugs (because, they reasoned, such persons should feel free to refuse the officer's request).[21] The Court also decided that no "seizure" occurs (and therefore the Fourth Amendment does not apply) when a police officer is in a foot pursuit (here, a juvenile being chased by an officer threw down an object, later determined to be crack cocaine).[22]

Searches and Seizures With and Without a Warrant

As with arrests, the best means by which the police can search a person or premises is with a search warrant that has been issued by a neutral magistrate. A warrant will be issued only if a judge finds, after reviewing a police officer's sworn affidavit, that there is probable cause to believe the named person possesses sought-after evidence (e.g., a weapon or stolen goods), or that the evidence is present at a particular location. Again, as with an arrest with a warrant, typically it is the detectives who have the "luxury" of searching and seizing with a warrant after interviewing victims and witnesses, perhaps obtaining analyses at a forensics laboratory, and gathering enough available evidence to request a search warrant. Figure 6.1 shows the pertinent parts of a search and seizure warrant form for persons or property that is used by the U.S. district courts, for execution by agents of the federal government.

Patrol officers rarely have the opportunity to perform a search with a warrant, as the flow of events typically requires quick action to prevent escape and to prevent evidence from being destroyed or hidden. Five types of searches may be conducted without a warrant: (1) searches incident to lawful arrest, (2) searches during field interrogation (stop-and-frisk searches), (3) searches of automobiles under special conditions, (4) seizures of evidence that is in "plain view," and (5) searches when consent is given.

Searches Incident to Lawful Arrest

In *Chimel v. California* (1969), officers arrested Chimel without a warrant in one room of his house and then searched his entire three-bedroom house, including the garage, attic, and workshop. The U.S. Supreme Court said that searches incident to a lawful arrest are limited to the area within the arrestee's immediate control or that area from which he or she might obtain a weapon, emphasizing that such searches are justified in large part to ensure officer safety. Therefore, if the police are holding a person in one room of the house, they cannot search and seize property in another part of the house, away from the arrestee's immediate physical presence.[23]

Another advantage given the police was the Court's allowing a warrantless, in-home "protective sweep" of the area in which a suspect is arrested to reveal the presence of anyone else who might pose a danger. Such a search, if justified by the circumstances, is not a full search of the premises and may include only a cursory inspection of those spaces where a person could be hiding.[24] In April 2009, the Supreme Court held in *Arizona v. Gant* that, when an individual has been arrested and is in police custody away from his or her vehicle, unable to access the vehicle, officers may not then search the vehicle without a warrant. Here, the officers did so, and discovered a handgun and a plastic bag of cocaine; the Court overturned nearly three decades of such a police practice, saying it is a violation of the Fourth Amendment's protection against unreasonable searches and seizures.[25] In essence, the Court is saying that police may search the passenger compartment of a vehicle incident to a recent occupant's arrest only if it is reasonable to believe that the arrestee might access the vehicle at the time of the search or that the vehicle contains evidence of the offense of arrest.

Figure 6.1 /// Federal Court Search and Seizure Warrant Form, U.S. District Courts

AO 93 (Rev. 11/13) Search and Seizure Warrant

UNITED STATES DISTRICT COURT

for the

In the Matter of the Search of)
(Briefly describe the property to be searched)
or identify the person by name and address)) Case No.
)
)
)

SEARCH AND SEIZURE WARRANT

To: Any authorized law enforcement officer

An application by a federal law enforcement officer or an attorney for the government requests the search of the following person or property located in the ________________ District of ____________________________
(identify the person or describe the property to be searched and give its location):

I find that the affidavit(s), or any recorded testimony, establish probable cause to search and seize the person or property described above, and that such search will reveal *(identify the person or describe the property to be seized)*:

YOU ARE COMMANDED to execute this warrant on or before ______________________ *(not to exceed 14 days)*
❒ in the daytime 6:00 a.m. to 10:00 p.m. ❒ at any time in the day or night because good cause has been established.

Unless delayed notice is authorized below, you must give a copy of the warrant and a receipt for the property taken to the person from whom, or from whose premises, the property was taken, or leave the copy and receipt at the place where the property was taken.

The officer executing this warrant, or an officer present during the execution of the warrant, must prepare an inventory as required by law and promptly return this warrant and inventory to ____ .
(United States Magistrate Judge)

❒ Pursuant to 18 U.S.C. § 3103a(b), I find that immediate notification may have an adverse result listed in 18 U.S.C. § 2705 (except for delay of trial), and authorize the officer executing this warrant to delay notice to the person who, or whose property, will be searched or seized *(check the appropriate box)*
❒ for ____ days *(not to exceed 30)* ❒ until, the facts justifying, the later specific date of ______________________ .

Date and time issued: __________________________ ____________________________________
Judge's signature

City and state: __________________________ ____________________________________
Printed name and title

AO 93 (Rev. 11/13) Search and Seizure Warrant (Page 2)

Return		
Case No.:	Date and time warrant executed:	Copy of warrant and inventory left with:
Inventory made in the presence of :		
Inventory of the property taken and name of any person(s) seized:		

Certification

I declare under penalty of perjury that this inventory is correct and was returned along with the original warrant to the designated judge.

Date: ______________

____
Executing officer's signature

____
Printed name and title

Practitioner's Perspective

Special Agent

Name: Dan Burke
Position: Special Agent, Food and Drug Administration's Office of Criminal Investigations

What is your career story? I started off doing internships. That's the best advice I can give college students: get in there, get internships, and get to know someone. I was lucky in that I got the job right out of college, but I did a co-op. That's when the government gives you a paid internship, and after you graduate, they agree to hire you as an agent. I came on as an agent in the Treasury Department, and from there I worked my way up.

What are some challenges and misconceptions you face in this position? There are a lot of challenges when it comes to law enforcement, and particularly federal law enforcement. You're dealing with crimes that are national, and oftentimes international in scope, and when you're dealing with that, you have to deal with a lot of different jurisdictions and laws, so it takes a while to understand the nuances of the jurisdictions that you're working in. The investigations also tend to take a lot longer.

The Internet is also a big challenge. It's the digital, cyber Wild West and it takes a lot of training and experience to understand how to investigate in that environment. But at the end of the day, it's satisfying when the long arm of the law can reach beyond your borders and reach through your screen. I've had cases where I never really had to get up from my desk; I was able to do almost entire cases sitting in front of the computer.

Some of the myths definitely do come from television. Cases are not made in hour- or half-hour-long segments. I've spent days and weeks on surveillance. Wire taps are not easy to get, contrary to popular belief, and they're very expensive. In general, managers don't like you spending a lot of money to pursue an investigation. It's difficult to write up a search warrant. You have to spend time on that, and you can't just go kicking in doors.

What role does diversity play in this position? In my career in federal law enforcement, I've arrested doctors, lawyers, even a CEO of a major corporation. So those cases can be quite interesting and challenging. You just never know how someone of that high status who's put a lot of their life into their career is going to react when it all comes tumbling down when I come through the door with an arrest warrant or search warrant. I've also been party to arrests involving religious people, and those are very sensitive circumstances. One thing I have learned is that it really runs the gamut in terms of the diversity of people who chooses to commit crimes.

What directions do you envision your department going in the future? I think there definitely is a future in law enforcement. It's a frustrating job, and students need to know that going into it. You're not always going to be thought of as the good guy. What you do are controversial things; you serve subpoenas and search warrants, you arrest people. That's sometimes difficult to understand and get your head around. There will always be positions because there will always be crime. So job security–wise, you're going to be okay. But you have to understand what you're going to be doing. You're putting yourself out there as the tip of the spear, if you will. You're that line between the good guys and the bad guys, and you have to make sure you stay on the good side of that line. You have to try to do the right thing at all times, even if that means losing a case, even if that means the bad guy is going to walk.

To learn more about Dan Burke's experiences as a Special Agent, watch the Practitioner's Perspective video in SAGE Vantage.

Finally, in mid-2013, in a decision that Justice Samuel Alito described as "the most important criminal procedure case this court has heard in decades," the Supreme Court held in *Maryland v. King* that police can collect DNA from people arrested but not yet convicted. The majority compared taking a DNA swab to fingerprinting, viewing it as a reasonable search that can be considered a routine part of a booking procedure.[26]

The David Riley Case: The San Diego police officer searched Riley's cell phone "incident to a lawful arrest," almost as an extension of Riley's person—like his wallet or a bag he may have been carrying. Was the search as invasive as taking a cheek swab for DNA? Or is it more like looking for weapons or contraband on an arrestee?

In 2014, the U.S. Supreme Court agreed with Riley's attorneys that the police search of Riley's cell phone was illegal. Speaking for the majority, Chief Justice John Roberts wrote:

> Digital data stored on a cell phone cannot itself be used as a weapon to harm an arresting officer or to effectuate the arrestee's escape. Law enforcement officers remain free to examine the physical aspects of a phone to ensure that it will not be used as a weapon—say, to determine whether there is a razor blade hidden between the phone and its case. Once an officer has secured a phone and eliminated any potential physical threats, however, data on the phone can endanger no one.[27]

Roberts also confirmed the realities of our modern digital lives and how the Court must interpret the Constitution with contemporary issues in mind, especially when reviewing police power:

> Modern cell phones are not just another technological convenience. With all they contain and all they may reveal, they hold for many Americans "the privacies of life." The fact that technology now allows an individual to carry such information in his hand does not make the information any less worthy of the protection for which the Founders fought.

Stop and Frisk

In 1963, Cleveland detective McFadden, a veteran of 19 years of police service, first noticed Terry and another man who appeared to be "casing" a retail store. McFadden observed the suspects making several trips down the street, stopping at a store window, walking about a half-block, turning around and walking back again, pausing to look inside the same store window. At one point they were joined by a third party, who spoke with them and then moved on. McFadden claimed that he followed them because he believed it was his duty as a police officer to investigate the matter further.

Soon the two rejoined the third man; at that point McFadden decided the situation demanded direct action. The officer approached the subjects and identified himself, then requested that the men identify themselves as well. When Terry said something inaudible, McFadden "spun him around so that they were facing the other two, with Terry between McFadden and the others, and patted down the outside of his clothing." In a breast pocket of Terry's overcoat, the officer felt a pistol. McFadden found another pistol on one of the other men. The two men were arrested and ultimately convicted for concealing deadly weapons. Terry appealed on the grounds that the search was illegal and that the evidence should have been suppressed at trial.

The U.S. Supreme Court disagreed with Terry, holding in *Terry v. Ohio* that the police have the authority to detain a person briefly for questioning even without probable cause if they have "**reasonable suspicion**" that the person has committed a crime or is about to commit a crime.

Reasonable suspicion: suspicion that is less than probable cause but more than a mere hunch that a person may be involved in criminal activity.

©New York Daily News Archive/New York Daily News/Getty Images

Despite officers being trained in mock scenarios, as seen here in the Bronx, the NYPD's aggressive stop-and-frisk program was abandoned after being deemed unconstitutional.

Terry Stop: also known as a "stop and frisk"; when a police officer briefly detains a person for questioning and then frisks ("pats down") the person if the officer reasonably believes he or she is carrying a weapon.

Reasonable suspicion is a lower standard than probable cause but more than a mere hunch that someone may be involved in criminal activity. Officers must be able to articulate a reasonable factual basis for a stop. Such detention—known as a **Terry Stop**—does not constitute an arrest, and a person may be frisked for a weapon if an officer reasonably suspects the person is carrying one and fears for his or her life. The *Terry* decision ushered in a new era for police officers, allowing them to stop and frisk suspects based only on reasonable suspicion, and not the higher threshold of suspicion required under a probable cause standard.[28]

An important extension of the *Terry* doctrine—known as the "plain feel" doctrine—was handed down in 1993 in *Minnesota v. Dickerson*, in which a police officer observed a man leave a notorious crack house and then try to evade the officer.[29] The man was eventually stopped and patted down, or frisked, during which time the officer felt a small lump in the man's front pocket that was suspected to be drugs. The officer removed the object—crack cocaine wrapped in a cellophane container—from the man's pocket. Although the arrest and conviction were later thrown out as not being allowed under *Terry*, the Court also allowed such seizures in the future when officers' probable cause is established by the sense of touch.

Another important Supreme Court decision in February 1997 took officer safety into account. In *Maryland v. Wilson*, the Court held that police may order passengers out of vehicles they stop, regardless of any suspicion of wrongdoing or threat to the officer's safety.[30] Citing statistics showing officer assaults and murders during traffic stops, the Court noted that the "weighty interest" in officer safety is present whether a vehicle occupant is a driver or a passenger.

In January 2013, the Supreme Court heard arguments in a landmark Fourth Amendment case involving nearly 50 years of uncertainty over whether or not police securing blood tests (for people who might be driving under the influence) without a suspect's consent is constitutional. Tyler McNeely, of Cape Girardeau, Missouri, was pulled over for speeding. McNeely refused an on-scene breath test, so the trooper took him to a hospital, where McNeely again refused a test. The trooper told the lab technician to take a blood sample anyway, without a warrant. McNeely's blood-alcohol level was almost double the legal limit. However, the Supreme Court ruled in *Missouri v. McNeely* (2013) that police must obtain a warrant in order to subject a drunk-driving suspect to a blood test.[31] The Court found that the natural metabolism of alcohol in the bloodstream did not constitute an exigency that would justify a warrantless blood test. Further in 2016, the Supreme Court held in *Birchfield v. North Dakota* that, while the Fourth Amendment permits states to require warrantless breath tests prior to an arrest for drunk driving, it does not permit states to criminalize the refusal of warrantless blood tests by suspected drunk drivers.[32]

Searches of Automobiles

The third general circumstance allowing a warrantless search is when an officer has probable cause to believe that an automobile contains criminal evidence. The U.S. Supreme Court gives the police greater latitude in searching automobiles, because a vehicle can be used to effect a getaway and to hide evidence. In *Carroll v. United States* (1925), officers searched the vehicle of a known bootlegger without a warrant but with probable cause, finding 68 bottles of illegal booze. On appeal, the Court ruled that the seizure was justified. *Carroll* established two rules, however: First, to invoke the *Carroll* doctrine, the police must have enough probable cause that if there had been enough time, a search warrant would have been issued; second, urgent circumstances must exist that require immediate action.[33]

The Court also allows the police to enter an impounded vehicle following a lawful arrest in order to inventory its contents,[34] and has stated that a person's general consent to a search of the interior of an automobile also justifies a search of any closed container found inside the car that might reasonably hold the object of the search.[35] And, when an officer has probable cause to search a vehicle, the officer may search objects belonging to a passenger in the vehicle, provided the item the officer is looking for could reasonably be in the passenger's belongings[36] (such as finding drugs in a passenger's purse). Finally, motorists have no expectation of privacy during a traffic stop if contraband is hidden in a vehicle and detected by a drug-sniffing dog,[37] although police cannot extend the length of a routine traffic stop to allow a drug-sniffing dog to arrive on scene if the original purpose of the stop has been addressed (e.g., a ticket has been issued), unless the officer can demonstrate reasonable suspicion of additional criminal drug activity.[38]

The courts have held that motorists have no expectation of privacy during a traffic stop if contraband hidden in a vehicle is detected by a drug-sniffing dog.

In 2012, the Supreme Court ruled in *U.S. v. Jones* that police violated the Constitution when they attached a global positioning system (GPS) device to a suspect's vehicle without a search warrant.[39] Police had followed a drug trafficking suspect for a month and eventually found nearly 100 kilograms of cocaine and $1 million in cash when raiding the suspect's home in Maryland. Justice Antonin Scalia noted that the Fourth Amendment's protection of "persons, houses, papers, and effects, against unreasonable searches and seizures" extends to automobiles as well, and that even a small trespass, if committed in "an attempt to find something or to obtain information," constitutes a "search" under the Fourth Amendment. More recently, the Supreme Court in *Byrd v. United States* (2018) ruled that police violated Terrence Byrd's Fourth Amendment rights when they searched his car and found heroin and body armor in the trunk.[40] The police conducted a warrantless search, noting that Byrd was driving a rental car (with the renter's permission) but was not listed on the rental agreement. The Supreme Court ruled unanimously that Byrd, even though not listed on the rental agreement, had a reasonable expectation of privacy against a warrantless government vehicle search.

Plain-View and Open-Field Searches

Essentially, police do not have to search for items that are seen in plain view. If police are lawfully on the premises and the plain-view discovery is inadvertent, then they may seize the contraband. For example, if an officer has been admitted into a home with an arrest or search warrant and sees drugs and paraphernalia on a living room table, he or she may arrest the occupants on drug charges as well as the ones related to the warrant. Or, if an officer during a lawful traffic stop observes drugs in the backseat of the car, he or she may arrest for that as well. When an officer found a gun under a car seat while looking for the vehicle identification number, the Court upheld the search and the resulting arrest as being a plain-view discovery.[41] Furthermore, fences and the posting of "No Trespassing" signs afford no expectation of privacy and do not prevent officers from viewing open fields without a search warrant.[42] Nor are police prevented from making a naked-eye aerial observation of a suspect's backyard or other curtilage (the grounds around a house or building).[43]

However, in *Collins v. Virginia* (2018), the Supreme Court ruled that police may not make a warrantless entry of curtilage in order to conduct a search, even if it involves a vehicle search.[44] Officer David Rhodes, while investigating two traffic incidents, observed a motorcycle under a tarp on a residential property. Suspecting that this was the motorcycle involved in the traffic incidents, Rhodes entered the property without a warrant, lifted the tarp, and confirmed that the vehicle identification number (VIN) matched the VIN of a stolen motorcycle. When Ryan Austin Collins returned to the property, he was arrested and the key to the motorcycle was found in his possession. While lower courts ruled that the warrantless search was justified under the Fourth Amendment's automobile exception, the Supreme Court ruled that this exception does not permit the warrantless entry of a home *or its curtilage* in order to search a vehicle.

The Supreme Court has ruled that residential porches are also considered part of the home itself and cannot be searched without a warrant. Officers from the Miami-Dade Police Department received a tip that marijuana was being grown in a house. The officers approached the front door (porch) with a drug dog. The Labrador retriever alerted officers to the presence of marijuana in the house; subsequently, officers obtained a search warrant and discovered the plants inside. The Supreme Court ruled, in *Florida v. Jardines* (2013), that the police can use residential porches to communicate with occupants, but a search that involves more intrusive measures (in this case, the use of a drug dog) constitutes an illegal search.[45] Thus, searches conducted from porches (and that go beyond the basic senses of a police officer) are not subject to the plain-view or open-field search exception.

Consent to Search

Another permissible warrantless search involves situations in which citizens consent to a search of their persons or effects—provided that the person's consent is given voluntarily. However, police cannot deceive people into believing they have a search warrant when they in fact do not.[46] Nor can a hotel clerk give a valid consent to a warrantless search of the room of one of the occupants; hotel guests have a reasonable expectation of privacy, and that right cannot be waived by hotel management.[47] Further, absent consent from hotel operators, police must have a warrant or administrative subpoena to inspect hotel guest records.[48]

What if one occupant of a home consents to a search while the other occupant refuses? In a recent case, a wife gave police permission to search for her husband's drugs, but the husband refused to give such permission; the officers went ahead and searched and found cocaine. The Supreme Court reversed the husband's conviction, saying the Constitution does not ignore the privacy rights of an individual who is present and asserting his rights.[49] However, in a feature of the law that should be important to college students with roommates, an occupant may still give police permission to search when the other resident is absent or does not protest. In fact, in 2014, the Supreme Court expanded police powers to seek consent from a co-occupant, ruling that even after an occupant has refused consent to search and the police then legally arrest that occupant for some other reason, the police may return and seek consent from a co-occupant who then has the right to voluntarily consent to a search of the entire premises.[50]

ROBYN BECK/AFP/Getty Images

Looking at the plain-view doctrine, assume police officers are lawfully inside a home to execute an arrest warrant on a suspected bank robber. If, by chance, they happen to observe this illegal marijuana operation, the officers could then also arrest the occupant on drug charges.

Electronic Surveillance

Katz v. United States (1967) held that any form of electronic surveillance, including wiretapping, is a search and violates a reasonable expectation of privacy.[51] The case involved a public telephone booth, deemed by the Court to be a constitutionally protected area where the user has a reasonable expectation of privacy. This decision expressed the view that the Constitution protects people, not places. Thus, the Court has required that warrants for electronic surveillance be based on probable cause, describe the conversations that are to be overheard, be for a limited period of time, name subjects to be overheard, and be terminated when the desired information is obtained.[52] However, the Court has also held that electronic eavesdropping (i.e., when an informant wears a "bug," or hidden microphone) does not violate the Fourth Amendment.[53]

THE FIFTH AMENDMENT

> No person shall be held to answer for a capital, or otherwise infamous crime, unless on a presentment or indictment of a Grand Jury, except in cases arising in the land or naval forces, or in the Militia, when in actual service in time of war or public danger; nor shall any person be subject for the same offense to be twice put in jeopardy of life or limb; nor shall be compelled in any criminal case to be a witness against himself, nor be deprived of life, liberty, or property, without due process of law; nor shall private property be taken for public use, without just compensation.

Application

Today, the **Fifth Amendment** applies not only to criminal defendants but also to any witness in a civil or criminal case and anyone testifying before an administrative body, a grand jury, or a congressional committee. However, the privilege does not extend to blood samples, handwriting exemplars, and other such items not considered as testimony.[54]

Fifth Amendment: in the Bill of Rights, among other protections, it guards against self-incrimination and double jeopardy.

The right against self-incrimination is one of the most significant provisions in the Bill of Rights. Basically it states that no criminal defendant shall be compelled to take the witness stand and/or give evidence against himself or herself. This right is expressed in the so-called *Miranda* warning (also known as "Mirandizing" someone), which requires police, at the point of having someone in custody and before questioning him or her about a crime or criminal activity, to inform the suspect "you have the right to remain silent. Anything you say can and will be used against you in a court of law. You have the right to an attorney. If you cannot afford an attorney, one will be provided for you."

The *Miranda* warning stems from a case involving a 23-year-old man, Ernesto Miranda, who raped 18-year-old Barbara Ann Johnson. Johnson was walking to a Phoenix, Arizona, bus stop on the night of March 2, 1963, when Miranda shoved her into his car, tied her hands and ankles, and drove her to the edge of the city, where he raped her. He then drove Johnson to a street near her home, letting her out of the car and asking that she pray for him. The police subsequently picked up Miranda and included him in a police station lineup. Miranda was identified by several women; one identified him as the man who had robbed her at knifepoint a few months earlier, and Johnson thought he was the rapist.

Miranda was a 23-year-old eighth-grade dropout with a police record dating back to age 14, and he had also served time in prison for driving a stolen car across a state line. During questioning, the police told Miranda that he had been identified by the women; Miranda then made a statement in writing that described the rape incident. He also noted that he was making the confession voluntarily and with full knowledge of his legal rights. He was soon charged with rape, kidnapping, and robbery.

Gerald L. Nino, CBP, U.S. Dept. of Homeland Security

The *Miranda* warning must be given after a suspect is in custody and before interrogation begins; otherwise, any statements might be thrown out in court.

At trial, officers admitted that during the interrogation the defendant was not informed of his right to have counsel present and that no counsel was present. Nonetheless, Miranda's confession was admitted into evidence. He was convicted and sentenced to serve 20–30 years for kidnapping and rape.

On appeal, the U.S. Supreme Court overturned Miranda's conviction, stating that

> the current practice of incommunicado interrogation is at odds with one of our Nation's most cherished principles—that the individual may not be compelled to incriminate himself. Unless adequate protective devices are employed to dispel the compulsion inherent in custodial surroundings, no statement obtained from the defendant can truly be the product of free choice.[55]

There are two legal triggers for the *Miranda* warning: custody and interrogation. Once a suspect has been placed under arrest, the police must give the suspect the *Miranda* warning before an interrogation is conducted for a suspected offense, whether a felony or a misdemeanor. If an officer does not give the warning when required, a defense attorney may ask the court to exclude the suspect's confession or statements. Once a Mirandized suspect invokes his or her right to silence, however, interrogation must cease. The police also may not readminister *Miranda* and interrogate the suspect later unless the suspect's attorney is present. If, however, the suspect initiates further conversation, any confession he or she provides is admissible.[56] Moreover, after an accused has invoked the right to counsel, the police may not interrogate the same suspect about a different crime.[57]

An exception to the *Miranda* requirement is the brief, routine traffic stop, but a custodial interrogation of a DUI (driving under the influence) suspect requires the *Miranda* warning.[58] And the police may question a suspect who poses a public-safety threat—known as the "public-safety exception." For example, when a clearly armed suspect runs into a public place and hides the gun, officers can legally ask the suspect about the location of the gun without violating *Miranda*.[59] In a more extreme example under the same exception, police were able to question Boston Marathon bomber Dzhokhar Tsarnaev for 16 hours without a *Miranda* warning because of the potential threat of co-conspirators and more planned bombings.[60]

Decisions Eroding *Miranda*

It has been held that a second interrogation session held after the suspect had initially refused to make a statement did not violate *Miranda*.[61] The Court also decided that when a suspect waived his or her *Miranda* rights, believing the interrogation would focus on minor crimes, but the police shifted their questioning to a more serious crime, the confession was valid; there was no police deception or misrepresentation.[62] And when a suspect invoked his or her right to assistance of counsel and refused to make written statements, then voluntarily gave oral statements to police, the statements were admissible.[63] Finally, a suspect need not be given the *Miranda* warning in the exact form as it was outlined in *Miranda v. Arizona*. In one case, the waiver form said the suspect would have an attorney appointed "if and when you go to court." The Court held that as long as the warnings on the form reasonably convey the suspect's rights, they need not be given verbatim.[64] Further, a suspect must clearly invoke the *Miranda* right. The Supreme Court has ruled that even when a suspect remained silent during three continuous hours of police questioning, the suspect had not invoked his rights under *Miranda*—suspects must specifically say that they are choosing to remain silent.[65]

Lineups and Other Pretrial Identification Procedures

Lineup: a procedure in which police ask suspects to submit to a viewing by witnesses to determine the guilty party, based on personal and physical characteristics; information obtained may be used later in court.

A police **lineup** or other face-to-face confrontation after the accused has been arrested is considered a critical stage of criminal proceedings; therefore, the accused has a right to have an attorney present. If counsel is not present, the evidence obtained is inadmissible.[66] However, the suspect is not entitled to the presence and advice of a lawyer before being formally charged, or when the police are using the more typical modern technique of a photo lineup, by showing a witness individual photos or a group of several photos typically referred to as a "six-pack."[67]

Faulty eyewitness identifications are the leading cause of wrongful convictions, and as a result, the law regarding their use has changed significantly in recent years (see Chapter 9 for

You Be the . . .

Judge

In 2013, the U.S. Supreme Court reviewed a case that focused on issues related to the Fifth Amendment, *Miranda* rights, and refusal to answer law enforcement questioning. *Salinas v. Texas* involved the questioning of Genovevo Salinas, who investigators suspected might have information pertaining to a double homicide. The investigators interviewed Salinas at his parents' home, and Salinas' father voluntarily gave the police his son's shotgun. Salinas voluntarily answered questions later at the police station for about an hour until an officer asked him if ballistics analysis would show that his shotgun was used in the killings; at that point Salinas remained silent and began behaving in a suspicious manner (e.g., staring at the floor, biting his lip, clenching his hands, shuffling his feet).

Salinas was later arrested and found guilty of murder. During his trial, the prosecutor told the jury that Salinas refused to answer the ballistics question and argued that this was proof of his guilt. His lawyers appealed and argued that he was punished for exercising his Fifth Amendment right against self-incrimination.

1. Does *Miranda* apply during voluntary police interviews (outside of a custodial interrogation)?
2. Can prosecutors use a suspect's silence during police interviews to infer guilt during a criminal trial, or do Fifth Amendment protections apply?

The Supreme Court's decision for this case is provided in the Notes section at the end of the book.[68]

more on wrongful convictions). Lineups—whether in person or through photos—must be fair and cannot be unreasonably suggestive, for example, by the suspect's being much taller than the others in the lineup or being the only person wearing a leather jacket similar to that worn by the perpetrator.[69] If identification procedures violate a suspect's due process rights to fairness, police risk having identifications made during these procedures excluded from evidence at trial. Further, in light of extensive social science research on faulty eyewitness identifications, many law enforcement agencies have changed their identification practices to include safeguards such as informing the witness that the suspect may or may not be in the lineup or photo set, preventing detectives working the case from being present to avoid giving inadvertent supportive or other cues to witnesses, and having witnesses record their level of certainty at the time of identification.[70] Some courts have instituted much stricter legal standards for admitting eyewitness evidence.[71]

Poole

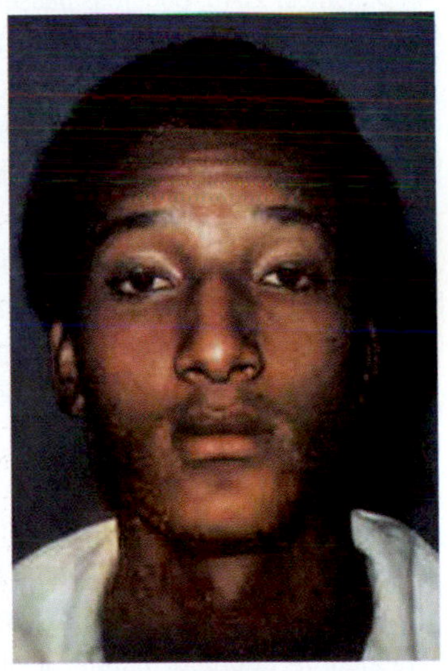

Cotton

Explore the case of Ronald Cotton (right), who served 11 years in prison for a sexual assault he did not commit. Despite studying her attacker closely, the victim misidentified Cotton in live and photo lineups, mistaking him for the real perpetrator, Bobby Poole (left).

THE SIXTH AMENDMENT

> In all criminal prosecutions the accused shall enjoy the right to a speedy and public trial, by an impartial jury of the State and district wherein the crime shall have been committed, which district shall have been previously ascertained by law, and to be informed of the nature and cause of the accusation; to be confronted with the witnesses against him; to have compulsory process for obtaining witnesses in his favor, and to have the assistance of counsel for his defense.

Juvenile Justice Journal

Law and Legislation

Given the difference in philosophy and treatment of juvenile offenders, as explained in earlier chapters and espoused by legislators, the courts, and the general public, the laws must reflect those differing viewpoints. Furthermore, policy makers believe that juveniles have more potential for rehabilitation than their adult counterparts. Next we discuss some of the more salient aspects of juvenile law and legislation.

The Juvenile Justice and Delinquency Prevention Act (JJDPA) was first authorized in 1974, to ensure that all states and territories met certain standards for how youth across the country were treated in the juvenile justice system. It established two principles: a prohibition on the incarceration of youth charged with status offenses (discussed in Chapter 1), and a requirement that youth have sight and sound separation from adult inmates. Later, in 2002, a prohibition was added against housing young people in adult facilities while awaiting trial, and it was mandated that states address disproportionate minority contact.

In December 2018, Congress passed H.R. 6964, which adds a requirement that states collect and analyze data on racial and ethnic disparities, determine at which points racial and ethnic disparities occur, and develop a plan to address such disparities. Second, states are required to ensure sight and sound separation and jail removal for youth awaiting trial as adults (this protection previously applied only to youth being held on juvenile court charges). Third, youth who are found in violation of a valid court order may be held in detention, but for no longer than seven days, if the court finds that such detention is necessary.

The Centerpiece: *In Re Gault*

The U.S. Supreme Court held in 1967 that the Fourteenth Amendment is not "for adults alone." The Court held that juveniles were entitled to the same basic procedural safeguards afforded therein, including advance notice of charges, right to counsel and to confront and cross-examine witnesses, and the privilege against self-incrimination.

Right to the *Miranda* Warning

In 2011, the U.S. Supreme Court decided *J. D. B. v. North Carolina*. In that case, a 13-year-old North Carolina boy was questioned by police without an attorney or guardian present concerning some burglaries, and he eventually confessed. The Supreme Court said that age must be considered in determining whether a suspect is aware of his or her rights. This decision tells police they cannot avoid giving a youth the *Miranda* warning simply by questioning the youth at school, away from his or her parents or guardians; "when in doubt, give the Miranda warnings."

Right to Counsel

Kent v. United States (1966) involved a 16-year-old boy who was arrested for robbery, rape, and burglary. The juvenile court transferred him to a criminal court, whereupon he was tried and convicted as an adult. Kent argued the transfer violated his right to due process. The Supreme Court agreed and also decided that there must be a meaningful right to representation by counsel and that the court must also provide reasons for the transfer.

Burden of Proof Standard

In 1970, *In re Winship* involved a 12-year-old boy who, at his juvenile court trial, had the "preponderance of the evidence" standard of proof used against him rather than the more demanding "beyond a reasonable doubt" standard used in adult courts. The U.S. Supreme Court reversed Winship's conviction on grounds that the "beyond a reasonable doubt" standard had not been used.

Trial by Jury and Double Jeopardy

In *McKeiver v. Pennsylvania* (1971), two boys arrested for robbery were denied a jury trial by the juvenile court; the Supreme Court said that while juveniles do not have an absolute right to trial by jury, such a trial is left to the discretion of state and local authorities. Then, in *Breed v. Jones* (1975), the Court held that juveniles, like adults, are protected from double jeopardy (being tried twice for the same offense).

Juvenile Life Without Parole Sentences

In *Roper v. Simmons* (2005), the U.S. Supreme Court banned the juvenile death penalty; later the Court banned life without parole sentences for youth convicted of nonhomicide crimes as well as mandatory sentences of life without parole for youth convicted of homicide crimes. However, in some states a youth may still be sentenced to discretionary life without parole in homicide cases if the sentencing court, after a hearing, determines that the youth is permanently incorrigible and incapable of rehabilitation. Twenty-eight states still allow such a sentencing option for juveniles.

Many people believe that the **Sixth Amendment** right of the accused to have the assistance of counsel before and during trial is the greatest right we enjoy in a democracy.

Sixth Amendment: in the Bill of Rights, it guarantees the right to a speedy and public trial by an impartial jury, the right to effective counsel at trial, and other protections.

Powell v. Alabama (1932) established that in a capital case, when the accused is poor and illiterate, he or she enjoys the right to assistance of counsel for his or her defense and due process.[72] In *Gideon v. Wainwright* (1963), the Supreme Court mandated that all indigent people charged with felonies in state courts be provided counsel.[73] *Gideon* applied only to felony defendants, but *Argersinger v. Hamlin* (1973) extended the right to counsel to indigent people charged with *misdemeanor* crimes if they face the possibility of incarceration (however short the incarceration may be).[74]

For defendants with resources, who represents them in court is a matter of their own choice. In 2012, a home health care agency owner, Sila Luis, was indicted on Medicare fraud charges. Prosecutors alleged that Luis paid illegal kickbacks to patients and billed the federal government for services not rendered by her companies. Following existing federal law, the courts decided to freeze all her assets, including the $45 million she was accused of fraudulently procuring, as well as another $2 million contained within her accounts. In 2016, the U.S. Supreme Court ruled in *Luis v. United States* that the defendant's Sixth Amendment right to retain her counsel of choice was violated since she was unable to hire an attorney using her own assets.[75] As a result, pretrial restraint of assets unrelated to the crime(s) in which a defendant is charged is now deemed unconstitutional.

Another landmark decision concerning the right to counsel is *Escobedo v. Illinois* (1964).[76] Danny Escobedo's brother-in-law was fatally shot in 1960; Escobedo was arrested and questioned at police headquarters; and his request to confer with his lawyer was denied, even after the lawyer arrived and asked to see his client. The questioning of Escobedo lasted several hours, during which time he was handcuffed and forced to remain standing. Eventually, he admitted being an accomplice to murder. At no point was Escobedo advised of his rights to remain silent or to confer with his attorney. Escobedo's conviction was ultimately reversed by the U.S. Supreme Court, based on a violation of his right to counsel. However, the real thrust of the decision was his Fifth Amendment right not to incriminate himself and to be informed of his rights; when a defendant is scared, flustered, ignorant, alone, and bewildered, he or she is often unable to effectively make use of protections granted under the Fifth Amendment without the advice of an attorney. Note that *Miranda*, decided two years later, simply established the guidelines for the police to inform suspects of all of these rights.

The Sixth Amendment's provision for a speedy and public trial is discussed in Chapter 7.

/// IN A NUTSHELL

- The Fourth Amendment protects the right of the people to be secure in their persons, papers, and effects, against unreasonable searches and seizures; and no search or arrest warrants shall be issued without probable cause, describing the place to be searched and disclosing the persons or things to be seized.
- The standard for a legal arrest, search, and seizure is probable cause, meaning that for an officer to make an arrest or conduct a search of someone's person or effects, he or she must have a reasonable suspicion that a crime has been or is about to be committed by that individual (which, in the case of a search, can include the person's merely possessing some form of illegal contraband).
- The exclusionary rule forbids evidence obtained in violation of the Fourth Amendment from being used against the defendant in a criminal trial.
- There are two ways for the police to make an arrest: with a warrant, and without a warrant. It is always best for a police officer to arrest someone with a warrant, because a warrant means that a neutral magistrate has examined the facts and determined that the individual should be arrested in order to make an accounting of the charge(s).
- Police officers must obtain a warrant when making a felony arrest, unless exigent circumstances are present (immediate action being required) and the officer possesses probable cause to make the arrest.

- As with arrests, the best means by which the police can search a person or premises is with a search warrant issued by a neutral magistrate. The process for obtaining a search warrant is the same as for an arrest warrant; in this case, a determination is made after the magistrate hears evidence from an affiant about whether probable cause exists to believe a person possesses evidence of a crime, such as stolen property or a weapon believed to have been used in committing the crime.
- The police may detain a person briefly for questioning, even without probable cause, if they believe that the person has committed a crime or is about to commit a crime, and the person may be frisked for a weapon if an officer reasonably suspects the person is carrying one and fears for his or her life. An extension of that rule involves the "plain feel" doctrine, according to which police, having reasonable cause to believe a person possesses drugs and feeling what resembles such an object on his or her person, may remove the object and make an arrest.
- The police have greater latitude in searching automobiles, because they can be used to conduct a getaway and to hide evidence.
- The police may search incident to lawful arrest and do not need a search warrant for items that are seen in plain view or when citizens give police consent. The police must, however, obtain a search warrant to search a suspect's cell phone or conduct a blood test.
- Police may use drug-sniffing dogs at traffic stops but not near homes without a warrant; they may also collect DNA swabs from arrestees as part of a routine booking procedure.
- The right against self-incrimination, under the Fifth Amendment, states that no criminal defendant shall be compelled to take the witness stand and give evidence against himself or herself. Once a suspect is arrested, the *Miranda* warning must be given before he or she is interrogated for any offense, be it a felony or a misdemeanor.
- The Sixth Amendment guarantees the accused the right to have the assistance of counsel during custodial interrogation and during trial.

/// KEY TERMS & CONCEPTS

Review key terms with eFlashcards at **edge.sagepub.com/peakbrief**.

Affidavit 127
Exclusionary rule 126
Exigent circumstance 127
Fifth Amendment 137
Fourth Amendment 125
Lineup 138
Probable cause 125
Reasonable suspicion 133
Search and seizure 125
Sixth Amendment 141
Terry Stop 134
Warrant, arrest 125
Warrant, search 125

/// REVIEW QUESTIONS

Test your understanding of chapter content. Take the practice quiz at **edge.sagepub.com/peakbrief**.

1. What protections are afforded citizens by the Fourth, Fifth, and Sixth Amendments?
2. What is an example of probable cause?
3. What, from both police and community perspectives, are the ramifications of having or not having the exclusionary rule?
4. How would you distinguish between arrests and searches and seizures with and without a warrant? Which method for arresting or searching is always best and why?
5. In what significant ways has the *Miranda* decision been eroded, and what future challenges might arise given the continuing tension between law enforcement powers and individual civil rights?
6. What limitations are placed on the police in their ability to use high-tech electronic equipment in order to listen in on conversations? Search for drugs?
7. What are two major court decisions concerning right to counsel, and how do they apply in everyday life?

/// LEARN BY DOING

1. Your criminal justice professor has assigned a class project wherein class members are to determine which amendment in the Bill of Rights—the Fourth, Fifth, or Sixth—contains the most important rights that are protected under a democracy. You are to analyze the three amendments and present your findings as to which one is the most important.

2. As part of a criminal justice honor society exercise, you are debating which period of legal history was the most important: the so-called due process revolution of the Warren Court (when the Supreme Court granted many additional rights to the accused through *Gideon, Miranda, Escobedo,* and so forth), or the more conservative era that followed under the Rehnquist Court (during which time many Warren Court decisions were eroded, and more rights were given to the police). Choose a side, and defend your opinion.

3. From the time of his confirmation in 1969, Chief Justice Warren Burger viewed the exclusionary rule as an unnecessary and unreasonable intrusion on law enforcement. Assume that as part of a group project, you are to prepare a pro-con paper that examines why there should or should not be an exclusionary rule as a part of our system of justice. What will be your arguments, pro and con?

/// STUDY SITE

Get the tools you need to sharpen your study skills. SAGE edge offers a robust online environment featuring an impressive array of free tools and resources.

Access practice quizzes, eFlashcards, video, and multimedia at **edge.sagepub.com/peakbrief**

PART 3

THE COURTS

This part consists of three chapters. **Chapter 7** examines court structure and functions at the federal, state, and trial court levels; included are discussions of pretrial preparations, the actual trial process, the jury system, and some court technologies.

Chapter 8 looks at the courtroom work group, including the judges, prosecutors, and defense attorneys, and legal defenses that are allowed under the law.

Chapter 9 discusses sentencing, punishment, and appeals. Included are the types and purposes of punishment, types of sentences convicted persons may receive, federal sentencing guidelines, victim impact statements, and capital punishment.

©istockphoto.com/stone18

CHAPTER 7

COURT ORGANIZATION

Structure, Functions, and the Trial Process

The place of justice is a hallowed place.

—Francis Bacon

Courts are at the center of life's important moments.

—Howard Conyers, former Washoe County, Nevada, district court administrator

LEARNING OBJECTIVES

As a result of reading this chapter, you will be able to

1. Explain the methods and purposes of the adversarial court system
2. Explain court jurisdiction and how it is determined
3. Identify the purpose and process of state appeals courts
4. List the functions of the various tiers of the federal court system, including appeals courts
5. Relate the activities that occur during the pretrial process as attorneys prepare for trial
6. Delineate the trial process, from opening statements through conviction and appeal
7. Discuss the impact of new technologies on the courts

ASSESS YOUR AWARENESS

Test your knowledge of court structure and functions as well as the trial process and jury system by responding to the following six true-false items; check your answers after reading this chapter.

1. America's court system relies on the adversarial system, which includes, among other things, the cross-examination of witnesses.
2. America's court system consists of a national system of federal courts as well as 50 state court systems, plus those in the District of Columbia and U.S. territories.
3. Federal judges are nominated by the president and confirmed by the Senate, and they have a lifetime appointment.
4. All courts generally have unlimited jurisdiction to hear civil and criminal trials as well as appeals.
5. The Sixth Amendment gives the accused the right to a trial by an impartial jury of peers—meaning 12 people who are "similar in nearly all regards" to the defendant.
6. A convicted person may quickly and easily—and at any time—leave the state court appellate system and appeal in a federal court.

<< Answers can be found on page 293.

In Orlando, Florida, 25-year-old Casey Anthony was charged with the murder of her 2-year-old daughter Caylee. It was Anthony's mother who called to report Caylee missing after she learned her granddaughter had been gone for 31 days. Anthony told police that Caylee was with a babysitter and that she worked at Universal Studios; both of these claims (and other statements she made) were found to be untrue. Police and hundreds of volunteers looked for Caylee before her decomposed body was found 6 months later in a wooded area not far from Anthony's home. Prosecutors alleged that Anthony used chloroform (traces were found in the trunk of her car) and suffocated her daughter by placing duct tape over the girl's mouth and nose. Before, during, and after the trial, the case captivated the nation.

Judges who preside over these types of trials must determine whether they believe that jurors can impartially weigh evidence presented in trial. Given the extent of media coverage surrounding Anthony's case, Judge Belvin Perry, Jr., faced the difficult task of finding jurors who could set aside their preformed judgments about Anthony's guilt or innocence. Motions for a change of venue (which are discussed as part of the pretrial motions that you will learn about in this chapter) are often filed by defense attorneys in high-profile cases. Yet Judge Perry decided to hold the trial where Anthony was charged (in Orlando) and import a jury from Clearwater, Florida, 100 miles away. This decision was costly and caused many hardships for jurors, who were sequestered for two months in a hotel far away from home. Jurors were not permitted to use personal computers or cell phones, and they were unable to watch news media channels.

Do you agree with critics that a traditional change of venue would have been a better course of action for this trial? Do you believe that these jurors were more impartial than jurors who might have been selected from the jurisdiction in which the murder took place?

As you read this chapter, consider these questions and the role of our nation's courts, which deal not only with high-profile cases like Anthony's but also with the much more minor and numerous cases that overload our state trial courts.

INTRODUCTION

Courts have existed in some form for thousands of years. The court system has survived the dark eras of the Inquisition and the Star Chamber (which enforced unpopular political policies in England during the 1500s and 1600s and, without a jury or public view, meted out severe punishment, including whipping, branding, and mutilation). The U.S. court system developed rapidly after the American Revolution and led to the establishment of law and justice on the western frontier.

In this chapter, we first consider the adversarial system the courts employ and how the courts influence our lives as policy-making bodies. Then the focus shifts to the several levels of courts found in the United States. Your attention is drawn to the case of Barry Kibbe, presented near the beginning of the section on federal courts; this case provides a rare look at the entire criminal trial and appeals process, all the way to the U.S. Supreme Court.

Colonial Courts: An Overview

Our nation's colonial courts began modestly. Thomas Olive, deputy governor of West Jersey in 1684, described the prevailing court system when he wrote that he was "in the habit of dispensing justice sitting in his meadow."[1] As the population of the colonies grew, however, formal courts of law appeared in Virginia, Massachusetts, Maryland, Rhode Island, and Connecticut, among other venues. The colonies drew on the example set by the English Parliament in that the colonial legislatures became the highest courts.[2]

Beneath the legislatures (those who make and change laws) were the superior courts, which heard both civil and criminal cases. Over time, some colonies established trial courts headed by a chief justice and several associate justices. Appeals from the trial courts were heard by the governor and his council in what were often called "courts of appeals."[3] Courts established at the county level played a key role in both the government and the social life of the colonies. In addition to having jurisdiction in both civil and criminal cases, county courts fulfilled many administrative duties, including setting and collecting taxes, supervising the building of roads, and licensing taverns.

Dual court system: the state and federal court systems of the United States.

State court system: civil or criminal courts in which cases are decided through an adversarial process; typically including a court of last resort, an appellate court, trial courts, and lower courts.

Federal court system: the four-tiered federal system that includes the Supreme Court, circuit courts of appeal, district courts, and magistrate courts.

Eventually, however, the founders of the new republic had profound concerns about the distribution of power between courts and legislatures, and between the states and the federal government. Coming from the experience of living under English rule, they feared the tyranny that could flow from the concentration of governmental power. At the same time, they were living with the problems associated with a weak centralized government. This conflict prompted the delegates to the Constitutional Convention of 1787 to create a federal judiciary that was separate from the legislative branch of government. As a result, today we have the **dual court system**—one implemented by the **state court system** (inherited directly from the Crown of England) and the other created by Congress and entrusted to the **federal court system**.[4]

Our Adversarial System

Ralph Waldo Emerson stated that "every violation of truth . . . is a stab at the health of human society."[5] Certainly, most people would agree that the traditional, primary purpose of our

Practitioner's Perspective

Attorney

Name: Jeff Mason
Position: Attorney
Location: Colorado

What is your career story? I became interested in the law at an early age. My dad was a cop growing up, so I grew up on cop stories and probably still get my fair share of them. That's what interested me in getting my bachelor's degree in criminal justice and criminology. I also went on to get a master's degree. And it was really during my undergraduate studies that I developed an interest in writing and, specifically, in persuasive writing. I really found that I enjoyed and began to excel at writing and trying to persuade an audience to adopt my viewpoints. So I'm not entirely sure how it happened, but I decided that based on those interests that I might do well in a legal career.

What are some challenges and misconceptions you face in this position? I think there are a number of challenges from the legal side. The law is a big thing, it's always changing and no two cases are the same. How the law applies to a particular case is always different from that of the previous case. It's also a sensitive area, dealing with the criminal system. It can be a very emotionally challenging field. It's an area ripe with challenges and a lot of that has to do with the nature of the work itself.

I'm the first to appreciate a good crime drama or movie, but it certainly does create a challenge as far as the public perception of the judicial process. Everything isn't exactly how it's portrayed, and I often find myself watching these shows and saying well this is incorrect. There's a public perception that the judicial process is more action packed than it might really be at trial.

What role does diversity play in this position? Diversity is something I see all the time. There is a diverse body of lawyers these days. I even notice that in law school, that there are men and women of every background. The same is true of the parties as well; the courts are unique in that they deal with cases of every type, civil and criminal. So you see corporations, individuals, men and women of a variety of ethnic backgrounds.

What are some challenges you face in your position? Technology has become the biggest challenge. The Constitution and laws are evolving as technology evolves and changes, so that is a battle that legislatures and judges have to deal with on a daily basis. These laws have been around for some time now, and we have to find ways to apply them, given the different and unique circumstances in which these crimes and offenses are committed. I know that, for example, technological advancements such as GPS, cell phones, even night vision goggles, interplay with constitutional rights against unreasonable searches and seizures. These are things that courts have to wrestle with and address on a daily basis as law enforcement tactics also advance.

To learn more about Jeff Mason's experiences as an Attorney, watch the Practitioner's Perspective video in SAGE Vantage.

courts is to provide a forum for seeking and obtaining the truth. Indeed, the U.S. Supreme Court declared in 1966 in *Tehan v. U.S. ex rel. Shott* that "the basic purpose of a trial is the determination of truth."[6]

Adversarial system: a legal system wherein there is a contest between two opposing sides, with a judge (and possibly jury) sitting as an impartial arbiter, seeking truth.

Our American court system relies on the **adversarial system**, which uses several means to get at the truth. First, evidence is tested under this approach through cross-examination of witnesses. Second, power is lodged with several different people; each courtroom actor is granted limited powers to counteract those of the others. If, for example, the judge is biased

Justin Sullivan/Getty Images News/Getty Images

California significantly reduced prison populations by nearly 30,000 inmates over the course of 15 months following the U.S. Supreme Court ruling in *Brown v. Plata*.

or unfair, the jury can disregard the judge and reach a fair verdict. If the judge believes the jury has acted improperly, he or she can set aside the jury's verdict and order a new trial. Furthermore, both the prosecuting and defense attorneys have certain rights and authority under each state's constitution. This series of checks and balances is aimed at curbing misuse of the criminal courts.

The Influence of Courts in Policy Making

Determining what the law says and providing a public forum involve the courts in policy making. **Policy making** can be defined as choosing among alternative choices of action. The policy decisions of the courts affect virtually all of us in our daily living. In recent decades, the courts have been asked to deal with issues that previously were within the purview of the legislative and executive branches. Because many of the Constitution's limitations on government are couched in vague language, the judicial branch must help interpret the law and deal with potentially volatile social issues, such as those involving prisons, abortion, and schools.[7]

Policy making: the act of creating laws or setting standards to govern the activities of government; the U.S. Supreme Court, for example, has engaged in policy making in several areas, such as affirmative action, voting, and freedom of communication and expression.

U.S. Supreme Court decisions have dramatically changed race relations, resulted in the overhaul of juvenile courts, increased the rights of the accused, prohibited prayer and segregation in public schools, legalized abortion, and allowed for destruction of the U.S. flag. State and federal courts have together overturned minimum residency requirements for welfare recipients, equalized school expenditures, and prevented road and highway construction from damaging the environment. They have eliminated the requirement of a high school diploma for a firefighter's job and ordered increased property taxes to desegregate public schools. More recently, the federal courts blocked travel bans that sought to keep foreign nationals from Middle Eastern nations from entering the United States. Cases in which courts make policy determinations usually involve government, the Fourteenth Amendment, and the need for equity—the remedy most often used against governmental violations of law. Recent policy-making decisions by the judicial branch have been based not on the Constitution but, rather, on federal statutes concerning the rights of the disadvantaged and consumers and the environment.[8]

Perhaps nowhere have the nation's courts had more of an impact than in the prisons—from which tens of thousands prisoner petitions are filed each year in the U.S. district courts (approximately 80 percent of them filed by state prisoners).[9] Judicial intervention has extended to prisoners the recognition of the constitutional rights of free speech, religion, and due process; abolished the South's plantation model of prisons; reinforced the adoption of national standards for prisons; and promoted increased accountability and efficiency of prisons. And while judicial intervention previously failed to prevent the explosion in prison populations and costs,[10] recent judicial decisions have ordered reductions in the numbers of inmates housed in correctional facilities. In 2011, the U.S. Supreme Court ruled that court-mandated inmate population limits are necessary to protect prisoners' constitutional rights and prevent **Eighth Amendment** violations involving cruel and unusual punishments (in *Brown v. Plata*, serious overcrowding in California's 33 prisons was the "primary cause" for violations of the Eighth Amendment).[11]

Eighth Amendment: in the Bill of Rights, it contains the protection against excessive bail and fines, as well as cruel and unusual punishment.

It may appear that the courts are overbroad in their review of issues. However, judges "cannot impose their views . . . until someone brings a case to court, often as a last resort after complaints to unresponsive legislators and executives."[12] Plaintiffs must be truly aggrieved and have a legal right to bring their case to court. The independence of the judicial branch, particularly at the federal court level, where judges enjoy lifetime appointments, allows the courts to champion the causes of the underclasses: those with fewer financial resources or votes (by virtue of, say, being a minority group) or without a positive public profile.[13] Also, the judiciary is considered to be the "least dangerous branch," having no enforcement powers. Moreover, the decisions of the courts can be overturned by legislative action. Thus, the judicial branch depends on a perception of legitimacy surrounding its decisions.[14]

AMERICAN COURTS: A DUAL COURT SYSTEM

To better understand the U.S. court system, it is first important to know that this country has a dual court system: both a national system of federal courts and 50 state court systems. Although often sharing similar names, they operate under different constitutions and laws, as outlined in the sections that follow.

Before exploring these different courts, we take up the critical issue of **jurisdiction**, which is best defined as a court's legal authority to hear and decide a particular type of case. Jurisdiction is set forth in state and federal law, and it is typically based on geography (i.e., where the case is physically located) and subject matter (what the case is about: criminal, civil, juvenile, etc.). Courts hear only cases that fall under their jurisdiction and authority; our courts are organized along state and federal lines and then further along subject matter lines (see Table 7.1). At the lower end are the limited-jurisdiction state trial courts, handling traffic cases, misdemeanors, and juvenile matters, for example. At the highest level is the U.S. Supreme Court, which hears a limited number of appeals on federal and constitutional legal issues. The great majority of the nation's judicial business occurs at the state—not the federal—level, and our discussion begins there.

Jurisdiction, court: the authority of a court to hear a particular type of case, based on geography (city, state, or federal) and subject matter (e.g., criminal, civil, probate).

STATE COURTS

Where Most Cases Begin: State Trial Courts

At the lowest level of state courts are trial courts of limited jurisdiction, also known as inferior courts or lower courts. There are more than 13,500 trial courts of limited jurisdiction in the United States, staffed by about 18,000 judicial officers. The lower courts constitute 85 percent of all judicial bodies in the United States.[15]

Table 7.1 /// State and Federal Courts

State Courts	Federal Courts
Appellate Jurisdiction—Courts of Last Resort	
State supreme court	U.S. Supreme Court
Appellate Jurisdiction—Intermediate Courts of Appeals	
State courts of appeals	Circuit courts of appeals
Trial Courts	
State trial courts	U.S. district courts
Limited Jurisdiction Trial Courts	
Traffic, juvenile, justice of the peace (for example)	Tax, admiralty, bankruptcy (for example)

Variously called district, justice, justice of the peace, city, magistrate, or municipal courts, the lower courts decide a restricted range of cases. These courts are created and maintained by city or county governments and therefore are not part of the state judiciary. The caseload of the lower courts is staggering—more than 84 million cases a year, an overwhelming number of which are traffic cases (more than 45 million).[16] The workload of the lower courts can be divided into felony criminal cases, nonfelony criminal cases, and civil cases. In the felony arena, lower court jurisdiction typically includes the preliminary stages of felony cases. Therefore, after an arrest, a judge in a trial court of limited jurisdiction will hold the initial appearance, appoint counsel for indigents, and conduct the preliminary hearing. Later, the case will be transferred to a trial court of general jurisdiction for trial (or plea) and sentencing.[17]

General Jurisdiction: Major Trial Courts

State trial courts of general jurisdiction are usually referred to as the major trial courts. There are an estimated 2,000 major state trial courts, staffed by more than 11,000 judges.

Each court has its own support staff consisting of a clerk of the court, a bailiff, and others. In most states, the trial courts of general jurisdiction are also grouped into judicial districts or circuits. In rural areas, these districts or circuits encompass several adjoining counties, and the judges are true generalists who hear a variety of cases and literally ride the circuit; conversely, larger counties have only one circuit or district for the area, and the judges are often specialists assigned to hear only certain types of cases.[18] The term *general jurisdiction* means that these courts have the legal authority to decide all matters not specifically delegated by state law to the lower courts of limited jurisdiction. The most common names for these courts are district, circuit, and superior.[19] On the criminal justice front, most serious criminal violations, including many drug-related offenses, are heard in these trial courts of general jurisdiction. Although the courts are overburdened with heavy case filings, most criminal cases do not go to trial, and the dominant issue in the courts of general jurisdiction is not guilt or innocence, but what penalty to apply to someone who has entered a guilty plea through plea bargaining (described later in this chapter). Figure 7.1 shows an organizational structure for a county district court serving a population of 300,000. Note the variety of functions and programs that exist, in addition to the basic court role of hearing trials and rendering dispositions.

Figure 7.1 /// Organizational Structure for a District Court Serving a Population of 300,000

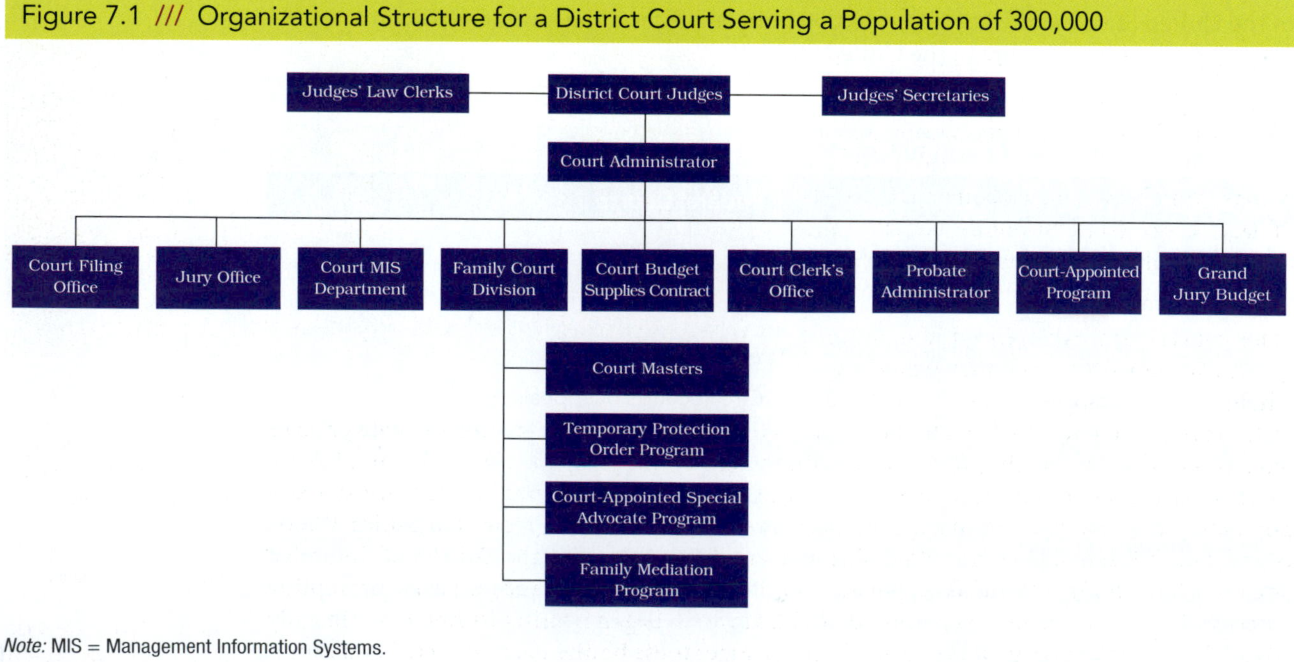

Note: MIS = Management Information Systems.

Figure 7.2 /// State Court Incoming Caseloads, 2007–2016

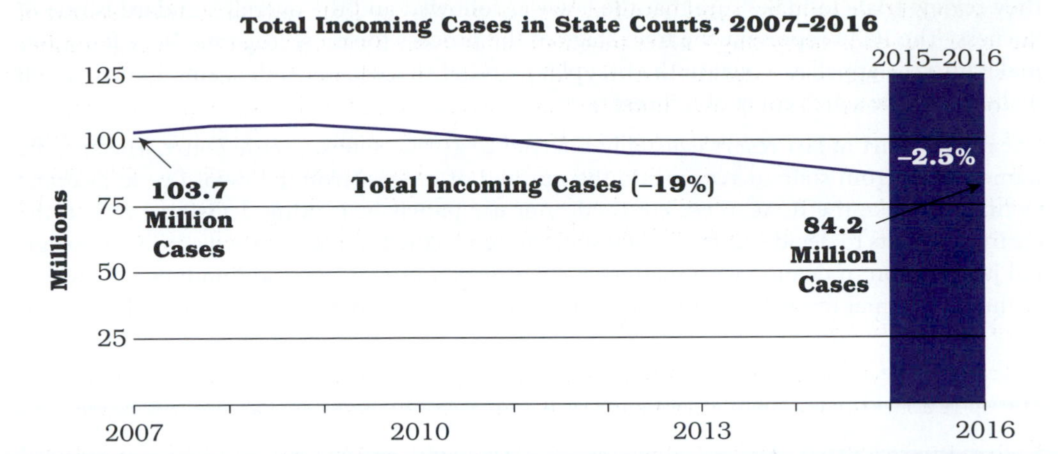

Total incoming caseloads in state courts reached an apex of just over 106.1 million cases at the onset of the Great Recession in 2008. Between 2009 and 2015, aggregate caseloads declined at an average annual rate of nearly 3.5 percent. However, in 2016, that rate fell to a 2.5 percent decrease overall.

Source: www.courtstatistics.org.

Intermediate court of appeals: a state court that stands between a trial court and a court of last resort; it typically has appellate jurisdiction only.

Within the state trial court system, a number of specialty courts or problem-solving courts, such as drug, family, mental health, veterans, and domestic violence courts, have become more common. They are used to divert criminal defendants into treatment programs rather than jail or prison (see "Diversion Programs/Problem-Solving Courts," later in this chapter).

Figure 7.2 shows the caseload trends for state criminal trial courts in the United States. As with crime in general, caseloads are declining in the trial courts.

Appeals Courts

After a conviction in a criminal case or a judgment in a civil case, the first stop for an appeal in the state courts is known as an **intermediate court of appeals** (ICA). These courts stand between trial courts and courts of last resort (state supreme courts), and they typically have appellate jurisdiction only; that is, they hear only appeals.

©AP Photo/Bebeto Matthews

Appellate court judges Sandra Sgroi, left, John Leventhal, center, and Mark Dillon, right, three members of a four-judge panel, listen to oral arguments from attorneys appealing for the public release of records from the grand jury in the videotaped chokehold death of Eric Garner, Tuesday, June 16, 2015, in New York.

State courts have experienced a significant growth in appellate cases that would overwhelm a single appellate court such as a state's supreme court. To alleviate the caseload burden on courts of last resort, state officials have responded by creating ICAs, which must hear all properly filed appeals. As of 2019, 41 states had established permanent ICAs. The only states not having an ICA are sparsely populated with low volumes of appeals: Delaware, Maine, Montana, New Hampshire, Rhode Island, South Dakota, Vermont, West Virginia, and Wyoming (also, the District of Columbia does not have ICAs). Nevada's population and caseloads grew in recent years, prompting voters in 2014 to approve the creation of an ICA, which began hearing its first cases in early 2015.[20] The structure of the ICAs varies. In most states these bodies hear both civil and criminal appeals, and these courts typically use rotating three-judge panels. Also, the state ICAs' workload is demanding: According to the National Center for State Courts, state ICAs report that more than 185,000 cases are filed annually.[21] ICAs engage primarily in error corrections; they review trials to make sure that the law was followed and the overall standard is one of fairness. The ICAs represent the final stage of the process for most litigants. Very few cases make it to the appellate court in the first place, and of those cases, only a small portion will be heard by the state's court of last resort.[22]

Court of last resort: the last court that may hear a case at the state or federal level.

Causation: a link between one's act and the injurious act or crime, such as one tossing a match in a forest and igniting a deadly fire.

The state's **court of last resort** is usually referred to as the state supreme court. The specific names differ from state to state, as do the number of judges (from a low of five to as many as nine). Unlike the ICAs, these courts do not use panels in making decisions; rather, the entire court sits to decide each case. All state supreme courts have a limited amount of original jurisdiction in dealing with matters such as disciplining lawyers and judges.[23] Nowhere is the policy-making role of state supreme courts more apparent than in deciding death

You Be the . . .

Judge

On a very cold night in Rochester, New York, Barry Kibbe and a friend met Stafford at a bar. Stafford had been drinking so heavily that the bartender refused to serve him more alcohol, so Kibbe offered to take him barhopping elsewhere. They visited other bars, and at about 9:30 p.m., Kibbe and his friend drove Stafford to a remote point on a highway and demanded his money. They also forced Stafford to lower his trousers and remove his boots to show he had no money hidden. Stafford was then abandoned on the highway, in the cold. A half-hour later, a man driving his truck down the highway saw Stafford standing in the highway, waving his arms for him to stop. The driver didn't see Stafford soon enough to stop, and Stafford was struck and killed. Even though the driver of the truck killed Stafford, Kibbe was arrested and charged with robbery and second-degree murder, raising a challenging question of **causation**—a necessary element in holding someone criminally responsible for a death.

State Court Actions: In *State v. Kibbe,* Kibbe was tried and convicted of robbery and murder in the second degree. At trial, the judge did not instruct the jury on the subject of causation (e.g., that the government had to prove Kibbe had actually caused Stafford's death). On appeal to New York's appeals court, in *Kibbe v. Henderson,* Kibbe argued that the judge should have given the jury such an instruction so that they would have to determine whether the state had proved this element beyond a reasonable doubt, but the appellate court affirmed his conviction. Then, on appeal to New York's supreme court, the conviction was also affirmed. Both courts found that the judge did not err in failing to instruct the jury about causation.

Federal Court Actions: Having exhausted all possible state remedies, Kibbe then sought redress in the federal court, filing a writ of *habeas corpus* with the U.S. district court having jurisdiction and arguing that the trial judge had violated his due process rights by not giving the jury instructions regarding causation. The district court denied the habeas petition, saying that no constitutional question had been raised.

Next, Kibbe appealed to the Second U.S. Circuit Court of Appeals, making the same argument. This court, however, reversed his conviction, saying that Kibbe had been deprived of due process because of the trial judge's failure to instruct the jury on causation. Next, the government appealed, this time to the U.S. Supreme Court; in *Henderson v. Kibbe,* the Supreme Court, issuing a writ of certiorari (defined later in this chapter) to the lower court, decided to hear the case.

1. How do you believe the U.S. Supreme Court should rule—for or against Kibbe? Why?
2. Who was Henderson (Kibbe's adversary) in this case?
3. What is meant by "exhausting all possible state remedies"?
4. What is meant by *habeas corpus?*

Answers to these questions are provided in the Notes section at the end of the book.[24]

penalty cases—which, in most states, are appealed automatically to the state's highest court, thus bypassing the ICAs. The state supreme courts are also the ultimate review board for matters involving interpretation of state law.[25] In states not having ICAs, the state supreme court has no power to choose which cases will be placed on its docket. However, the ability of most state supreme courts to choose which cases to hear makes them important policy-making bodies. Whereas ICAs review thousands of cases each year, looking for errors, state supreme courts handle a hundred or so cases that present the most challenging legal issues arising in that state.

FEDERAL COURTS

The accompanying "You Be the . . . Judge" box describes a case that uniquely demonstrates the entire appellate process through the dual state and federal court systems, going through the state courts and all the way to the U.S. Supreme Court.

Federal Trial Courts: U.S. District Courts

The U.S. **district courts**, like their counterparts in the state court system, may be fairly described as the "workhorses" of the federal judiciary, because nearly all civil or criminal cases heard in the federal courts are initiated at the district court level. Indeed, there are about 359,000 annual combined filings for civil cases and criminal defendants in the U.S. district courts.[26]

District courts: trial courts at the county, state, or federal level with general and original jurisdiction.

The locations of the U.S. district courts are outlined in Figure 7.3 (sometimes designated by region—"northern" or "eastern," for example). Congress created 94 U.S. district courts, 89 of

Figure 7.3 /// Geographic Boundaries of U.S. Courts of Appeals and District Courts

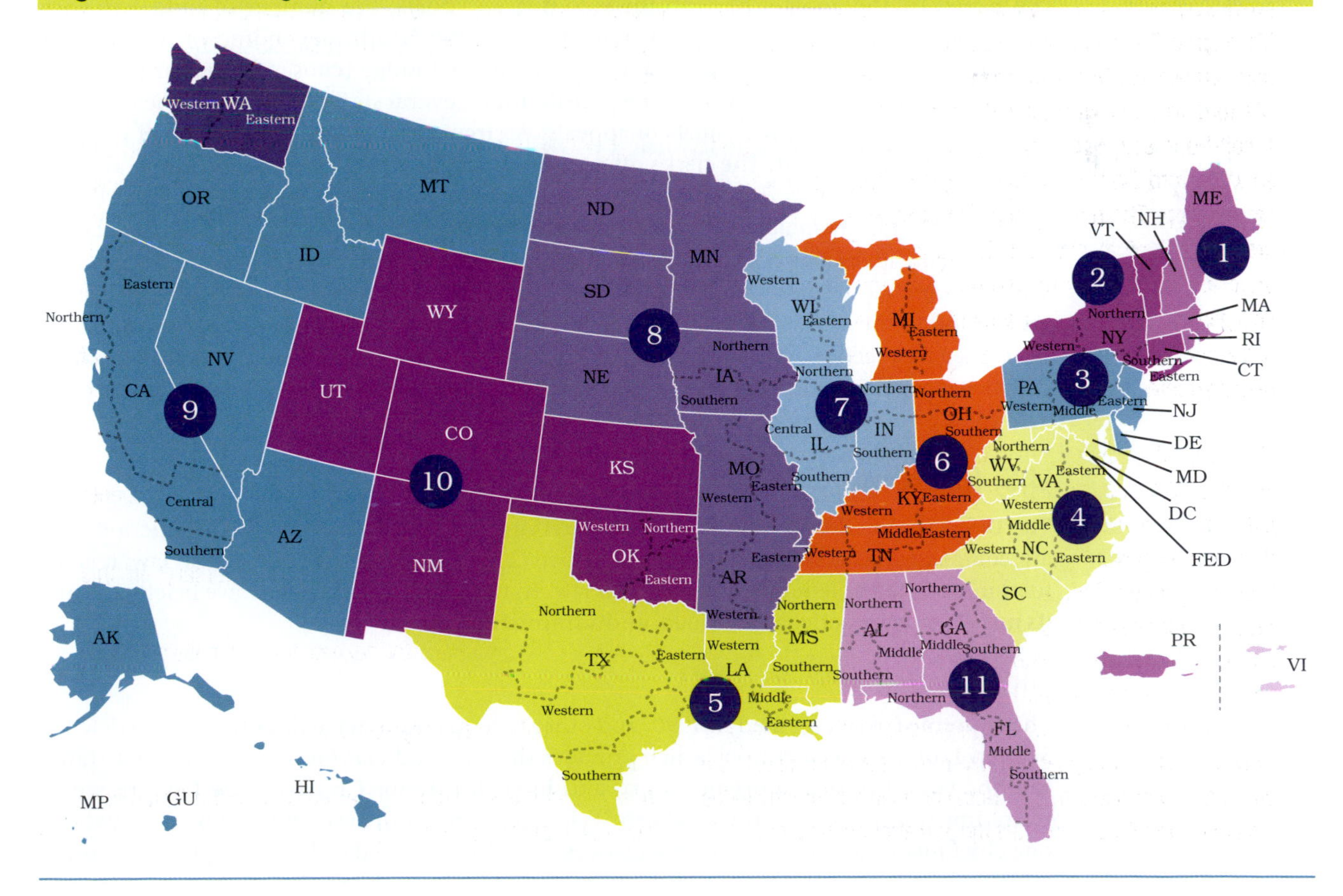

which are located within the 50 states. There is at least one district court in each state (some states have more, such as California, New York, and Texas, each of which has four). Congress has created 678 district court judgeships for the 94 districts. The president nominates district judges, who must then be confirmed by the Senate; they then serve for life unless removed for cause. In the federal system, the U.S. district courts are the federal trial courts of original jurisdiction for all major violations of federal criminal law (some 500 full-time magistrate judges hear minor violations).[27]

District court judges are assisted by an elaborate supporting cast of clerks, secretaries, law clerks, court reporters, probations officers, pretrial services officers, and U.S. marshals. The larger districts also have a public defender. Another important actor at the district court level is the U.S. attorney—the federal prosecutor. There is one U.S. attorney in each district. The work of district judges is significantly assisted by 352 bankruptcy judges, who are appointed for 14-year terms by the court of appeals in which the district is located.

U.S. Courts of Appeals: Circuit Courts

Circuit courts: originally courts wherein judges traveled a circuit to hear appeals, now courts with several counties or districts in their jurisdiction; the federal court system contains 11 circuit courts of appeals (plus the District of Columbia and territories), which hear appeals from district courts.

Much like the state court system, the federal system provides intermediate courts of appeals—intermediate between U.S. district courts and the U.S. Supreme Court. These courts are known as the **circuit courts** of appeals, referring to the 13 geographic areas—or circuits—where these courts are seated and over which they exercise jurisdiction. Eleven of the circuits are identified by number, and two others are called the D.C. Circuit and the Federal Circuit (see Figure 7.3). A court of appeals hears appeals from the U.S. district courts located within its circuit, as well as appeals from decisions of federal administrative agencies. These courts, receiving about 28,000 appellate filings per year, are the last stop for an appeal before it reaches the U.S. Supreme Court.[28]

The judges who sit on the circuit courts of appeals are nominated by the president and confirmed by the Senate. As with the U.S. district courts, the number of judges in each circuit varies, from 6 in the First Circuit to 28 in the Ninth, depending on the volume and complexity of the caseload. Each circuit has a chief judge (chosen by seniority or on a rotating basis) who has supervisory responsibilities. Several staff positions aid the judges in conducting the work of the courts of appeals. A circuit executive assists the chief judge in administering the circuit. The clerk's office maintains the records. Each judge is also allowed to hire three law clerks. In deciding cases, the courts of appeals may use rotating three-judge panels. Or, by majority vote, all the judges in the circuit may sit together to decide a case or reconsider a panel's decision. Such *en banc* hearings are rare, however.[29] Because the U.S. Supreme Court hears so few cases (see next section), the U.S. circuit court decisions are very often the last word on legal issues and, as such, their decisions have great impact in setting legal precedent and in shaping the law for the many people who live within a circuit court's geographic jurisdiction.

U.S. Supreme Court

U.S. Supreme Court: the court of last resort in the United States, also the highest appellate court; it consists of nine justices who are appointed for life.

The **U.S. Supreme Court**, as the highest court in the nation, has ultimate jurisdiction over all federal courts, as well as over state courts in cases involving issues of federal law; it is the final interpreter of federal constitutional law. In the sections that follow, we briefly discuss its jurisdiction, practices, workload, and administration.

Judges and Advocacy

The Supreme Court, formed in 1790, and other federal courts have their basis in Article III, Section 1, of the Constitution, which provides that "the judicial Power of the United States, shall be vested in one supreme Court, and in such inferior Courts as the Congress may from time to time ordain and establish."[30] The Supreme Court is composed of nine justices: one chief justice and eight associate justices. As with other federal judges appointed under

Article III, they are nominated to their post by the president and confirmed by the Senate, and they serve for life.[31] Each new term of the Supreme Court begins, by statute, on the first Monday in October.

The U.S. Supreme Court, in Washington, D.C., is where the court's nine justices meet, deliberate, and render the law of the land.

Not just any lawyer may advocate a cause before the high court; all who wish to do so must first secure admission to the Supreme Court bar. Applicants must submit an application form that requires applicants to have been admitted to practice in the highest court of their state for a period of at least three years (during which time they must not have been the subject of any adverse disciplinary action), and they must appear to the Court to be of good moral and professional character. Applicants must also swear or affirm to act "uprightly and according to law, and . . . support the Constitution of the United States."[32]

Conferences and Workload

The Supreme Court does not meet continuously in formal sessions during its nine-month term. Instead, the Court divides its time into four separate but related activities. First, some amount of time is allocated to reading through the thousands of petitions for review of cases that come annually to the Court—usually during the summer and when the Court is not sitting to hear cases. Second, the Court allocates blocks of time for oral arguments—the live discussion in which lawyers for both sides present their clients' positions to the justices. During the weeks of oral arguments, the Court sets aside its third allotment of time, for private discussions of how each justice will vote on the cases they have just heard. Time is also allowed for the justices to discuss which additional cases to hear. These private discussions are usually held on Wednesday afternoons and Fridays during the weeks of oral arguments. The justices set aside a fourth block of time to work on writing their opinions.[33]

The Court has complete discretion to control the nature and number of the cases it reviews by means of the *writ* (order) *of certiorari*—an order from a higher court directing a lower court to send the record of a case for review. The Court considers requests for *writs of certiorari* according to the *rule of four.* If four justices decide to review a case—to "grant cert," the Court will hear the case. Several criteria are used to decide if a case requires action: First, does the case concern an issue of constitutional or legal importance? Does it fall within the Court's jurisdiction (the Court can hear only cases that are mandated by Congress or the Constitution)? Does the party bringing a case have **standing**—a strong vested interest in the issues raised in the case and in its outcome?[34] The Court hears only a tiny fraction of the thousands of petitions that come before it. When it declines to hear a case, the decision of the lower court stands as the final word on it. The Court's caseload has increased over time; today the Court has 7,000–8,000 cases on the docket per term, and formal written opinions are delivered in about 70–80 cases.[35]

Standing: a legal doctrine requiring that one must not be a party to a lawsuit unless he or she has a personal stake in its outcome.

Administration

The chief justice orders the business of the Supreme Court (a description of the chief justice's role is provided in Chapter 8) and administers the oath of office to the president and vice president upon their inauguration. According to Article I, Section 3, of the Constitution of the United States, the chief justice is also empowered to preside over the Senate in the event that it sits as a court to try an impeachment of the president.

The clerk of the Court serves as the Supreme Court's chief administrative officer, supervising a staff of 30 under the guidance of the chief justice. The marshal of the Court supervises all building operations. The reporter of decisions oversees the printing and publication of the

Court's decisions. Other key personnel are the librarian and the public information officer. In addition, each justice is entitled to hire four law clerks, almost always recent top graduates of law schools, many of whom have served clerkships in a lower court the previous year.[36]

MAKING PREPARATIONS: PRETRIAL PROCESSES

In Chapter 1, we briefly discussed the sequence of major events that compose the U.S. criminal justice system, from the point of entry of the offender into the system and through the related police, courts, and corrections components. Here, we focus on the pretrial process, the events that occur prior to the trial itself. (We discuss judges as well as the roles and strategies of prosecution and defense attorneys in Chapter 8.)

Booking, Initial Appearance, Bail, and Preliminary Hearing

As stated in Chapter 1, the criminal justice process is engaged when one is arrested. Following that, the accused will then proceed through the steps detailed in the sections that follow.

Booking

Booking: the clerical procedure that occurs after an arrestee is taken to jail, during which a record is made of his or her name, address, charge(s), arresting officers, and time and place of arrest.

Basically a clerical function, **booking** usually involves taking the suspect to a police station or sheriff's office, where he or she may be fingerprinted, photographed, questioned, read his or her rights, and possibly be given a bail amount that must be paid in order to gain release until the next stage in the process. Meanwhile, the prosecutor will be sent a copy of the offense report written by the police, and will be considering whether or not enough probable cause (discussed in Chapter 6) exists to believe the suspect committed the offense.

Initial Appearance

Initial appearance: a formal proceeding during which the accused is read his or her rights and informed of the charges and the amount of bail required to secure pretrial release.

Within a reasonable time after arrest, the suspect has his or her **initial appearance** in court, where the judge gives the defendant formal notice of the charge(s), advises the suspect of his or her rights—including the right to a government-provided attorney if the defendant cannot afford one, and sets bail in appropriate cases. The judge also performs a cursory review of the evidence, and if he or she does not believe such evidence establishes probable cause that the defendant may have committed the crime charged, the case will be dismissed. Notice that defendants typically do not enter a plea at this stage because they have not yet been appointed an attorney, so they have not yet had a chance to consult anyone about their case.

Bail

Bail: surety (e.g., cash or paper bond) provided by a defendant to guarantee his or her return to court to answer to criminal charges.

Bail is also known as "pretrial release," allowing defendants to remain out of jail while awaiting trial and affording them time to help with their defense and to maintain ties with family and their job. But bail is not a right, nor is it guaranteed under the U.S. Constitution. In fact, most state laws provide that bail cannot be granted in cases involving certain violent felonies. If a judge does grant bail, however, the Constitution's Eighth Amendment prohibits the bail amount from being "excessive." To ensure a fair decision-making process, a bail hearing is conducted (usually as part of the initial appearance), and the judge must consider a variety of factors, including

- The risk that the defendant will flee before trial
- The defendant's criminal record
- The seriousness of the charges
- The safety of the community

There are several ways to make bail, all of which ensure that the accused will later appear in court and most of which require a financial commitment from defendants or their family or friends. The court typically requires either cash or a bond, the latter of which is basically an insurance policy the defendant buys insuring his or her appearance. Defendants can "make bail" directly with some courts, but in other jurisdictions the defendant will have to get a bond through a bail bond company, which typically requires a nonrefundable payment of 10 percent of the bail amount. Either way, if the defendant fails to appear in court, the court issues a warrant for the defendant's arrest, and he or she loses any money posted with the court.

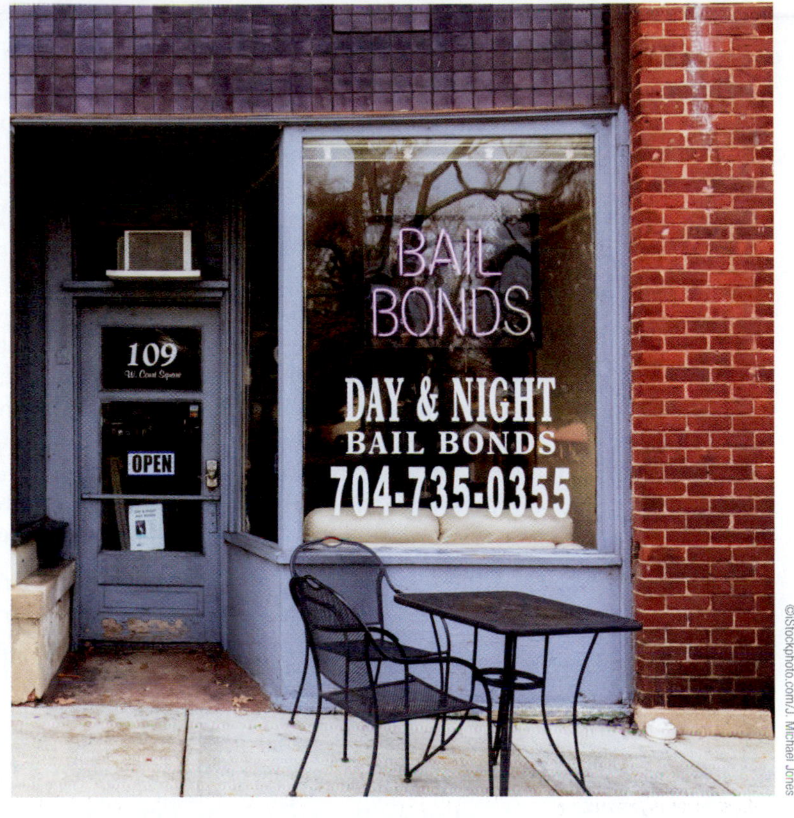

©iStockphoto.com/J. Michael Jones

Bail bonds operations are commonly located near county jails to serve criminal defendants seeking freedom on bail while their cases are pending. Judges and bail bond agents use similar criteria in making decisions about bail.

The bail decision can be critical for criminal defendants. Again, pretrial release affords defendants the chance to keep their jobs and maintain family and community ties, and to assist with their defense. Being denied bail—pretrial detention—can be devastating on all of these fronts, and it can provide prosecutors and police better access to the defendant for continued questioning and investigation (subject to certain *Miranda* limitations). So it is important to consider what types of defendants are detained and whether the process is truly fair. In 2015, researchers released findings showing that the bail decision is unfairly affecting the poor and racial minorities, with African American defendants being detained four times more often than whites.[37] The same study concluded that 75 percent of people held in pretrial detention are there for nonviolent offenses, but they simply cannot afford bail and so they are incarcerated in county jails until their cases are resolved. Indeed, 65 percent of jail populations are unconvicted defendants awaiting trial.[38] Therefore, the bail decision affects not only defendants but also our already-overcrowded jails and the criminal justice professionals tasked with keeping those populations secure.

Preliminary Hearing

The **preliminary hearing** (which the defendant may choose to waive in some jurisdictions) allows a judge (no jury is present) to decide whether or not probable cause is sufficient against the person charged to proceed to trial. The prosecutor will offer physical evidence and testimony to try to get the accused "bound over" for trial while the defense offers counterevidence. If the judge finds enough probable cause, he or she will order the accused to appear at trial to answer the state's formal charges, which the prosecutor will file in the form of an "information" (versus an "indictment," discussed in the next section). The preliminary hearing can help the accused and his or her counsel to prepare for trial, because they are able to hear much of the state's case. But as discussed in the next section, in some states, the prosecutor avoids this effect by presenting the case before a secretive grand jury.

Preliminary hearing: a stage in the criminal process conducted by a magistrate to determine whether a person charged with a crime should be held for trial based on probable cause; does not determine guilt or innocence.

Grand Juries

The primary function of the modern **grand jury** is to review the evidence presented by the prosecutor and to determine whether there is probable cause to return an indictment. Although all states have some form of the grand jury, only about half of all states use grand juries routinely to bring formal charges, rather than relying solely on a prosecutor's decision (or after a preliminary hearing).[39] The prosecutor presents evidence to the grand jury, and, much like the judge in a preliminary hearing, the grand jury must find probable cause to

Grand jury: a body that hears evidence and determines probable cause regarding crimes and can return formal charges against suspects; use, size, and functions vary among the states.

charge the defendant with a crime. If that happens, the grand jury issues an "indictment"—a formal charge—and trial (or plea bargaining) will ensue.

The importance of the grand jury function cannot be overstated because, without an indictment, the state cannot move forward with criminal charges against a defendant. Consider the 2014 case from Ferguson, Missouri, where a white Ferguson police officer, Darren Wilson, shot and killed unarmed, 18-year-old Michael Brown during an encounter in the street. The case sparked intense media coverage and debate over police-citizen race relations and the appropriate use of police force in encounters with unarmed suspects, with many people assuming Wilson had acted criminally because Brown was unarmed. The case against Wilson went to a St. Louis County grand jury, which—after 25 days of hearing testimony and reviewing evidence—determined there was not enough probable cause to conclude that Wilson acted criminally. But the Brown-Wilson proceedings were different from a typical grand jury review in several ways: The prosecutor did not recommend charges against Wilson, more than 60 witnesses testified (versus the typical three or four), and Officer Wilson himself testified for more than four hours.[40] Some observers believe that, in cases like Wilson's, the state effectively "tries" the case but without the public oversight of a typical criminal trial, and, in that way, the grand jury role can be significant.

Although the use of grand juries and their roles vary across states, the Fifth Amendment to the U.S. Constitution requires a grand jury indictment for *all* federal criminal charges. For federal cases involving complex and long-term investigations (such as those involving organized crime, drug conspiracies, or political corruption), "long-term" grand juries will be impaneled. In most jurisdictions, grand jurors are drawn from the same pool of potential jurors as are any other jury panels, and in the same manner. But unlike potential jurors in regular trials, grand jurors are not screened for biases or other improper factors.

The Federal Rules of Criminal Procedure provide that the prosecutor, grand jurors, and the grand jury stenographer are prohibited from disclosing what happened before the grand jury, unless ordered to do so in a judicial proceeding. Secrecy prevents the escape of people whose indictment may be contemplated, ensures that the grand jury can deliberate without outside pressure, prevents witness tampering prior to trial, and encourages people with information about a crime to speak freely.

A prosecutor can obtain a subpoena to compel anyone to testify before a grand jury, without showing probable cause and, in most jurisdictions, without even showing that the person subpoenaed is likely to have relevant information.

In the federal system, a witness cannot have his or her lawyer present in the grand jury room, although witnesses may interrupt their testimony and leave the grand jury room to consult with their lawyer. A few states do allow a lawyer to accompany the witness. A witness who refuses to appear before the grand jury risks being held in contempt of the court.

If the grand jury refuses to return an indictment, the prosecutor can try again; double jeopardy does not apply to a grand jury proceeding. No judge is present in the grand jury room when testimony is being taken.[41]

The Casey Anthony Case: Florida uses grand juries to bring formal charges against a suspect and issue indictments in first-degree murder cases. In the case against Anthony, 19 grand jurors—10 women and 9 men—heard testimony from a variety of witnesses, including police officers, an FBI agent, a cadaver dog handler, and Anthony's father. At the time of the grand jury hearing, Anthony's daughter was still missing; her body had not yet been found. Still, the grand jury deliberated for only about 30 minutes before issuing a criminal indictment for first-degree murder.

Arraignment

After being formally charged, the accused will again be advised of his or her rights, and is asked to enter a formal plea. A defendant may plead "not guilty," "guilty," or "no contest" (also known as *nolo contendere*). A "not guilty" plea means the defendant claims innocence and is forcing the state to prove its case at trial, whereas a "guilty" plea relieves the state of that burden and the defendant is convicted without trial. The *nolo* plea is unique in that it allows a defendant to plead guilty for purposes of avoiding trial, but the plea cannot be used against him or her later in a civil case (because technically he has not admitted guilt). The court must always approve a plea of *nolo contendere*.

If the plea at **arraignment** is "not guilty," a trial must occur (unless plea bargaining takes place).

BILL GREENBLATT/UPI/Newscom

Michael Brown was fatally shot by Officer Darren Wilson on August 9, 2014, in Ferguson, Missouri. On November 24, 2014, the St. Louis County grand jury decided not to indict Wilson, finding that he acted within the limits of the lethal-force law.

Plea Negotiation

When defendants enter a "guilty" or "no contest" plea, they often do so through a **plea negotiation**, or **plea bargaining**. In fact, the great majority of criminal convictions—typically more than 90 percent—are obtained through plea deals, without any courtroom fact-finding. In essence, in exchange for the defendant's plea of guilty, the government is willing to give the defendant certain concessions.

There are several forms of plea bargaining. The accused can engage in *charge bargaining* (offering to plead guilty to a lesser offense than the one charged, thus hoping for a lighter sentence), *count bargaining* (pleading guilty to, say, three of the charged counts and having the remaining six counts thrown out), or negotiating how he or she will serve an imposed sentence (e.g., two five-year terms to be served concurrently, as opposed to consecutively).

As the defendant benefits from plea bargaining, so does the government. First, criminal court dockets in most jurisdictions are seriously overburdened and the state simply cannot take every case to trial, so there is great pressure to resolve cases out of court. Further, the prosecutor "wins" a conviction while reducing the time, money, and uncertainty involved in taking the case to trial, whereas a jury may acquit because the defense produces unanticipated evidence or witnesses. The process also eases the burden on witnesses and prospective jurors, and can reduce the overcrowding of jails by more quickly funneling offenders out of pretrial detention and into prison or some form of community corrections (discussed in Chapter 11).

Arraignment: a criminal court proceeding during which a formally charged defendant is informed of the charges and asked to enter a plea of guilty or not guilty.

Plea negotiation (or bargaining): a preconviction process between the prosecutor and the accused in which a plea of guilty is given by the defendant, with certain specified considerations in return—for example, having several charges or counts tossed out, and a plea by the prosecutor to the court for leniency or shorter sentence.

Certainly, there are many reasons why a high percentage of cases are bargained out of the system, and one can only imagine the havoc that would be caused in the court system if 9 of 10 cases that are plea bargained had to be tried. However, by signing the deal and pleading guilty, the defendant waives a number of constitutional protections. But consider the realities of a typical barroom killing. There might be evidence of premeditation and malice that is sufficient to justify a jury verdict of murder in the first degree. Or, the defendant's longtime status as an alcoholic might convince the jury that he was unable to form the necessary intent to be heavily punished. Perhaps the defendant may indicate that he acted in the "heat of passion," pointing to a verdict of manslaughter, or even that he acted in self-defense. When such cases are given to the jury to decide, a variety of outcomes are possible. Therefore, plea negotiations allow the prosecutor and the defense to arrive at some middle ground of what experience has shown to be "justice," without the defense running the risk of heavy punishment for the defendant, and the government not having to devote many days to trial—with the risk of the defendant's being acquitted.[42]

Some authors believe plea bargaining reduces the courthouse to something akin to a Turkish bazaar, where people barter over the price of copper jugs. They see it as justice on the cheap. Others believe that plea bargaining works to make the job of the judge, the prosecutor, and the defense attorney much easier, while sparing the criminal justice system the expense and time needed to conduct many more trials.

No matter where one stands on the issue, however, it is ironic that, in general, both police and civil libertarians oppose plea bargaining. Police and others in the crime control camp view plea bargaining as undesirable because defendants can avoid conviction and responsibility for crimes they committed when allowed to plead guilty to (and be sentenced for) lesser and/or fewer charges; police, in the crime control camp, would prefer to see the defendant convicted for the crime committed.

Civil libertarians and other supporters of the due process model also oppose plea bargaining, but for different reasons: When agreeing to negotiate a plea, the accused forfeits a long list of legal protections afforded under the Bill of Rights: the presumption of innocence; the government's burden of proof (beyond a reasonable doubt); and the right to face one's accuser, testify, and present witnesses in one's defense, the right to have an attorney and a trial by jury (except for lesser offenses), and the right to an appeal if convicted. Another concern is that an innocent defendant might be forced to enter a plea of guilty because of the threat of trial or police and prosecutorial coercion.

It is important to consider both positions. In our quest for justice, we must ask whether plea bargaining sacrifices too many of the defendant's rights. We must also ask if plea bargaining impedes justice by giving too many benefits to guilty persons.

Jury Trials

Most civilizations—even the most primitive in nature—have used some means to get at the truth: to tell right from wrong, guilt from innocence, and so forth. In Burma, each suspected party to a crime had to light a candle, and the person whose candle burned the longest was not punished.[43] In Borneo, suspects poured lime juice on a shellfish; whoever's shellfish squirmed first was the guilty party.[44] The "trial by ordeal" method was also used around the world, with people's guilt or innocence determined by subjecting them to a painful task (often using fire and water); the idea was that God would intercede and help the innocent by performing a miracle on their behalf.[45]

The Sixth Amendment to the U.S. Constitution ensures that our method is more civilized, guaranteeing the defendant a trial by an impartial jury of his or her peers. Many people view the jury as the most sacred aspect of our criminal justice system, because it is where common

Figure 7.4 /// Jury Trials and Service in United States

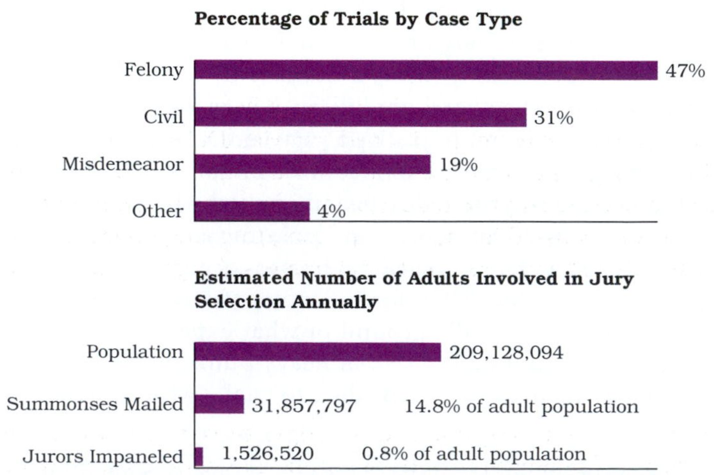

Note: This estimate was extrapolated using survey results for 1,546 counties representing 70 percent of the U.S. population.

Source: Reprinted with permission from The National Center for State Courts, Center for Jury Studies.

citizens determine the truth and assess punishment. Not all criminal defendants are guaranteed a right to trial by jury (i.e., if charged with a *lesser misdemeanor*—one that has a penalty of less than six months in jail); furthermore, the defendant may waive the right to a jury trial and be tried by a judge alone (known as a *bench trial*). There are advantages and disadvantages to each, and a wise defendant will want to discuss them with an attorney.

Criminal trials in the United States often involve a jury of one's peers to hear the evidence; then, if rendering a conviction, the same jury may be used to determine the proper form and extent of punishment.

The method of selecting citizen *peers* to hear the evidence is important. First, in most states a questionnaire is mailed to people (whose names were obtained from voter, taxpayer, driver, or other lists) to determine who is qualified to serve, with certain exemptions given to specific groups. Those who are qualified to serve are then sent a *summons* to appear and form a jury pool, from which a smaller number of prospective jurors is selected for a process known as *voir dire* ("to speak the truth"), where the prosecutor and defense attorney question and screen pool members to determine whether they can be fair and impartial and decide the case based on the evidence presented. With the judge overseeing the selection process, both the prosecution and the defense can challenge and have removed an unlimited number of jurors *for cause*, meaning for some reason one is prejudiced against their side. But both the prosecution and the defense also receive a limited number of *peremptory challenges* (usually set by statute), which allow them to remove jurors without any reason or explanation.

Generally, 12 jurors and 2 alternates are selected for a criminal trial, but that number is not required by the Constitution. In *Williams v. Florida,* the U.S. Supreme Court observed that the decision to fix the size of a jury at 12 "appears to have been a historical accident" and that a 6-member jury satisfied the constitutional requirement.[46] Nor is a unanimous verdict by the jury required by either the U.S. Constitution or the U.S. Supreme Court, so in some states a majority of "votes" from the jurors will support a verdict.[47] Figure 7.4 shows several aspects of jury service and trials in the United States as reported by the Center for Jury Studies: the estimated number of jury trials, the percentage of actual trials by case type, and the estimated number of adults involved in jury service each year.

Pretrial Motions

Prior to the trial, either the defendant or the prosecutor may file **pretrial motions/processes** with the court in order to be better positioned for trial. Defense motions include requests to suppress evidence (e.g., the defense believes that a police search, physical evidence, or a confession was obtained illegally), to reduce bail (if the accused is still in jail awaiting trial), to conduct discovery (discussed in the next section), to change venue (to move the trial to another city, if a highly publicized or emotional crime is charged), and to delay trial (known as a "continuance").

Pretrial motions/processes: any number of motions filed by prosecutors and defense attorneys prior to trial, for instance, to quash evidence, change venue, conduct discovery, challenge a search or seizure, raise doubts about expert witnesses, or exclude a defendant's confession.

Discovery

No member of the court work group—including judges, prosecutors, and defense attorneys—likes major surprises or "bombshell" evidence coming to light in the courtroom. **Discovery** is simply the exchange of information between prosecution and defense, in order to promote a fair adversarial contest between the two sides and help the truth come to light. Essentially, each side is entitled to learn the other's strengths and weaknesses as well as the evidence and theories on which each will rely.

Discovery: a procedure wherein both the prosecution and the defense exchange and share information as to witnesses to be used, results of tests, recorded statements by defendants, or psychiatric reports, so that there are no major surprises at trial.

You Be the . . .

Defense Attorney

In 2001, Michael Vick rose to national fame as a star quarterback with the NFL's Atlanta Falcons—the first African American player to be selected in the first round of the NFL draft. Vick then had six successful seasons with the Falcons, going to the playoffs twice and being selected to three Pro Bowls. But in 2007, Vick was implicated in an illegal dog-fighting operation in his home state of Virginia. Federal and state law enforcement authorities found evidence of an extensive dog-fighting venue at Vick's sprawling rural compound, along with evidence of illegal interstate activities, including dog trafficking, online betting, and drug sales. Authorities also found evidence of dog torture and killings, and Vick was soon charged with breaking a host of both Virginia state laws and federal laws.

With his NFL career in shambles, Vick faced prosecution in both venues of our dual court system, and many observers asked where Vick would first answer for his alleged crimes and would he—could he—be brought to justice twice, raising questions of double jeopardy under the Fifth Amendment to the U.S. Constitution. The U.S. government took the lead in prosecuting Vick first. He was granted bail with certain conditions, including drug testing. He was indicted by a federal grand jury, meaning the government could proceed with formal charges against him. Facing such charges and the prospect of extensive evidence of his illegal activities and brutality against the dogs used in his fighting operations, what would you have recommended as Vick's defense attorney?

1. Would you have recommended that he seek a plea deal (meaning, a plea negotiation) rather than take the case before a jury? Why?
2. What would have been the pros and cons of pursuing a jury trial, given the facts of this case?

More about the outcome of this case is provided in the Notes section at the end of the book.[48]

Source: CNN Library, *Michael Vick Fast Facts*, November 10, 2014, http://www.cnn.com/2013/06/24/us/michael-vick-fast-facts/.

Discovery has become quite controversial in recent years, with prosecutors often being accused of withholding evidence that should have been provided to the defense, to the point that many jurisdictions have individuals—typically attorneys, skilled in the laws of evidence—serving as "discovery masters" to ensure fair exchange of information by both sides. Generally, the prosecution has a higher burden of providing "exculpatory" evidence (that which tends to support the defendant's innocence).[49] However, because the U.S. Supreme Court has required the prosecution to disclose only evidence that is both material and exculpatory, this has become a confusing area of law and formal/informal policy—with some states adopting conservative, others liberal, and still others "middle ground" rules of discovery[50]—so that the question of what is to be exchanged is not always clear-cut.

Diversion Programs/Problem-Solving Courts

Diversion program: a sentencing alternative that removes a case from the criminal justice system, typically to move a defendant into another treatment program or modality.

The increasing number of criminal cases in our already-overburdened courts, together with the high recidivism (reoffending) rate, has spawned a number of alternative courts and **diversion programs** around the country, including drug courts, mental health courts, veterans courts, and even some courts dedicated to offenders with gambling addictions. By 2015, more than 3,000 drug courts were operating in every U.S. state and territory,[51] and today more than 300 mental health courts operate in most U.S. jurisdictions.[52] A 2016 report by the National Drug Court Institute suggests that many more adult drug courts, veterans treatment courts, family drug courts, and hybrid drug/DUI courts will be established over the next few years.[53] These "problem-solving" courts allow eligible defendants—typically first-time, nonviolent offenders—to move their cases to a court where specialized court professionals (prosecutors, defense attorneys, judges, social workers, physicians, and treatment professionals) can better address the unique features of these defendants and their cases.

©AP Photo/Rich Pedroncelli

Amanda Nagel hugs her 9-month-old daughter, Alexis, while waiting to appear in drug court in Placer County, California. Nagel, who had been arrested on a methamphetamine charge, told the judge she wants to go straight and stay out of jail to take care of her daughter. Drug courts, mental health courts, and veterans courts are just a few of the "specialty courts" operating across the country to divert low-risk offenders out of the criminal justice system.

The model for these specialty courts is fairly similar across jurisdictions, utilizing either a pre-plea or post-plea process. In the former instance, a defendant demonstrates eligibility for the court's services and is not required to plead guilty but is diverted to the specialty court's program. In the latter instance, the defendant is diverted to the court's program after an initial guilty plea. In both models, defendants are then given the opportunity to complete a program to treat the problems that landed them in criminal trouble in the first place: drug rehabilitation for the drug offender; mental health counseling/medication for the mentally ill offender; and specialized debriefing/counseling for military veterans, most of whom are suffering from posttraumatic stress disorder (PTSD) and other postcombat issues. If the defendant successfully completes the court-ordered "program," the charges are dropped (for a pre-plea case) or the guilty plea is vacated (in a post-plea case). If the defendant fails to carry out his or her end of the court agreement, the state can pursue the original criminal charges and the defendant will risk a conviction and a possible jail or prison sentence.

Research indicates that many of these courts, drug courts in particular, are succeeding in providing treatment and reducing recidivism. Many drug courts report an average of 8 to 14 percent lower recidivism rates than other justice system responses, with some of the best courts yielding a 35 to 80 percent lower rate. The long-term effects of such programs appear to be promising as well, with positive effects (nonrecidivism, or remaining "clean") lasting from 3 years to 14 years in some cases.[54]

THE TRIAL PROCESS

After all pretrial processes have been addressed, the next challenge is to get the case into the courtroom in a timely manner and then see that certain rules are followed and defendants' rights protected. This **trial process** is described in the sections that follow.

Trial process: all of the steps in the adjudicatory process, from indictment or charge to conviction or acquittal.

Right to a Speedy Trial: "Justice Delayed . . ."

Swift justice is a term that is fairly well emblazoned in our collective psyche—and has even been the title for a number of books, movies, and even some television series. Bringing offenders to justice in a timely manner is felt to be essential to sending a meaningful message to offenders, to convey a message of deterrence to the general public, to maintain public confidence in the judicial process, and generally to help the criminal justice system better do its job.

Similarly, the adage that "justice delayed is justice denied" says much about the long-standing goal of processing court cases with due deliberate speed. Charles Dickens condemned the practice of slow litigation in 19th-century England.[55] Dickens was considerably harsh toward England's Chancery Courts in his novel *Bleak House,*[56] and Shakespeare mentioned "the law's delay" in *Hamlet.*[57] Most important, even our founding fathers saw fit to hasten the movement of criminal matters into the courtroom: The Sixth Amendment to the Constitution states in part, "In all criminal prosecutions, the accused shall enjoy the right

Delay (trial): an attempt (usually by defense counsel) to have a criminal trial continued until a later date.

to a speedy and public trial." As a result, the consequences of **delay** to society are potentially severe. The U.S. Supreme Court has ruled that if the defendant's right to a speedy trial has been violated, then the indictment must be dismissed and/or the conviction overturned.[58] Certainly some criminal defendants want their trial dates delayed (or "continued") as long as possible—giving time for the community's emotions surrounding the crime to subside, the memories of its victims and witnesses to fade, and the defense to uncover further evidence. Others, however—particularly those who cannot post bail and are awaiting trial in a jail, and/or have jobs and family to return to—want their "day in court" to arrive as soon as possible.

Speedy Trial Act of 1974: later amended, a law originally enacted to ensure compliance with the Sixth Amendment's provision for a speedy trial by requiring that a federal case be brought to trial no more than 100 days following the arrest.

But what does a "speedy trial" mean in practice, and how does an appellate court know if the right has been denied? Those questions have been addressed at the *federal* level, with Congress enacting the **Speedy Trial Act of 1974**.[59] This act mandates a 30-day limit from the point of arrest to indictment and 70 days from indictment to trial. Thus, federal prosecutors have a total of 100 days from the time of arrest until trial.

However, there is very little in the way of fixed, enforced time limits at the *state* level, and the U.S. Supreme Court has refused to give the concept of a "speedy trial" any precise time frame.[60] Also, where they exist, most state laws fail to provide the courts with adequate and effective enforcement mechanisms; furthermore, if a prosecutor has clearly taken an excessively long amount of time to bring a case to trial, existing time limits may be waived due to the court's own congested dockets. As a result, there are no "teeth" in state statutes concerning time limits, so state-level speedy trial laws are often not followed in practice.

The concern, then, is with *unnecessary* delay. Where a court must determine whether or not the defendant's right to a speedy trial was violated, the Supreme Court in *Barker v. Wingo* established the following test:[61]

- *Length of delay*: A delay of a year or more from the date of arrest or indictment, whichever occurs first, was termed "presumptively prejudicial"; however, as noted earlier, the Supreme Court has never explicitly ruled that any absolute time limit applies.
- *Reason for the delay*: The prosecution may not excessively delay the trial for its own advantage; however, a trial may be delayed for good reason, such as to secure the presence of a key witness.
- *Time and manner in which the defendant has asserted his right*: If a defendant agrees to the delay when it works to his own benefit, he cannot later claim that he has been unduly delayed.
- *Degree of prejudice to the defendant that the delay has caused*: Delays can affect evidence and impair the defense; for example, witnesses might die, move away, or their memories fade; records or evidence can be lost or destroyed.[62]

The Casey Anthony Case: Anthony was arrested for her daughter's murder on July 16, 2008, and was officially charged (by a grand jury) with first-degree murder on October 14, 2008. Jury selection for Anthony's trial didn't begin until May 8, 2011. On July 3, 2011, the trial's closing arguments began. On July 5, 2011, nearly three years after she was first arrested, Anthony was found not guilty of murdering her daughter.[63]

Trial Protocols

After the judge has given the jury its preliminary instructions—emphasizing that the defendant is presumed innocent until proven guilty—and other pretrial issues have been settled, typically the pattern of the trial process is as follows:

1. *Opening statements:* The prosecutor goes first, as he or she has the burden of proof (and must prove every element of the crimes charged—beyond a reasonable doubt), followed by the defense (although in many jurisdictions the defense can opt to defer making its opening statements until later, when it presents its main case, or waive it altogether). The purpose of this step is to succinctly outline the facts they will try to prove during the trial—and it is *not* a time to argue with the other side.
2. *Prosecution's case:* The prosecution will present its side of the case, presenting and questioning its witnesses and admitting relevant evidence. The defense may cross-examine these prosecution witnesses. A "redirect" allows the prosecution to reexamine its witnesses. Once the prosecution has finished presenting its evidence, it will rest its case.
3. *Motion to dismiss:* As a formality, at this point during a criminal trial the defense will often make a motion to dismiss all charges, arguing that the state has not proved its case and, that being so, there is no need for the defense to put on its case. This request is generally denied by the judge, opening the way for the defense case.
4. *Defense's case:* Next, the defense presents its main case through direct examination of its chosen witnesses. The prosecution is then given an opportunity to cross-examine the defense witnesses, and, during redirect, the defense may reexamine its witnesses. The defendant cannot be compelled to testify against himself or herself but has the right to testify in his or her own defense if so desired. The defense then rests. Because a defendant is presumed innocent until proven guilty, the defense is not required to put on a case at all. If the defense does not have good evidence or bases for cross-examination of the state's witnesses, the defense can simply rest, and the jury or judge will have to decide if the state has proven its case beyond a reasonable doubt.
5. *Prosecution rebuttal:* The prosecution may offer evidence to refute the arguments made by the defense.
6. *Closing arguments:* This is a time for both sides to review the evidence so that it is clear to the jury before they begin their deliberations. The order of closing arguments varies by jurisdiction. In some jurisdictions, the state always argues first, but in others the defense does.[64] The prosecution will offer reasons why the evidence proves the defendant's guilt, and the defense will explain why the defendant should be acquitted. This is *not* the time for the prosecutor to offer personal opinion, make inflammatory or discriminatory remarks, or comment on the defendant's failure to testify. Such misconduct may result in a reversal of a conviction on appeal.
7. *Jury instructions:* The judge's instructions to the jury are important and, if improper, may later be grounds for a reversal and new trial. Also known as "charging the jury," the judge will explain the law that is applicable to that particular criminal case and the possible verdicts, and will typically include general comments concerning the presumption of the defendant's innocence, that guilt be proved beyond a reasonable doubt, and that the jury may not draw inferences from the fact that the defendant did not testify in his or her own behalf.
8. *Jury deliberations and verdict:* The jury will deliberate for as long as it takes to reach a verdict. In most states, unanimous agreement must be met for a verdict to be reached (however, as noted earlier, a unanimous verdict is not required by the Constitution). Once the jury has determined its verdict, either guilty or not guilty for each crime in question, the verdict will be read to the court. Then, either side, or the judge, may "poll the jury," asking all jurors individually if the verdict as read is theirs; if not unanimous, the jury may be returned to its room to deliberate again, or be discharged. If the jury acquits the defendant, the case is over. The prosecutor cannot appeal an acquittal because of the Fifth Amendment protection against double jeopardy. The jury may, instead, convict the defendant of some charges while acquitting him or her of others.

9. *Posttrial motions:* If the jury delivers a guilty verdict, the defense will usually ask the judge to override the jury's decision and acquit the defendant or grant him or her a new trial. This motion is almost always denied.
   - *Sentencing*: If the defendant was convicted of the crime(s), sentencing will be determined by the judge immediately after the verdict is read or at a later court date. In arriving at a sentence, the judge generally orders a presentence investigation report (PSI) by probation or court services personnel, to look at the history of the person convicted, any extenuating circumstances, a review of his or her criminal record, and a review of the specific facts of the crime. The person or agency preparing the PSI makes a recommendation to the court about the type and severity of the sentence. The judge may, however, be limited by federal and state sentencing guidelines (discussed in Chapter 9); some crimes carry a mandatory minimum sentencing requirement, while other sentences may be based largely on the discretion of the judge.
   - *Punishments*: After a conviction, and upon receiving the PSI or using sentencing guidelines, the judge may opt for one of the following punishments:
     - Incarceration
     - Probation
     - Fines
     - Restitution
     - Community service
10. *Appeal:* After conviction, the defendant may challenge the outcome. Potential grounds for appeal in a criminal case include legal error (e.g., improperly admitted evidence, improper jury instructions), juror misconduct, ineffective counsel, or lack of sufficient evidence to support a guilty verdict. If this error affected the outcome of the case, the appeal is granted—the conviction is overturned—and, in some instances, the case is remanded back to the trial court and a retrial is ordered (the prosecutor may choose to drop charges if facing a retrial without key evidence, for example). If the appeals court finds that the error(s) would not have affected the outcome, then the errors are considered harmless, and the appeal is denied.[65] Note also that, for the *first* appeal, the U.S. Supreme Court has said that if the person convicted is indigent, then free, appointed counsel must be provided.[66] However, the Court ruled later that, after losing the initial appeal, the convicted person is *not* entitled to free appointed counsel for any subsequent appeals.[67]

AP Photo/Glenn C. Silva

Michelle Carter, 20, encouraged her boyfriend through text messages and phone calls to kill himself, which he did. At trial, Carter's defense team argued for acquittal, as there were no laws against encouraging suicide. Carter was convicted of involuntary manslaughter in June 2017. After Massachusetts's highest court upheld her conviction, Carter's attorneys filed a final appeal with the U.S. Supreme Court in July 2019.

The Casey Anthony Case: Following Anthony's trial, the five women and seven men on the jury took less than 11 hours to decide that Anthony was not guilty of murdering her 2-year-old daughter (they did not convict her of either first-degree murder or manslaughter). The jury did find Anthony guilty of four misdemeanors related to providing false information to police.

TECHNOLOGIES IN THE COURTS

Like the police (particularly in the forensics area, described in Chapter 4), the courts are unveiling new technologies that it is hoped will provide more efficient and effective operations. Some of those technologies are described in the sections that follow.

Achieving Paper on Demand

©AP Photo/Star Ledger, Pool

Raymond Dargan, 20, of New Brunswick, New Jersey, is arraigned on burglary and robbery charges at the Hunterdon County Jail via videoconference. Conducting initial appearances and arraignments via videoconference is increasingly common, especially in metropolitan areas where the courts and jails process record numbers of offenders.

A major goal for all courts in the United States is to go "paper on demand" (POD)—denoting an environment in which the routine use of paper no longer exists in general. Rather, paper documents may be used for court business only rarely, and as a last resort. The ultimate goal is that there be no more lost files, all receipts be issued electronically, all police citations be issued electronically, all filing formats and forms be standardized, all judges use POD, and there be no more folders in the courtroom.

Toward that end, electronic case filing has been possible for many years and allows courts to realize dramatic increases in efficiency and reductions in related costs—in clerical staff alone. Electronic filing also enables some court services—such as the payment of fines and fees, collection of fines and penalties, provision of case information and documents to the public, and jury management—to be centralized or regionalized for improved efficiency and service. In addition, an electronic case file enables a court to better distribute its workload across the system.

Emerging Technologies

Following are five other areas in which court technologies are emerging or have already been put in place:

- *Digital recording*: Significant savings can be realized by replacing court stenographers with digital audio- or video-recording equipment. Many states have used digital recording extensively, and some states have used digital recording exclusively for many years without experiencing significant issues.
- *Conducting hearings via videoconferencing*: Videoconferencing has improved rapidly in both cost and quality over the past few years. Prices for basic capabilities have been reduced considerably, and the quality of the networks has improved steadily.
- *Use of tablets and apps to present evidence*: The Federal Judicial Center[68] highlights the Chambers Online Automation Training (COAT) program. This program provides online training modules to teach judges and other court staff how to use court technology effectively. Currently available training includes information on connecting to chambers from remote locations, computer security, courtroom technology, and the use of tablet devices (e.g., iPads).[69] The proliferation and relatively low cost of tablet devices are changing the way in which attorneys present evidence in the courtroom. For example, prosecutors in the San Diego County District Attorney's Office use TrialPad, an iPad application, that allows jurors to examine photos, videos, audio files, and transcripts on large courtroom screens and replaces expensive, single-use exhibit boards.[70]
- *Annotation monitors*: In addition to widely-adopted large, flat-screen, high definition monitors that attorneys use to display images to jurors (instead of using traditional projectors and screens), annotation monitors allow witnesses to mark an exhibit with notations. These marked exhibits can be preserved for later viewing during jury deliberations. More recently, these systems are being replaced by tablet devices with annotation capabilities.
- *Evidence cameras*: During in-court proceedings, an evidence camera can instantaneously convert a paper document or physical exhibit to an electronic image for monitor display. An evidence camera can enlarge small physical items (e.g., a four-inch-by-six-inch photograph or wristwatch face) for all courtroom participants to see.[71]

Juvenile Justice Journal

Underlying Principles of the Juvenile Court

In this chapter, which discusses court structure, functions, and processes, those structures and functions of the juvenile court should also be taken into account. Most states' juvenile court decisions and legislation contain the following three underlying principles:

The *presumption of innocence* is one of the hallmarks of our criminal justice system. It places the burden on the state to prove that the accused has committed an offense. The state cannot force accused persons to testify against themselves, cannot use illegally seized evidence, and must use a process consistent with due process standards to establish guilt.

The principle of *least involvement with the system* assumes that minors, like adults, have liberty interests that include the right to be left alone or the right to live in a family situation without state interference. The state has the burden of showing that intervention is necessary for the protection of either the minor or society. Diversion should be considered before a formal petition is filed, and probation should be considered before commitment to an institution. In the detention situation, many codes require that a child not be held unless a probable cause exists to believe that he or she has committed a crime and an immediate and urgent necessity exists to admit the child.

The primary purpose of juvenile justice is to operate in the *best interest of the child*. This interest must be balanced against the interests of society. Society benefits from programs that help minors mature into law-abiding citizens, and children benefit by being held accountable and developing responsibility.

Source: David W. Roush, *A Desktop Guide to Good Juvenile Detention Practice* (Washington, D.C.: Office of Juvenile Justice and Delinquency Prevention, 1996), pp. 26–27.

Educational institutions are helping to introduce new courtroom technologies. The National Judicial College in Reno, Nevada, offers its students access to its "Model Courtroom," which offers hands-on experience with some of the latest courtroom technologies. Persons can plug in their laptops to refer to notes; retrieve documents, charts, and photographs; and forward evidentiary material digitally to the presiding judge's monitor. Evidence can be shown on the LCD displays where court participants are sitting: the jury box/room, the attorneys' lectern, or the witness stand. Video/audio feeds may also be relayed to the media room for reporters covering the trial, the attorney conference room (where a victim may choose to view the trial away from the defendant), and a remote-site language interpreter. During trials, attorneys and witnesses may employ the LCD's touch screen technology to provide annotations on evidence. Cameras, evidence presentation tools, monitors, and computer hardware and software serve to introduce students to new technologies in a dynamic learning center. Instructional sessions can be viewed in real time by registrants with a computer and Internet capability, and mock trials can be streamed. The National Judicial College fields questions from court personnel across the country seeking input about incorporating technology into their own courtrooms.[72]

/// IN A NUTSHELL

- As the population of the U.S. colonies grew, formal courts of law appeared based on the English system; however, fearing tyranny from this concentration of governmental power, a federal judiciary was created that was separate from the legislative branch of government. We now have the dual court system—one implemented by the state courts, the other created by Congress and entrusted to the federal courts.

- The policy decisions of the courts affect virtually all of us in our daily living. Perhaps nowhere have the nation's courts had more of an impact than in the prisons.
- The courts must appear to do justice—and provide rights that are embodied in the due process clause. Our court system relies on the adversarial system, using several means to get at the truth: Evidence is tested through cross-examination of witnesses, and power is lodged with several different people. This series of checks and balances is aimed at curbing misuse of the criminal courts.
- Each state has a court of last resort, all but 11 states have an appellate court, and there are trial courts of general jurisdiction that decide all matters not specifically delegated to lower courts.
- Lower state trial courts have limited jurisdiction, but after an arrest, the judge conducts the initial appearance, appoints counsel for indigents, and conducts the preliminary hearing.
- There are 94 U.S. district courts, which are trial courts of original jurisdiction for all major violations of federal criminal law.
- Federal judges are nominated by the president and confirmed by the Senate, and they serve for life. The U.S. Supreme Court has complete discretion to control the nature and number of the cases it reviews, and it hears only a tiny fraction of the thousands of petitions that come before it. The chief justice orders the business of the U.S. Supreme Court.
- There are 11 circuit courts of appeals plus the D.C. Circuit and the Federal Circuit; they hear appeals from the federal district courts located within their circuits, as well as appeals from decisions of federal administrative agencies.
- The criminal justice process is engaged when one is arrested. Following that, the accused will then proceed through a series of steps; at some point, the prosecutor will prepare an information setting forth the charge against the defendant; some jurisdictions use a grand jury to bring formal charges, rather than the prosecutor's doing so unilaterally.
- Discovery is the pretrial exchange of information between prosecution and defense, in order to promote a fair adversarial contest between the two sides and help the truth come to light.
- The Sixth Amendment gives defendants the right to a trial by an impartial jury of his or her peers. The jury system is felt by many to be the most sacred aspect of our criminal justice system, because it is where common citizens sit as a forum to determine the truth and assess the punishment to be meted out.
- The Sixth Amendment guarantees the accused the right to a speedy and public trial. Although there are fixed, enforced time limits at the federal level, the Supreme Court has never defined a "speedy trial" with precise time frames at the state level. Rather, courts must use a test to determine whether or not the defendant's right to a speedy trial was violated.
- The courts are unveiling new technologies that it is hoped will provide more efficient and effective operations. These include paper on demand, digital recording, videoconferencing, use of tablets and apps, annotation monitors, and evidence cameras in the courtroom.

/// KEY TERMS & CONCEPTS

Review key terms with eFlashcards at **edge.sagepub.com/peakbrief**.

/// REVIEW QUESTIONS

Test your understanding of chapter content. Take the practice quiz at **edge.sagepub.com/peakbrief**.

1. How is the adversarial system of justice related to the truth-seeking function of the courts?
2. How do the courts influence public policy making?

3. What are the types of, and reasons for, courts having specific types of jurisdiction?
4. What are the roles of the state court systems?
5. What is the structure and function of the trial courts of general and limited jurisdiction?
6. How does the U.S. Supreme Court decide to hear an appeal, and approximately how many cases does the Court hear per term?
7. What are some of the pretrial activities that occur?
8. Why does our jury system exist, and how is a jury formed?
9. What is meant by a right to a "speedy" trial? What are the ramifications of a defendant's being denied this right?
10. What are the major points of the trial process, from opening statements through appeal?

/// LEARN BY DOING

1. Your local League of Women Voters is establishing a new study group to better understand the court system as it relates to political affairs. You are asked to explain the dual (federal and state) court system. You opt to use *Kibbe v. Henderson* as a good—and rare—example of a convicted offender's flow through both systems. Prepare your presentation.
2. Some countries do not subscribe to the adversarial process as part of their court system, believing that it is too combative, slow, and cumbersome, and can lead to a "win at all cost" mentality among the lawyers. Rather, they use a nonadversarial or inquisitorial system, in which the court or a part of the court is actively involved in determining the facts of the case (as opposed to the court's being primarily an impartial referee, as in the adversarial system). Your instructor asks you to participate in a pro-con group project concerning the adversarial process. Choose a side and make your defense.
3. Your criminal justice professor has assigned the class to debate the pros and cons of plea negotiation. What do you believe will be the prominent arguments presented by each side?
4. Assume you are a court administrator and your chief judge has tasked you to "bring the courtrooms into the new decade" by making recommendations concerning technologies that should be acquired. Using information and descriptions of the technologies presented in this chapter, select and prioritize which new technologies you would recommend be obtained, and why.

/// STUDY SITE

Get the tools you need to sharpen your study skills. SAGE edge offers a robust online environment featuring an impressive array of free tools and resources.

Access practice quizzes, eFlashcards, video, and multimedia at **edge.sagepub.com/peakbrief**

CHAPTER 8

istockphoto.com/Chris Ryan

THE BENCH AND THE BAR

Those Who Judge, Prosecute, and Defend

Fiat justitia ruat coelum [Let justice be done, though heaven should fall].

—Emperor Ferdinand I, 1563

Four things belong to a Judge:

To hear courteously,

To answer wisely,

To consider soberly, and

To decide impartially.

—Socrates

LEARNING OBJECTIVES

As a result of reading this chapter, you will be able to

1. Explain the five methods of judicial selection, and why the subject of judicial selection has come under scrutiny
2. List some of the benefits, training, and challenges of judges
3. Describe why courtroom civility is important, as well as the meaning of "good judging"
4. Relate the major duties of prosecutors and defense attorneys (to include their roles in plea negotiation)
5. List the six main actors who participate in the courtroom work group and describe the courtroom subculture
6. Discuss the various defenses that defendants may use in criminal cases to reduce or eliminate their criminal liability

ASSESS YOUR AWARENESS

Test your knowledge of the duties of judges, prosecutors, and defense attorneys by responding to the following seven true-false items; check your answers after reading this chapter.

1. Studies indicate that there is no difference in terms of how judges are selected; in all states, they are simply elected.
2. In recent years, people involved with courtroom matters have become much less friendly and less well-behaved.
3. The prosecutor may be fairly said to be the single most powerful person in the American criminal justice system.
4. A criminal defense attorney's primary role is to help the defendant escape punishment, even if the defendant is indeed guilty as charged.
5. A prosecutor's primary duty is not to convict, but to see that justice is done.
6. One's transition from public or private attorney to the role of judge can involve a number of psychological problems and issues.
7. An individual can use one of only two defenses against being charged with a crime: "I didn't know it was a crime" or "I was too intoxicated to know what I was doing."

<< Answers can be found on page 293.

In early 2019, Tracie Hunter, former Hamilton County (Ohio) juvenile court judge, was waiting for a federal judge to decide whether she would serve a six-month sentence imposed for a judicial misconduct conviction. A grand jury initially indicted Hunter in 2014 on nine felony charges: two counts of tampering with evidence, two counts of forgery, two counts of theft in office, two counts of having unlawful interest in a public contract, and one count of misuse of credit cards. She was eventually convicted on only one charge related to using her judicial position to give her brother, a juvenile court employee, confidential documents.[1] She was allowed to remain free while pursuing appeals.

The judge who initially sentenced Hunter, Judge Norbert Nadel (now retired), stated that jail time was warranted in this case because Hunter breached the public's trust. After the jury's verdict, Nadel remarked, "The evidence showed that the conduct of Judge Hunter dealt a serious blow to public confidence in our judicial system, and could very well justify a jail sentence for Judge Hunter."[2]

After the federal judge who reviewed the case rejected her appeal, Common Pleas Judge Patrick Dinkelacker imposed the original six-month sentence in July 2019. At the hearing, Scott Croswell, a special prosecutor during Hunter's original trial remarked, "What she wants to do is play by her own set of rules. That's the very attitude and the very conduct that put her in the predicament that she's in and, frankly, has caused all this pain to her and caused all this turmoil to the community."[3] Upon being sentenced, Hunter refused to walk and a bailiff dragged her out of the courtroom.

Judicial misconduct represents a serious threat to public confidence in those we trust to carry out justice. High-profile cases include instances in which judges have been removed from office for the following:[4]

- Posting comments considered sexist, racist, perverse, and homophobic
- Having an affair (including having sex in judicial chambers) with a party involved in a case over which the judge was presiding
- Initiating a fistfight with a public defender just outside the courtroom
- Campaign misconduct
- Domestic violence

We often think of judges as all-powerful players in the criminal justice system. As you read this chapter, think about the role of each of the courtroom actors. How much power and discretion should each of the actors have? What ethical standards should these actors be held to, both in the courtroom and in their personal lives?

INTRODUCTION

In Chapter 7, we looked at the general nature of courts and judges; this chapter expands that discussion, focusing more on judges and other key personnel who are involved in the courts and their operation.

It is a part of our human nature that we hate losing. Therefore, even though in theory attorneys in a criminal courtroom are engaged in a truth-seeking process, make no mistake: They are *competing* from beginning to end—trying to persuade the judge to include or exclude evidence or witnesses, to persuade the judge or jury of the guilt or innocence of the defendant, to sway the judge or jury that the convicted person should or should not be severely punished, and so on. This adversarial legal process is what drives our criminal justice system. Indeed, renowned defense attorney Percy Foreman is said to have remarked, "The best defense in a murder case is that the deceased should have been killed."[5] In a murder case where a woman was charged with shooting her husband, Foreman so slandered the victim that "the jury was ready to dig up the deceased and shoot him all over again."[6] As will be seen in this chapter, the challenges (and criticisms) facing today's judges are several. They must successfully serve many masters and occupy many roles. As criminologist Abraham Blumberg noted:

> The "grand tradition" judge, the aloof brooding charismatic figure in the Old Testament tradition, is hardly a real figure. The reality is the working judge who must be politician, administrator, bureaucrat, and lawyer in order to cope with a crushing calendar of cases.[7]

The chapter opens by considering the means by which judges ascend to the bench. This once-simple task has come under intense scrutiny—particularly in relation to the partisan election of judges, for which they must often solicit campaign contributions. Then we discuss the benefits and problems that occur when one becomes a judge; following that is an examination of the need for courtroom civility and a look at judicial misconduct. The roles and strategies of two other very important court figures—prosecutors and defense attorneys—are also reviewed.

THOSE WHO WOULD BE JUDGES: SELECTION METHODS AND ISSUES

The manner in which state and local court judges assume the bench matters—and it differs widely from state to state. The method used to select judges is important for at least four reasons: The type of judicial selection system affects judges' experience level; it determines the ability of qualified, but less politically connected, individuals to serve; it affects the gender and racial diversity of the judiciary; and it affects the public's perception of judicial impartiality and independence.[8] Across the United States at least five methods of **judicial selection** are used, but no two states use exactly the same selection method. In many states, more than one method of selection is used—for judges at different levels of the court system and even among judges serving at the same level. And when the same method is used, there are still variations in how the process works in practice.

Judicial selection (methods of): means by which judges are selected for the bench, to include election, a nominating commission, or a hybrid of these methods.

Methods of Selection in State Courts

As noted in Chapter 7, all federal judges are nominated by the president and confirmed by the Senate, and they serve for life (unless they resign or are impeached). When a vacancy (due to death, retirement, or resignation of a judge)[9] occurs at the state level, however, candidates are as likely as not to face an election as part of their selection process. This becomes particularly important given that 97 percent of the cases heard in the United States are handled by state judges. Furthermore, every year, millions of Americans find themselves in state courts, whether called for jury service, to address a minor traffic offense, as a crime victim, or in a small claims case.[10]

The following five methods of selection used in state courts[11] are also depicted in Figure 8.1:

- *Commission-based appointment* (also known as **merit selection** or the Missouri Plan): Judicial applicants are evaluated by a nominating commission, which then sends the names of the best-qualified candidates to the governor, who appoints one of those nominees. In most commission-based appointment systems, judges run unopposed in periodic retention elections, where voters are asked whether the judge should remain on the bench.
- *Partisan election*: In a partisan election, multiple candidates may seek the same judicial position. Voters cast ballots for judicial candidates as they do for other public officials, and candidates run with the official endorsement of a political party. The candidate's party affiliation is listed on the ballot.
- *Nonpartisan election*: In a nonpartisan election, a judicial candidate's party affiliation, if any, is not designated on the ballot.
- *Gubernatorial appointment*: A judge is appointed by the governor (without a judicial nominating commission). The appointment may require confirmation by the legislature or an executive council.
- *Legislative appointment/election*: This is the process by which judges are nominated and appointed or elected by legislative vote only.

Merit selection: a means of selecting judges whereby names of interested candidates are considered by a committee and recommendations are then made to the governor, who then makes the appointment; known also as the Missouri Plan.

Figure 8.1 /// Initial Selection of State Judges (Trial Courts of General Jurisdiction)

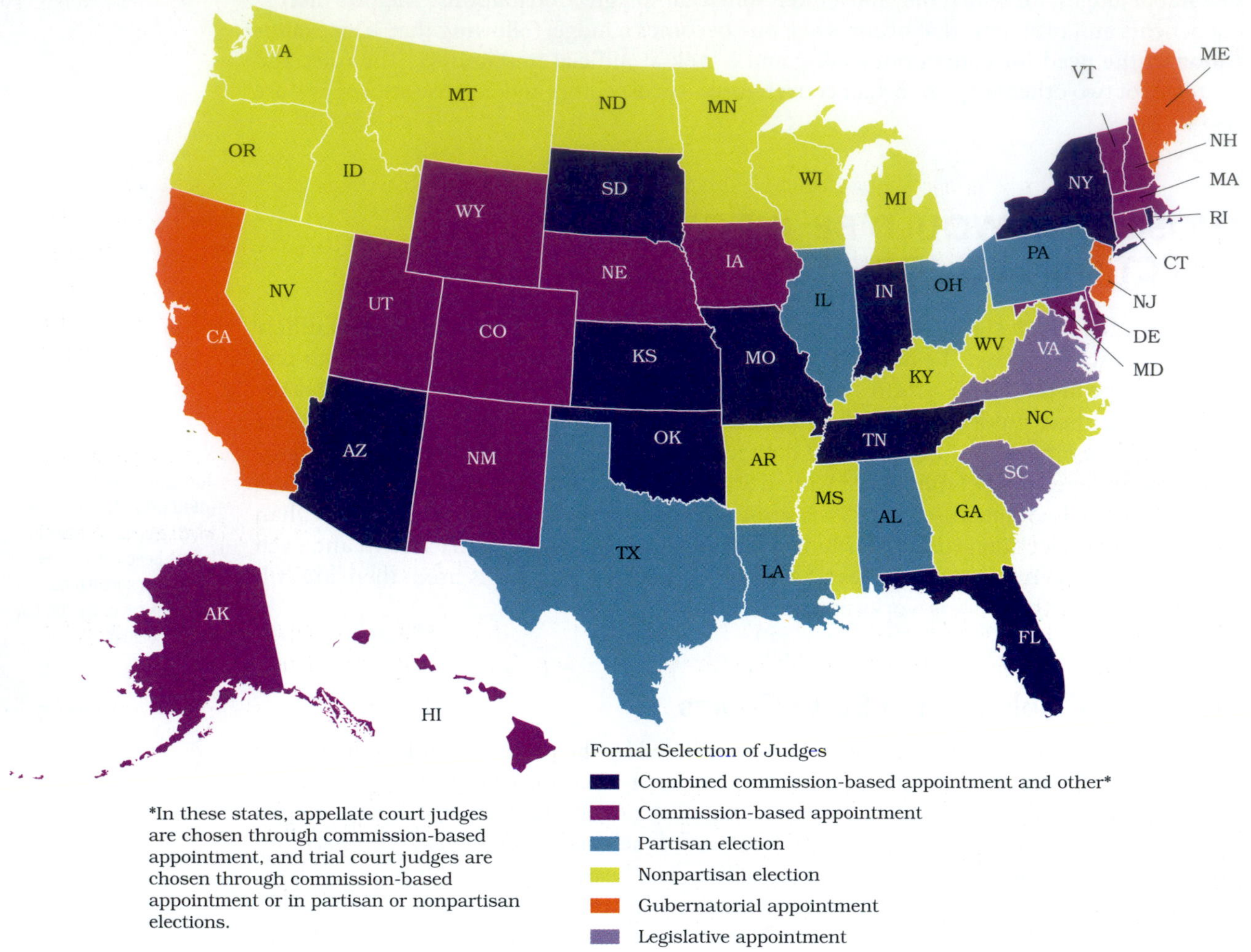

Formal Selection of Judges
- Combined commission-based appointment and other*
- Commission-based appointment
- Partisan election
- Nonpartisan election
- Gubernatorial appointment
- Legislative appointment

*In these states, appellate court judges are chosen through commission-based appointment, and trial court judges are chosen through commission-based appointment or in partisan or nonpartisan elections.

Source: Reprinted with permission from Institute for the Advancement of the American Legal System and American Judicature Society.

Debating Judges and Politics

"They're awful. I hate them." Thus spoke former U.S. Supreme Court justice Sandra Day O'Connor (the first female appointed to the U.S. Supreme Court) concerning her views of judicial elections in May 2009 at an American Bar Association summit.[12] O'Connor added that the public is growing increasingly skeptical of elected judges in particular, with surveys showing that more than 70 percent of the public are considerably more distrustful of judges than they have been in the past. At risk, O'Connor said, is the public perception that judges are "just politicians in robes."[13] O'Connor also said in November 2007 that "if I could wave a magic wand, I would wave it to secure some kind of merit selection of judges across the country."[14]

Our courts make decisions every day that affect nearly every aspect of our lives. Therefore, to a large extent, the quality of justice Americans receive depends on the quality of the judges who dispense it. The debate over how America chooses its judges has escalated in the 21st century. Consider this: In March 2009, the U.S. Supreme Court considered a case concerning a newly elected West Virginia Supreme Court of Appeals justice, Brent Benjamin, who voted on a mining company dispute; the mining company contributed $3 million in an election campaign to help Benjamin get elected. Instead of removing himself from the vote (known in the courts as *recusal*), Benjamin instead possibly cast the deciding vote in

You Be the . . .

Judge

In 1984, 18-year-old Terrance Williams was convicted of murdering 56-year-old Amos Norwood in Philadelphia. The trial prosecutor sought the death penalty in this case and, upon conviction, Williams was indeed sentenced to death.

In 2012, Williams filed an appeal, his attorneys claiming that the trial prosecutor had engaged in misconduct by obtaining false testimony and suppressing exculpatory evidence (i.e., evidence favorable to the defendant). The appeals court found that the trial prosecutor committed these violations and stayed Williams's execution, while also ordering a new sentencing hearing. However, on further appeal the Pennsylvania Supreme Court reinstated Williams's death sentence.

In 2015, the U.S. Supreme Court agreed to review the state supreme court's ruling. At issue was the fact that the chief justice of Pennsylvania Supreme Court was former district attorney Ronald Castille, the person who earlier in his career had approved the trial prosecutor's request to seek the death penalty for Williams in the first place. Although Williams had filed a motion asking Chief Justice Castille to recuse himself (to avoid a conflict of interest), Castille denied the motion and voted in this case.

1. Under what circumstances should judges recuse themselves from a case?
2. Did Castille act inappropriately when he denied Williams's motion for recusal?

The U.S. Supreme Court's ruling in this case is provided in the Notes section at the end of the book.[15]

the 3–2 case—in favor of the mining company. There was no law in West Virginia saying a judge can't hear a case involving someone who financed his or her campaign. During oral arguments in the case, former justice David Souter said, "The system . . . is not working well."[16] The U.S. Supreme Court ruled in June 2009 that Benjamin's failure to recuse himself violated the Fourteenth Amendment's due process clause.[17]

Several states are now evaluating their judicial selection systems with a view to altering their current processes. And, by ruling in the West Virginia case, the Supreme Court certainly put a spotlight on this issue—one that has already been settled in about two dozen states by eliminating political fund-raising by their judicial candidates through the use of various merit selection systems.[18]

"Investing" in Judges?

Adding fuel to the fire over judicial selection is the amount of money now being spent to fund judges' elections. Expenditures on state supreme court elections has skyrocketed since the 1990s. For example, whereas 7 states had at least one sitting justice who had been involved in a race costing $1 million or more during his or her tenure in 1999, this number jumped to more than 20 states by the start of 2017.[19] Furthermore, special interest groups have ramped up their efforts to influence the composition of state courts, making contributions to candidates, funding television ads, and pressuring candidates to speak publicly about their political views. In state supreme court elections in 2015–2016, special interest group spending represented 40 percent of the total dollars spent in such races.[20]

JUDGES' BENEFITS, TRAINING, AND CHALLENGES

Judges enjoy several distinct benefits of office, including life terms for federal positions and in some states. Ascending to the bench can be the capstone of a successful legal career for a

Judges are elected in many U.S. jurisdictions, but their campaigns and the money used to fund them are under increasing scrutiny amid questions of how judges can possibly remain objective and resist influence on the bench.

lawyer, even though a judge's salary can be less than that of a lawyer in private practice. Judges warrant a high degree of respect and prestige as well; from arrest to final disposition, the accused face judges at every juncture involving important decisions about their future: bail, pretrial motions, evidence presentation, trial, and punishment.

Although it would seem that judges are the primary decision makers in the courts, such is not always the case. Judges often accept recommendations from others who are more familiar with the case—for example, bail recommendations from prosecutors, plea negotiations struck by prosecuting and defense counsels, and sentence recommendations from probation officers. Judges frequently accept such input in the kind of informal courtroom network that exists. Although judges run the court, if they deviate from the consensus of the courtroom work group, they may be sanctioned: Attorneys can make court dockets go awry by requesting continuances or by not having witnesses appear on time.

Newly elected judges are not simply "thrown to the wolves" and expected to immediately begin to conduct trials, listen to arguments, understand rules of evidence, render verdicts and sentences, and possibly write opinions, without the benefit of training or education. Many states mandate judicial education for new judges, sometimes even prior to assuming the role, as well as mandatory in-service or continuing education thereafter.

Other challenges can await a new jurist-elect or appointee. Judges who are new to the bench commonly face three general problems:

- *Mastering the breadth of law they must know and apply.* New judges would be wise, at least early in their careers, to depend on other court staff, lawyers who appear before them, and experienced judges for invaluable information on procedural and substantive aspects of the law and local court procedures. Through informal discussions and formal meetings, judges learn how to deal with common problems. Judicial training schools and seminars have also been developed to ease the transition into the judiciary.
- *Administering the court and the docket while supervising court staff.* One of the most frustrating aspects of being a presiding judge is the heavy caseload and corresponding administrative problems. Instead of having time to reflect on challenging legal questions or to consider the proper sentence for a convicted felon, trial judges must move cases. They can seldom act like a judge in the "grand tradition."[21] Judges are required to be competent administrators, a fact of judicial life that comes as a surprise to many new judges. One survey of 30 federal judges found that three-fourths had major administrative difficulties on first assuming the bench, while half complained of heavy caseloads. One judge maintained that it takes about four years to "get a full feel of a docket."[22]
- *Coping with the psychological discomfort that accompanies the new position.* Most trial judges experience psychological discomfort on assuming the bench. Three-fourths of new federal judges acknowledged having psychological problems in at least one of five areas: maintaining a judicial bearing both on and off the bench, dealing with the loneliness of the judicial office, sentencing criminals, forgetting the adversary role, and handling local pressure. One judge remembers his first day in court: "I'll never forget going into my courtroom for the first time with the robes and all, and the crier

tells everyone to rise. You sit down and realize that it's all different, that everyone is looking at you and you're supposed to do something."[23] Like police officers and probation and parole workers, judges complain that they "can't go to the places you used to. You always have to be careful about what you talk about. When you go to a party, you have to be careful not to drink too much so you won't make a fool of yourself."[24] And the position can be a lonely one:

> After you become a . . . judge some people tend to avoid you. For instance, you lose all your lawyer friends and generally have to begin to make new friends. I guess the lawyers are afraid that they will someday have a case before you and it would be awkward for them if they were on too close terms with you.[25]

Judges frequently describe sentencing criminals as the most difficult aspect of their job: "This is the hardest part of being a judge. You see so many pathetic people and you're never sure of what is a right or a fair sentence."[26]

THE ART OF JUDGING, COURTROOM CIVILITY, AND JUDICIAL MISCONDUCT

To fully understand judges' role in the criminal justice system, we must consider some of the less obvious facets of their work on and off the bench: the art and craft of judging, maintaining civility among the many courtroom players, and the critical issue of judicial misconduct.

"Good Judging"

What traits make for "good judging"? Obviously, judges should treat each case and all parties before them in court with absolute impartiality and dignity while providing leadership for court operations. But judging requires more than just those activities and roles. For example, a retired jurist with 20 years on the Wisconsin Supreme Court stated that the following qualities define the art and craft of judging:

- Judges are keenly aware that they occupy a special place in a democratic society. They exercise their power in the most undemocratic of institutions with great restraint.
- They are aware of the necessity for intellectual humility—an awareness that what we think we know might well be incorrect.
- They do not allow the law to become their entire life; they get out of the courtroom, mingle with the public, and remain knowledgeable of current events.[27]
- Other writers believe that judges should remember that the robe does not confer omniscience or omnipotence; as one trial attorney put it, "Your name is now 'Your Honor,' but you are still the same person you used to be, warts and all."[28]

Judges are responsible for regulating the behaviors of all courtroom participants and observers, including attorneys, defendants, witnesses, court personnel, jury members, and members of the public.

Tony Avelar/Christian Science Monitor/Getty Images

The Model Code of Judicial Conduct requires judges to be "patient, dignified, and courteous" to all persons engaged in court business.

Courtroom Civility

As if it weren't difficult enough to strive for and maintain humility, civility, and balance in their personal lives, judges must also enforce courtroom civility. Many persons have observed that we are becoming an increasingly uncivil society; the courts are not immune to acts involving misconduct.

Personal character attacks by lawyers, directed at judges, attorneys, interested parties, clerks, jurors, and witnesses, both inside and outside the courtroom, in criminal and civil actions have increased at an alarming rate.[29] For example, an attorney stated that opposing and other attorneys were "a bunch of starving slobs," "incompetents," and "stooges."[30] Such behavior clearly does not enhance the dignity or appearance of justice and propriety that is so important to the courts' public image and function. The Model Code of Judicial Conduct (discussed in Chapter 3) addresses these kinds of behaviors; Canon 3B(4) requires judges to be "patient, dignified, and courteous to litigants, jurors, witnesses, lawyers, and others with whom the judge deals in an official capacity" and requires judges to demand "similar conduct of lawyers, and of staff, court officials, and others subject to the judge's direction and control."[31] At a minimum, judges need to attempt to prevent such behavior and discipline offenders when it occurs. Among the means that judges have at their disposal to control errant counsel are attorney disqualifications, new trials, and reporting of attorneys to disciplinary boards.[32]

Judicial Misconduct

What types of **judicial misconduct** must the judiciary confront? Sometimes medications may affect a judge's cognitive process or emotional temperament, causing him or her to treat

Juvenile Justice Journal

Roles of the Juvenile Court Judge

Earlier chapters established that one of the fundamental duties of a democratic society is to nurture its children and to ensure their safe and successful development into adulthood. It may be fairly said that the ultimate legal authority in that duty falls to juvenile court judges. Juvenile court judges must, on a daily basis, assist families coming before them in such areas as adequately nurturing (both emotionally and physically), providing for, and educating their children; controlling delinquent youths' behavior; and protecting children from abuse. Conversely, harsh measures must be employed for those families failing to do so.

The juvenile judge—relying on accurate information provided by the police, attorneys, social services agencies, probation departments, and others—is the ultimate gatekeeper in making determinations such as whether a child should be removed from a parent, what types of services should be offered to the family, and whether the child should be returned to the family and the community or placed in another secure or nonsecure setting (e.g., foster care or a facility). All of this must be done while making certain that the legal and constitutional rights of all involved parties are preserved.

In some jurisdictions the juvenile court judge might also oversee a vast system that includes the juvenile probation department, juvenile court staff, other judges and attorneys, social services investigators, support personnel, psychologists, psychiatrists, and physicians—as well as the operation of foster homes, group homes, and detention facilities.

Source: Adapted from Leonard P. Edwards, "The Juvenile Court and the Role of the Juvenile Court Judge," *Juvenile and Family Court Journal* 43, no. 2 (1992), pp. 25–26.

parties, witnesses, jurors, lawyers, and staff poorly. Some stay on the bench too long; such judges will ideally have colleagues who can approach them, suggest retirement, and explain why this would be to their benefit. And sometimes, according to one author, judicial arrogance (sometimes termed "black robe disease" or "robe-itis") is the primary problem. This is seen when judges "do not know when to close their mouths, do not treat people with dignity and compassion, do not arrive on time, or do not issue timely decisions."[33]

Judicial misconduct: inappropriate behavior by a judge.

Some bar associations or judicial circuits perform an anonymous survey of a sample of local attorneys who have recently argued a case before a particular judge and then share the results with the judge. Sometimes these surveys are popularity contests, but a pattern of negative responses can have a sobering effect on the judge and encourage him or her to correct bad habits. Many judges will be reluctant to acknowledge they have such problems as those described here. In such cases, the chief judge may have to scold or correct a subordinate judge. Although difficult, it may be imperative to do so in order to maintain good relations with bar associations, individual lawyers, and the public. A single judge's blunders and behaviors can affect the reputation of the entire judiciary as well as the workloads of the other judges in his or her judicial district. Chief judges must therefore step forward to address such problems formally or informally.[34]

The Committee on Codes of Conduct of the Judicial Conference of the United States provides guidance to U.S. federal judges to help promote ethical behavior. In a recent attempt to curb judicial misconduct, the committee condemned the use of a wide variety of common social media behaviors. For example, they argue that judges must carefully word every personal social media comment, post, or blog in an effort to avoid any appearance of impropriety. They caution against using social media to engage in improper communications with lawyers or others (e.g., commenting on or "liking" posts that support particular people or causes); participating in activities that might adversely impact perceptions of impartiality (e.g., maintaining a blog that promotes particular political positions); using court email accounts for personal social activities (e.g., forwarding chain emails); or engaging in social media activities that could reveal any confidential, sensitive, or nonpublic information obtained through the court.[35] In 2017, an American Bar Association publication addressed similar ethical concerns associated with lawyers and judges using social media sites such as Facebook, Twitter, and YouTube.[36]

THE ATTORNEYS

Next we look at the roles and strategies of attorneys who serve on both sides of the courtroom aisle—as prosecutors and defense attorneys.

"Gatekeeper" of the Justice System: Prosecutor

The prosecutor may be fairly said to be the single most powerful person in the American criminal justice system—and to have tremendous discretion in what he or she does. As described by the Southern Poverty Law Center, "Prosecutorial discretion is a necessary and important part of our system of justice—it allocates sparse prosecutorial resources, provides the basis for plea-bargaining and allows for leniency and mercy in a criminal justice system that is frequently harsh and impersonal. They literally have unchecked power to decide who will stand trial for crimes."[37] Indeed, prosecutors have the authority and power to make all of the following decisions, at their sole discretion, all of which affect criminal defendants and the criminal justice system in general:

- The decision to charge
- Types of charges
- Whether to recommend granting or denying bail
- Plea agreements—whether to entertain such agreements and, if so, the terms
- Sentencing recommendations

Prosecuting attorney: one who brings prosecutions, representing the people of the jurisdiction.

Prosecutors represent the people, the victims in particular, and investigate crimes—often out in the field, having been called to the scene of a particularly heinous crime by the police. Still, remember that the primary role of the **prosecuting attorney**, as set forth by the U.S. Supreme Court (and as noted in Chapter 3), is "not that he shall win a case, but that justice shall be done."[38]

Once the police have completed a preliminary investigation, the prosecutor evaluates the arrest report and other documents to determine whether there is sufficient evidence to bring charges. Prosecutors also have the ability to scold officers who have not done their work properly—perhaps failing to have the requisite probable cause prior to making a search or an arrest—and can quash the arrest report. They have contact with the person suspected of the crime, the victim and witnesses, and the police. For them, the overarching question is, "Can I prove that a defendant committed a particular criminal act beyond a reasonable doubt?" If so, the prosecutor's office files charges and handles the case through pretrial negotiations (if any) and ultimately takes the case to trial. Other determining factors concerning how to handle a case are as follows:

- The type of crime charged (personal or property crime)
- The prior criminal record of the person accused
- The number of counts in the complaint (the more counts there are, the stiffer the sentence sought)
- Whether there are aggravating or mitigating circumstances in the case
- The victim's attitude—what he or she wants done with the case (this is particularly important in cases of violent crimes)

You Be the . . .

Victim Advocate

A relatively new member of the courtroom work group is the victim advocate. In our adversarial system of justice, the prosecutor represents the people of his or her state or county and seeks justice for them and the victim, but the prosecutor is not the victim's personal representative.

Autumn Driscoll

Michelle Cruz, the state of Connecticut's victim advocate, confers with Hakima Bey-Coon, an office attorney, during a hearing at New Britain Superior Court concerning grand jury testimony involving a missing person.

Despite best efforts, victims can often feel lost in the dizzying process that is a criminal case. Victim advocates seek to remedy that problem, providing victims with information about the criminal justice process, resources for recovery and counseling (especially for victims of violent crimes), legal rights, and a host of other services to help the victim navigate the uncertainties of postvictimization life.

If you were responsible for recruiting or hiring victim rights advocates in your jurisdiction:

1. What type of educational degree (e.g., field of study) and background experience would you want your candidates to possess?
2. How would you address burnout and psychological trauma that victim advocates (like many other first responders who assist victims) are likely to face?

To learn more about the role of victim advocates, visit the National Center for Victims of Crime at https://victimsof-crime.org.

Good prosecutors will also try to establish rapport with the victim prior to trial, to personalize the justice system. If possible, the prosecutor or a victim advocate might take the victim to the court and show him or her the courtroom and witness stand and where the offender will be seated, explaining the process along the way.

Victim advocates (both paid and volunteer) work in a variety of places along the criminal justice spectrum, often being called to crime scenes to comfort victims immediately after a crime. They also work at police stations and hospitals and are often in or around the court, to accompany victims who have been asked to testify, to work with prosecutors and defense attorneys as they negotiate plea deals or case outcomes, and to monitor the court process so that they can better inform the victim about what to expect. They also accompany victims' families to the morgue to claim personal effects.

Prosecutorial Immunity and Misconduct

Not only do prosecutors have nearly unfettered discretion and power, but they are also immune from prosecution for actions taken in their official capacity. In other words, defendants cannot sue prosecutors for civil damages for how they handled a case.[39] This "civil immunity" is unique in our system of justice. In fact, it's unique in the professional realm in general. Doctors, other lawyers, and most professionals (except judges) are subject to civil prosecution if they fail to maintain standards of conduct and performance that apply to their field. Probably the most dramatic contrasting example is that of police officers, who are not granted such immunity and yet are tasked with making a multitude of professional decisions in the field, many of them in a split second and in dynamic and dangerous situations.

Some scholars and observers believe prosecutorial immunity goes too far and unfairly insulates the most powerful player in the criminal justice system.[40] Proponents of this long-standing immunity counter that prosecutors, in serving the people and in seeking to do justice on behalf of communities, cannot be looking over their shoulders or second-guessing their decisions because they fear civil suits. As the U.S. Supreme Court has put it, prosecutorial immunity represents a "balance of evils," and that it is better "to leave unredressed the wrongs done by dishonest officers than to subject those who try to do their duty to the constant dread of retaliation."[41]

"Guiding Hand of Counsel": Defense Attorney

In many countries, a person mired in some stage of a legal proceeding might also find himself or herself standing alone in the courtroom, overwhelmed by fear and befuddled by the activities swirling around him or her. Not so in the United States, where the fundamental principles of liberty and justice require that all Americans, even the poorest among us, will be given the "guiding hand of counsel" at all critical stages of a criminal proceeding.

Duties and Strategies

As noted in Chapter 6, many people believe that the Sixth Amendment's provision for effective counsel is the most important right we enjoy in a democracy. The law is complicated, and by requiring the state to prove its case and helping defendants understand their options in the criminal justice system, **defense attorneys** can help ensure that the state does not commit innocent people to jail or prison. Furthermore, defendants have the right to counsel during all "critical stages" of the proceedings—those in which rights could be lost—which include interrogation, jury selection, arraignment, trial, sentencing, and first appeal of conviction (but not the initial appearance, where the judge simply informs the defendant of his or her charges and rights), as well as pretrial testing of fingerprints, blood samples, clothing, hair, and so on.[42]

Defense attorney: one whose responsibility it is to see that the rights of the accused are upheld prior to, during, and after trial; the Sixth Amendment provides for "effective" counsel, among other constitutionally enumerated rights that defense attorneys must see are upheld.

AP Photo/Rich Pendroncelli

Priscilla Prendez speaks with Public Defender Joe Cress during her first court appearance on September 1, 2017, in Sacramento County Superior Court. Prendez faced charges for vehicle theft and felony evasion stemming from a car chase that led to the fatal shooting of a Sacramento sheriff's deputy.

What does "ineffective" counsel mean in practice? Basically it means the attorney was deficient in his or her performance, and in being so, the resulting prejudice to the defendant was so serious as to bring the outcome of the proceeding into question.[43] This standard, often referred to as the Strickland test, was initially established by the U.S. Supreme Court in *Strickland v. Washington* (1984).[44] Examples would include one's failures to investigate an alibi defense, investigate prosecution witnesses, obtain experts to challenge the prosecution's physical evidence, or even attend or stay awake for hearings.[45] The burden of proving ineffective counsel is high, and is on the defendant to show that "a reasonable probability" exists that, but for counsel's unprofessional errors, the result of the proceeding would have been different.[46] But it is possible to do so: In one Texas case, a defense attorney claimed he did not believe he needed to go into "sleazy bars to look for witnesses." The appeals court essentially informed him that that's precisely what he would do, if doing so was required to locate witnesses for the defendant, persons to confirm the defendant's alibi, and so on.[47]

When someone is charged with a crime, his or her defense lawyer, either hired or court-appointed, should do the following:

- Explain the offense the accused is charged with, including the possible punishments and probation options
- Advise the accused of his or her rights, ensure those rights are upheld, and inform the accused of what to expect during the different stages of the criminal process
- Investigate the facts of the case
- Explain what is likely to happen if the case goes to trial
- If beneficial for the accused, attempt to negotiate a plea bargain with the prosecutor (bear in mind that more than 90 percent of criminal convictions come from negotiated pleas of guilty, which must be approved by the judge; therefore, less than 10 percent of criminal cases go to trial)[48]—this can involve arranging for reduced charges, a shorter sentence, sentences for different crimes to be served consecutively instead of concurrently, probation or a disposition that avoids certain imprisonment, and/or other consideration, in exchange for entering a plea of guilty
- If the case goes to trial, cross-examine government witnesses, object to improper questions and evidence, and present applicable legal defenses,[49] as described later in this chapter

Other defense strategies might include the following:

- Trying to make the victim appear to be the aggressor, or someone who "deserved" what happened to him or her (the general rule is that the defense attorney wants to try to overlook the victim's story, while the victim wishes to punish the defendant, and will also try to engender sympathy; particularly if the victim appears to have precipitated or participated in the crime, the defense will attack his or her faults at trial, within legal bounds)
- Coming up with ways to compensate the victim (e.g., determining whether he or she will accept restitution or be satisfied if the defendant attends counseling)
- Getting continuances (which might mean that key witnesses move away, emotions and local publicity surrounding the crime diminish, and so forth)

Practitioner's Perspective

Criminal Defense Attorney

Name: Ruth Moyer
Position: Criminal Defense Attorney
Location: Philadelphia, Pennsylvania

How long have you been a practitioner in this criminal justice position? I've been practicing law for six years.

What is your career story? I received my bachelor's degree in history, and I attended Temple Law School. While I was in law school, I took many public law elective courses. I took a course on federal courts and jurisdiction, an elective course on post-trial review in criminal cases, and I also audited a course on the death penalty, which I found very interesting. While I was in law school, I also interned in the U.S. attorney's office, which gave me a more prosecutorial background. Now, I work at a criminal defense firm, which is very small. So I definitely have gotten a sense of both sides of the criminal justice system.

What are the primary duties and responsibilities of a practitioner in this position? I write many, many trial motions and trial memorandums. I've served as court-appointed counsel for defendants charged with misdemeanors who are unable to afford their own counsel. I've also done a lot of work at the appellate and collateral review stages of litigation in which, after a person has been convicted, either at trial or through a guilty plea, he has a right to challenge his conviction through what's called an appellate review or collateral review.

What are some challenges you face in this position? One of the greatest challenges of criminal defense attorneys is really twofold: first is ensuring that an innocent person isn't convicted, and second is ensuring to the extent possible that even when a person is guilty the government doesn't obtain the conviction in some violation of a person's rights. One very specific challenge I've encountered is with defendants who have no prior criminal record. And I've encountered this a lot with my court-appointed misdemeanor clients. One conviction for an otherwise law-abiding person with no criminal history could have very detrimental consequences on their life. It could result in a loss of employment and a whole range of collateral consequences.

What advice would you give to someone either wishing to study, or now studying, criminal justice and wanting to become a practitioner in this position? One skill I think is important is the ability to listen. And listening is important whether it's interviewing clients or a potential witness, or in court during cross-examination. Another very important skill is the ability to ask the right questions. That can be either researching a legal issue or, in a courtroom, asking a witness the right question. Another very important skill, of course, is oral advocacy. Good oral advocacy develops through lots and lots of practice. It's not something that most people are innately born with. The ability to analyze law and to write well are also very important skills. Unfortunately, sometimes those skills aren't emphasized as much as they should be for criminal defense attorneys. But so much of being an effective advocate for one's client requires the ability to really analyze a statute, or take five different cases and the holding from each of those cases and make it into one coherent rule, or to write a brief that's going to persuade a judge on a close legal issue.

To learn more about Ruth Moyer's experiences as a Criminal Defense Attorney, watch the Practitioner's Perspective video in SAGE Vantage.

Indigent Services

In *Griffin v. Illinois* (1956), the U.S. Supreme Court observed that "there can be no equal justice where the kind of trial a man gets depends on the amount of money he has."[50] There are basically three systems for providing legal representation to indigent persons in criminal prosecutions: the public defender system, the assigned counsel system, and the contract system.[51]

Public defender: an attorney whose full-time job is to represent indigent defendants.

Public defenders, like prosecutors, are paid government employees—most commonly found in larger jurisdictions—whose sole function is to represent indigent defendants. Public defenders perform many of the duties of prosecutors, described earlier in this chapter. They provide representation to people who not only are indigent but also may be illiterate, uneducated, and uncooperative, while managing a large caseload. Public defenders might also represent juveniles charged with acts of delinquency (offenses that would be a felony if committed by an adult) as well as children in child abuse and neglect cases.

The assigned counsel system uses private attorneys appointed on an as-needed basis by the court. A primary problem with assigned counsel is that the attorney may have little or no experience handling the criminal matter at hand; indeed, it may have been long ago that the assigned counsel studied subjects such as criminal law, criminal procedure, rules of evidence, and so on, and such attorneys may have limited knowledge about their state's criminal statutes.

The contract system is one whereby an attorney, a law firm, or a nonprofit organization contracts for a certain dollar amount—often after engaging in competitive bidding—with a unit of government to represent its indigent defendants. Advantages of this system can include reduced and predictable costs, streamlining of the counsel appointment process, and greater expertise of the attorneys. A major disadvantage is that obtaining legal counsel from the "lowest bidder" may result in inadequate or ineffective legal services—which may, of course, interfere with the defendant's Sixth Amendment right to effective counsel and be the basis for appealing a conviction.

As with many things in life, it is said that with regard to legal representation, "You get what you pay for." Obviously, people with financial means to do so will typically "go to the marketplace" and hire the best-trained legal counsel they can afford to represent them for their particular criminal matter. At the opposite end of this continuum is the *pro se* defendant who chooses instead to represent himself or herself. In such cases, there is an old adage: "He who represents himself at trial has a fool for a client."

COURTROOM WORK GROUP

Like the police subculture you learned about in Chapter 5, courtroom actors are also part of a subculture. The courtroom subculture is influenced heavily by the need to process large numbers of cases. Although our trial courts are designed to promote adversarial processes, in which defense and prosecuting attorneys "battle" to win court cases, courtroom actors—including the prosecutor, defense attorney, and judge—must work together to keep cases moving as quickly as possible and prevent delays that could violate the accused's Sixth Amendment right to a speedy trial. Accordingly, interactions between courtroom actors, who make up the **courtroom work group**,[52] are more likely to be cooperative than combative in nature.

Courtroom work group: the criminal justice professionals who work together to move cases through the court system.

In addition to the judge, prosecutor, and defense attorney, there are three other key courtroom work group participants: the court reporter, the clerk, and the bailiff. The court reporter attends legal proceedings, including trials and depositions (in which witnesses answer attorneys' questions under oath prior to trial), in order to record what is said and create transcripts of the proceedings. According to the Bureau of Labor Statistics, the justice system employs almost 20,000 court reporters.[53] The clerk of the court performs a wide array of functions. In essence, the clerk organizes and catalogs all of the paperwork generated during a trial, including all exhibits introduced into evidence (e.g., photographs and transcripts). The clerk also manages the jury subpoena and selection processes, and administers oaths to witnesses and jurors. The bailiff is a law enforcement officer who is responsible for maintaining order and safety in the courtroom and the judge's chambers. Bailiffs assist with the movement of defendants and jury members in and out of the courtroom, and they remove any persons ordered out of the courtroom by the judge.

DEFENSES

As mentioned earlier in this chapter, defense attorneys present applicable legal defenses for defendants. In the U.S. system of justice, criminal defendants have the opportunity to defend their actions by asserting **affirmative defenses**, in which the defendant admits to the criminal conduct but offers his or her reasons for acting. Because the defendant asserts such defenses, he or she typically has the burden to prove them. These **defenses** fall into two categories: justifications and excuses. When using justification defenses, defendants argue they were justified in acting because, for example, they were defending themselves or others. A police officer who had to injure a fleeing felon could claim a justification defense. When using excuse defenses, defendants admit to the criminal act but claim they are not legally responsible—they are excused—because they are too young or are insane, for example.

Affirmative defense: the response to a criminal charge in which the defendant admits to committing the act charged but argues that for some mitigating reason he or she should not be held criminally responsible under the law.

Defense: the response by a defendant to a criminal charge, to include denial of the criminal allegations in an attempt to negate or overcome the charges.

Learning about defenses is another way to learn about the critically important element of *mens rea*. Most defenses are designed to negate or reduce criminal liability by showing that the accused did not act with the required criminal intent. The prosecution may be able to easily show that the criminal act—the *actus reus*—was committed (e.g., the killing, the assault, the breaking and entering), but the defense can then defeat the criminal charges by showing that the required *mens rea* simply was not present (killing in self-defense is not the same as killing intentionally because you are trying to save yourself, and breaking and entering to save yourself from freezing to death is not the same as burglarizing a home to steal things). As a result, examining these defenses helps us understand how *mens rea* operates in real cases.

Justification Defenses

Self-Defense

Self-defense is a justification defense rooted in legal doctrine that permits the use of force against others who pose a threat to one's person or interests. Under early common law of England, people had a "duty to retreat," also known as "retreat to the wall," prior to using force to defend themselves. In effect, a person could not respond to an attacker until he or she was "cornered" and had no other retreat option available. Today, most state laws do not impose a duty to retreat and, in fact, provide in many situations that people may "stand their ground." Under modern laws, one may use force, even deadly force, without first "retreating to the wall" against another person if he or she reasonably believes that an attack against him or her is imminent, but defensive actions must be proportionate to the threatened harm and not unreasonable for the circumstances. One cannot respond to an attack with a small tree branch by using a shotgun.

Necessity

If a criminal act is committed in the event of an emergency, and if the harm avoided outweighs the harm committed by the defendant, then a person might escape criminal liability by employing the necessity defense. For example, the necessity defense could be used by a person who trespassed on private property during a fire to save a child or unattended pet, someone who broke into a home to escape a life-threatening storm, or a person who drove at a reckless speed to transport a pregnant woman to the hospital. However, the necessity defense has not been successfully used by every defendant who has committed a crime with the intent to avoid a seemingly greater harm. For example, two people operated a needle-exchange program in Massachusetts to attempt to reduce the transmission of AIDS caused by contaminated hypodermic needles. They were charged with violating a state law prohibiting hypodermic needle distribution without a physician's prescription. Although the defendants were attempting to prevent harm, the court rejected the defendants' necessity defense on the grounds that the situation posed no clear and imminent danger.[54]

Duress

The duress defense involves defendants' claims that they committed the act only because they were not acting of their own free will. For example, the wife of a bank president calls her husband and informs him that someone has broken into their home and put a gun to her head, and if he does not bring money home immediately, she will be killed. The husband then removes the money from his bank to supply the ransom. The husband could argue that he acted under duress, only to save his wife. Other actual cases include a person who was forced by gangsters to commit certain criminal acts or be killed, a drug smuggler who argued that his family would have been killed if he did not do what he was told,[55] and a Texas prison inmate whose three cellmates planned an escape and threatened to slit his throat if he did not accompany them.[56] Again, the burden will be on the accused to convince the jury that he or she committed the act under duress. For the duress defense to be successful, a jury must conclude the following:

1. The threatening conduct was sufficient to create in the mind of a reasonable person the fear of death or serious bodily harm;
2. The conduct in fact caused such fear of death or serious bodily harm in the mind of the defendant;
3. The fear or duress was operating upon the mind of the defendant at the time of the alleged act; and
4. The defendant committed the act to avoid the threatened harm.[57]

Excuse Defenses

Age

The infancy defense excuses the acts of children ages 7 and under because they are too young to be criminally responsible for their actions—they are too young to form the requisite *mens rea*. Minors between ages 7 and 14 are presumed incapable of committing a crime, but prosecutors may challenge that assumption in certain cases. Minors over age 14 have no infancy defense, but those under 16 at the time of the crime are typically tried in juvenile court. Under some state statutes, however, serious felony cases are transferred automatically to adult court, or the prosecutor has the option to seek such a transfer if the juvenile is not a suitable candidate for the more lenient and protective philosophy and law of the juvenile court.

Entrapment

Entrapment: police tactics that unduly encourage or induce an individual to commit a crime he or she typically would not commit.

If the police induced a person to commit a crime that he or she would otherwise not have attempted, the defendant can claim the defense of **entrapment**.[58] But it is not always clear what constitutes entrapment. A state supreme court deemed that police officers posing as homeless persons with cash sticking out of their pockets was entrapment because it could tempt even honest persons who were not otherwise predisposed to committing theft. But the U.S. Supreme Court did not find entrapment where undercover drug agents provided an essential chemical to defendants who were already planning to manufacture illegal drugs.[59] Nor is it entrapment when a drug agent sells drugs to a suspected drug dealer, who then sells it to government agents. The defense of entrapment will fail where the government has merely set the scene for people to commit a crime they are predisposed to commit anyway, such as where a police officer positions himself on the route of a known working prostitute and offers her money for sex when she comes by.

Intoxication

The intoxication defense is rooted in the concept of *mens rea*, and defendants must show that they were operating under such "diminished capacity" that they could not know what they were doing and cannot be held responsible. The defense is not available in cases of voluntary intoxication (except in some cases of severe alcoholism where *mens rea* is impaired permanently) and is successful—albeit rarely—only in cases of involuntary intoxication (the spiked drink or slipped drug), where the intoxicant was ingested without awareness of its intoxicating nature or where the consumption was coerced. The burden on the defendant is high in these cases, and defense attorneys generally have a difficult time convincing juries that defendants should be excused (although diminished capacity can be useful for defense attorneys to seek reduced charges or punishment).

Double Jeopardy

The Fifth Amendment to the U.S. Constitution states that no person shall be "subject for the same offense to be twice put in jeopardy of life or limb," prohibiting the government from prosecuting someone for the same offense more than once (**double jeopardy**). Other than some specific exceptions (a mistrial, a reversal on appeal, or a situation in which the crime violates laws of separate jurisdictions such as civilian/military or federal/state), the government has only one attempt to obtain a conviction. If a defendant is acquitted, the prosecution may not appeal that conviction or retry the defendant.

Double jeopardy: the prosecution of an accused person twice for the same offense; prohibited by the Fifth Amendment except under certain circumstances.

Mental Illness/Insanity

The insanity defense is perhaps the most misunderstood area of the criminal law. Ask anyone on the street, and most people will say that the insanity defense is a way for criminals to "get off" by arguing that they were "crazy" at the time of the crime. Some crimes by their very nature seem to indicate that the perpetrator was indeed not thinking straight, and many people believe that those who commit especially gruesome crimes will, by default, plead insanity. These are just a few of the common misconceptions about this defense.

Several notorious trials in recent decades have contributed to the confusion. Most notably, in 1981, John Hinckley Jr. attempted to assassinate President Ronald Reagan but was found not guilty by reason of insanity after he claimed he had done so in an effort to impress actress Jodie Foster and after psychiatrists testified at length about his childhood.[60] Then, in 1986, Steve Roth hired two men who slashed model Marla Hanson's face with razors after Hanson rejected Roth's sexual advances. Her injuries required more than 100 stitches. Roth's insanity defense—based on the psychiatric effects of his short stature—failed, and all three men were convicted.[61] More recently, the movie *Milk* recalled the so-called Twinkie defense from the trial of Dan White, who in 1978 murdered Harvey Milk, a San Francisco gay rights activist and politician. Although the defense never even mentioned Twinkies during White's trial, the media coined the term following psychiatric testimony that White had been depressed and consuming junk food and sugar-laden soft drinks—allegedly "blasting sugar through his arteries and driving him into a murderous frenzy."[62] Despite these high-profile cases, the insanity defense is raised in less than 1 percent of felony cases and is successful in only a fraction of those.[63] The defense is really quite simple if you think of it as another way of examining the critical element of *mens rea:* Defendants must prove that they have a recognized, diagnosable mental illness—a disease of the brain (e.g., schizophrenia, bipolar disorder, psychosis)—and that because of that mental illness, they cannot be held criminally responsible for their actions.

State laws set forth the applicable test for legal insanity. Most states use the M'Naghten Rule, also known as the **right-wrong test**. Under this test, it must be "proved that, at the time of the committing of the act, the defendant suffered from a mental illness and because of that disease of the mind, was laboring under such a defect of reason that he did not know the nature and quality of the act he was doing, or if he did know it, that he did not understand what he was doing was wrong."[64] Another test is the "irresistible impulse test," requiring a showing that the defendant, because of a mental illness, could not control his or her impulses or volition. Also known as "the policeman at your elbow" test, this standard for legal insanity requires a showing that the defendant would have committed the crime even if a police officer had been on the scene, literally at the accused's elbow, thereby evidencing that the defendant had no impulse control.[65]

Right-wrong test: the test of legal insanity, asking whether the defendant understood the nature and quality of his or her act and, if so, if he or she understood it was wrong.

Contrary to another popular misconception, a finding of legal insanity does not mean the defendant walks free. Instead, the defendant will be committed to a psychiatric facility. In some states, in reaction to the Hinckley verdict, the jury can reach a "guilty but mentally ill" verdict, allowing mentally ill defendants to be found guilty but to receive psychiatric treatment while incarcerated or to be placed in a mental hospital until well enough to be moved to a prison to serve their sentences.[66] After Hinckley's trial, many states shifted the burden of proving insanity to the defense, requiring them to show either clear and convincing evidence or a preponderance of the evidence that the defendant was legally insane at the time of the crime. Consider the following notorious post-Hinckley cases in which the insanity defense failed:

- Jeffrey L. Dahmer, the serial killer who claimed that necrophilia drove him to murder and dismember/cannibalize 15 men and boys, was convicted in 1992.[67]
- David Berkowitz, known as the Son of Sam killer (he reported receiving messages from the devil through a neighbor's dog, Sam), murdered six people in New York in the mid-1970s and was deemed fit to stand trial (despite a psychiatric report that found him paranoid and delusional).[68]
- John Wayne Gacy, the Chicago-area so-called Killer Clown who murdered more than 30 youths, pleaded not guilty by reason of insanity but was convicted and executed in 1994.[69]

/// IN A NUTSHELL

- The debate over how we select our judges has escalated in the 21st century. Adding fuel to the fire over judicial selection is the amount of money now being spent by judges to fund their elections; across the nation, states use five basic methods of judicial selection. One of the more common methods is the merit plan (or "Missouri Plan").
- Judges enjoy several distinct benefits of office, including life terms for federal positions and a high degree of respect and prestige. But problems can await new judges as well: mastering the breadth of law they must know and apply; administering the court and the docket; supervising court staff; and coping with the psychological discomfort and loneliness that accompany the new position.
- The prosecutor is probably the most powerful person in our criminal justice system, controlling the floodgates in determining whether or not to file charges (and these attorneys also have the ability to rebuke officers who fail to do their work properly). The prosecutor also interacts with the person suspected of the crime, the victim, and witnesses.
- Defense attorneys require the state to prove its case, help defendants understand their options in the criminal justice system, and attempt to ensure that the entire slate of rights owed to the defendant is upheld. Like prosecutors, defense attorneys have a number of strategies at their disposal.
- The courtroom work group consists of individuals who play key roles during the trial process. The six main courtroom work group actors employed by the criminal justice system are the judge, prosecutor, defense attorney, court reporter, clerk, and bailiff.
- Our system of justice allows for persons charged with crimes to offer defenses for their behavior; one can argue that his or her acts were justified (e.g., self-defense, necessity, duress); or the accused can admit wrongdoing but argue that he or she is not deserving of blame due to circumstances surrounding the offense (e.g., age, entrapment, intoxication, mental illness).

/// KEY TERMS & CONCEPTS

Review key terms with eFlashcards at **edge.sagepub.com/peakbrief**.

/// REVIEW QUESTIONS

Test your understanding of chapter content. Take the practice quiz at **edge.sagepub.com/peakbrief**.

1. What are the five methods by which judges are selected?
2. Why is the partisan election method of selecting judges currently under severe criticism?
3. What are the key points of the merit selection plan for selecting judges?
4. What is meant by "good judging," and why is courtroom civility so important?
5. Why is the prosecutor believed to occupy the most powerful position in the criminal justice system?
6. What is prosecutorial immunity, and why is it both important and potentially dangerous in our adversarial system?
7. What is a defense attorney's primary responsibility under the Sixth Amendment?
8. Which individuals play prominent roles in the courtroom work group?
9. What is double jeopardy, why does this constitutional protection from the Fifth Amendment exist, and what are some examples of exceptions to the rule? Many years after O. J. Simpson was acquitted of murdering his ex-wife and her friend, he finalized a contract to publish a book entitled *If I Did It,* which detailed how he "could have" committed the murders. Could prosecutors retry Simpson for the murders based on this new "evidence"?
10. How can age, entrapment, intoxication, and duress each be used as a criminal defense?

/// LEARN BY DOING

1. Your local League of Women Voters is establishing a new study group to better understand merit selection, or the so-called Missouri Plan for selecting judges, so as to be better informed when the matter comes up for a referendum. You are asked to explain this system of selecting judges, including its pros and cons when compared with, say, judges running for election on a partisan ticket. Develop your presentation.
2. You have been asked by your criminal justice department chairperson to participate in the annual "Career Day" program that the faculty conducts. The focus is on different careers in law enforcement, courts, and corrections. Because the faculty members know you recently completed an internship with your local prosecutor's office, they ask you to make a presentation on the functions and challenges that exist for a prosecutor. Develop and organize into a 10-minute speech what you will say in your comprehensive presentation.
3. Following is a case study grounded in actual case facts and chapter materials concerning self-defense. Answer the following questions yourself before reviewing answers and/or outcomes for the case in the Notes section. For purposes of discussion, however, approach all of the questions as if there are no absolute, totally correct answers.
4. A woman is at home with her two children when her estranged husband arrives and begins threatening and assaulting her, following a pattern of alleged past conduct with this woman and prior women. Fearing for her and her children's safety, she flees to the garage but later claims she was unable to open the garage door. Instead, she gets a gun out of her car and returns to the house. Her husband tells her he will kill her, and their young son witnesses the threat. The woman fires a warning shot that hits the wall behind the husband and then deflects into the ceiling, injuring no one. The woman is charged with aggravated assault with a deadly weapon, but she lives in a "stand your ground" jurisdiction and argues that she acted in self-defense.
   - How can the prosecutor argue that the woman had no right to "stand her ground" and exercise her self-defense rights? What other options did she have in the situation?
   - How can the defense argue that this is precisely the type of case in which the wife should be able to argue self-defense? What other facts would you want to know about the husband-wife relationship?
   - What are the dangers if the wife is successful at using this defense? What are the dangers if she fails and is convicted of attempted murder? Think about broader social policy issues.[70]

/// STUDY SITE

Get the tools you need to sharpen your study skills. SAGE edge offers a robust online environment featuring an impressive array of free tools and resources.

Access practice quizzes, eFlashcards, video, and multimedia at **edge.sagepub.com/peakbrief**

istockphoto.com/AndreyPopov

CHAPTER 9

COURT METHODS AND CHALLENGES

Sentencing and Punishment

If you are going to punish a man retributively, you must injure him. If you are going to reform him, you must improve him. And men are not improved by injuries.

—George Bernard Shaw[1]

LEARNING OBJECTIVES

As a result of reading this chapter, you will be able to

1. Delineate the four goals of punishment and explain the factors that influence the type of punishment that a convicted person will receive
2. Describe the historical development of, and different philosophies regarding, crime and punishment from the colonial era to today, and how different types of prisons were built accordingly
3. Describe the differences between, and purposes of, both determinate and indeterminate sentences
4. Review federal and state-level sentencing guidelines
5. Explain the law and purposes surrounding the use of victim impact statements
6. Describe the fundamental arguments for and against capital punishment, including key Supreme Court decisions concerning its existence and application, methods of execution, and DNA exonerations from death sentences
7. Describe aggravating and mitigating circumstances as they apply to sentencing decisions
8. Explain the right to appeals by those who are convicted

ASSESS YOUR AWARENESS

Test your knowledge of criminal sentencing and punishment by responding to the following seven true-false items; check your answers after reading this chapter.

1. Historically, people have been punished for one purpose only: retribution.
2. A small number of offenders commit a disproportionately large number of offenses.
3. Today, U.S. society adheres to the rehabilitation model, which considers criminals to have been failed by society and emphasizes offender treatment.
4. Offenders' sentences can be served in determinate or indeterminate and concurrent or consecutive configurations.
5. Prosecutors and defense attorneys can influence judges' sentencing decisions.
6. Victims' families are not allowed to present impact statements in court at the time of sentencing.
7. Federal sentencing guidelines are to be merely advisory and not mandatory.

<< Answers can be found on page 293.

On November 30, 1983, Michael Conley shot and killed 21-year-old Marsy Nicholas, his neighbor and former girlfriend of three years, at his home. Nicholas was a model and award-winning equestrian. She was completing her last quarter at the University of California, Santa Barbara, and had plans to teach disabled children following graduation.

Conley became jealous after learning that Nicholas had been dating a mutual friend. Nicholas's mother said that Marsy went to see Conley because he had threatened suicide. Conley shot Nicholas in the head with a shotgun, later claiming that the gun went off by accident. Conley was eventually found guilty of second-degree murder.[2]

Nicholas's family became victim's rights advocates. Her brother said in an interview, "After [Marsy's] funeral service, we were driving home and stopped at a market so my mother could just run in and get a loaf of bread. And there in the checkout line was my sister's murderer, glowering at her."[3] Nicholas's family were key backers of a California Victims' Bill of Rights approved by voters in 2008. This proposition became known as "Marsy's Law" and called for increased protection and rights for victims. There are 17 Marsy's Rights, including victims' right to be heard at multiple stages of criminal proceedings (you will learn more about victim impact statements later in this chapter). Like accused persons are read their *Miranda* rights, victims in California who are contacted by police are now informed of their Marsy's Rights and provided with a "Marsy's Card" that lists these rights.

Similar laws have been passed by multiple states throughout the country, but they remain controversial. At least two state supreme courts have ruled that the laws are unconstitutional.

As you read this chapter and learn about the history and theories of punishment and sentencing in the United States, think about the moment when a court hands down a sentence to a criminal offender. What are we trying to achieve through the process? How can we best offer redress to victims and communities, if at all, through punishment and sentencing? Should victims or their families be allowed to testify, or is such testimony too inflammatory to juries? Similarly, should defendants be allowed to make statements about their crime and punishment? Finally, how can we possibly find an appropriate punishment for every new offender, or for the thousands of repeat offenders moving through our criminal justice system every year?

California Attorney General's Office

Contact Name: ______________________________

Phone No.: ______________________________

Police Report / Case No.: ______________________________

Notes: ______________________________

Marsy's Card and Resources

The California Constitution, Article 1, Section 28(b), confers certain rights to victims of crime. Those rights include:

1. **Fairness and Respect** – To be treated with fairness and respect for his or her privacy and dignity, and to be free from intimidation, harassment, and abuse, throughout the criminal or juvenile justice process.
2. **Protection from the Defendant** – To be reasonably protected from the defendant and persons acting on behalf of the defendant.
3. **Victim Safety Considerations in Setting Bail and Release Conditions** – To have the safety of the victim and the victim's family considered in fixing the amount of bail and release conditions for the defendant.
4. **The Prevention of the Disclosure of Confidential Information** – To prevent the disclosure of confidential information or records to the defendant, the defendant's attorney, or any other person acting on behalf of the defendant, which could be used to locate or harass the victim or the victim's family or which disclose confidential communications made in the course of medical or counseling treatment, or which are otherwise privileged or confidential by law.
5. **Refusal to be Interviewed by the Defense** – To refuse an interview, deposition, or discovery request by the defendant, the defendant's attorney, or any other person acting on behalf of the defendant, and to set reasonable conditions on the conduct of any such interview to which the victim consents.
6. **Conference with the Prosecution and Notice of Pretrial Disposition** – To reasonable notice of and to reasonably confer with the prosecuting agency, upon request, regarding, the arrest of the defendant if known by the prosecutor, the charges filed, the determination whether to extradite the defendant, and, upon request, to be notified of and informed before any pretrial disposition of the case.
7. **Notice of and Presence at Public Proceedings** – To reasonable notice of all public proceedings, including delinquency proceedings, upon request, at which the defendant and the prosecutor are entitled to be present and of all parole or other post–conviction release proceedings, and to be present at all such proceedings.
8. **Appearance at Court Proceedings and Expression of Views** – To be heard, upon request, at any proceeding, including any delinquency proceeding, involving a post– arrest release decision, plea, sentencing, post–conviction release decision, or any proceeding in which a right of the victim is at issue.
9. **Speedy Trial and Prompt Conclusion of the Case** – To a speedy trial and a prompt and final conclusion of the case and any related post–judgment proceedings.
10. **Provision of Information to the Probation Department** – To provide information to a probation department official conducting a pre–sentence investigation concerning the impact of the offense on the victim and the victim's family and any sentencing recommendations before the sentencing of the defendant.
11. **Receipt of Pre-Sentence Report** – To receive, upon request, the pre–sentence report when available to the defendant, except for those portions made confidential by law.
12. **Information About Conviction, Sentence, Incarceration, Release, and Escape** – To be informed, upon request, of the conviction, sentence, place and time of incarceration, or other disposition of the defendant, the scheduled release date of the defendant, and the release of or the escape by the defendant from custody.
13. **Restitution**

    A. It is the unequivocal intention of the People of the State of California that all persons who suffer losses as a result of criminal activity shall have the right to seek and secure restitution from the persons convicted of the crimes causing the losses they suffer.

    B. Restitution shall be ordered from the convicted wrongdoer in every case, regardless of the sentence or disposition imposed, in which a crime victim suffers a loss.

    C. All monetary payments, monies, and property collected from any person who has been ordered to make restitution shall be first applied to pay the amounts ordered as restitution to the victim.

14. **The Prompt Return of Property** – To the prompt return of property when no longer needed as evidence.

15. **Notice of Parole Procedures and Release on Parole** – To be informed of all parole procedures, to participate in the parole process, to provide information to the parole authority to be considered before the parole of the offender, and to be notified, upon request, of the parole or other release of the offender.

16. **Safety of Victim and Public are Factors in Parole Release** – To have the safety of the victim, the victim's family, and the general public considered before any parole or other post-judgment release decision is made.

17. **Information About These 16 Rights** – To be informed of the rights enumerated in paragraphs (1) through (16).

****

Additional Resources

The Attorney General does not endorse, have any responsibility for, or exercise control over these organizations' and agencies' views, services, and information.

Victim Compensation Board – Can help victims pay for: mental health counseling, funeral costs, loss of income, crime scene cleanup, relocation, medical and dental bills. **1-800-777-9229** *www.victims.ca.gov*

CA Dept. of Corrections and Rehabilitation, OVSRS – Provides information on offender release, restitution, parole conditions and parole hearings when the offender is incarcerated in prison. **1-877-256-6877** *www.cdcr.ca.gov/victim_services*

McGeorge School of Law – Victims of Crime Resource Center - Provides resources for victims by their geographic area along with information on victims' rights. **1-800-Victims (1-800-842-8467)** *www.1800victims.org*

National Domestic Violence Hotline – **1-800-799-7233** *www.thehotline.org*

Adult Protective Services County Information – (Elder abuse) 24 hour hotline numbers by county in California. *www.cdss.ca.gov/inforesources/County-APS-Offices*

National Child Abuse Hotline – Treatment and prevention of child abuse. **1-800-422-4453** *www.childhelp.org*

Rape, Abuse & Incest National Network – **1-800-656-4673** *www.rainn.org*

National Human Trafficking Resource Center Hotline – 24-hour hotline: **1-888-373-7888** *www.humantraffickinghotline.org*

The California Relay Service: For speech impaired, deaf or hard-of-hearing callers: Dial 711. TTY/HCO/VCO to Voice for English: 1-800-735-2929 and for Spanish: 1-800-855-3000. Voice to TTY/VCO/HCO for English: 1-800-735-2922 and for Spanish: 1-800-855-3000. Speech to Speech – English and Spanish: 1-800-854-7784.

Attorney General's Victims' Services Unit – Provides local victim/witness information, geographic resource information and appeal status to victims of crime. For more information, call **1-877-433-9069** or visit: *www.oag.ca.gov/victimservices* For local Human Trafficking information, visit: *www.oag.ca.gov/human-trafficking*

****

A 'victim' is defined under the California Constitution as "a person who suffers direct or threatened physical, psychological, or financial harm as a result of the commission or attempted commission of a crime or delinquent act. The term 'victim' also includes the person's spouse, parents, children, siblings, or guardian, and includes a lawful representative of a crime victim who is deceased, a minor, or physically or psychologically incapacitated. The term 'victim' does not include a person in custody for an offense, the accused, or a person whom the court finds would not act in the best interests of a minor victim." (Cal. Const., art. I, § 28(e).)

A victim, the retained attorney of a victim, a lawful representative of the victim, or the prosecuting attorney upon request of the victim, may enforce the above rights in any trial or appellate court with jurisdiction over the case as a matter of right. The court shall act promptly on such a request. (Cal. Const., art. I, § 28(c)(1).)

Funding is made possible through the United States Department of Justice, Victims of Crime Act, 2016-VA-GX-0057

VSU Rev 10/2017

INTRODUCTION

For what reasons and purposes are people punished? Do punishments always fit the crimes committed? Does capital punishment work? Punishing those who violate the right to life, liberty, and property of their fellow human beings is one of the primary functions of the American criminal justice system. However, many questions concerning the "how" and "how much" need to be addressed. As the chapter-opening quotation by George Bernard Shaw might indicate, an enigma for our time is what might best be done with people who must be punished for their transgressions. As will be seen in this chapter, sentencing and punishment are complicated issues, with both having financial and societal factors that must be included in these discussions.

Although violent crime rates have been declining in the new millennium, people are increasingly being victimized by intelligent white-collar criminals, identity thieves, and

Placing people in stocks was used internationally and during medieval, Renaissance, and colonial American times as a form of physical punishment as well as for public humiliation.

cybercriminals. How should our society deal with these new types of offenders? Into this complicated mix might also be included the adage that "it is better to let a hundred guilty people go free than to convict one innocent person." The federal and state sentencing guidelines discussed in this chapter indicate how much concern and effort have recently gone into sentencing and punishment. This chapter approaches sentencing and punishment from several perspectives, including their purposes, types, and methods, as well as the more recent influence of DNA evidence. Also examined are capital punishment and criminal appeals.

PURPOSES OF PUNISHMENT

The need to punish some of our fellow citizens has existed since the beginning of time—at least since biblical times, and likely much earlier. It would seem there have always been attempts—by a variety of methods and for a variety of reasons—to convince people that they should change their behavior and either obey the customs and laws of their society or suffer the consequences. Very often those attempts at changing behaviors meant that—by one means or another, such as imprisonment, banishment, or death—offenders would be removed from society in such a way that they were no longer in a position to do further harm to their fellow citizens. Here, we discuss the four goals of punishment—some or all of which are hoped to be achieved by all societies, even the most primitive. Furthermore, we will see that throughout history, crimes and criminals have been viewed and punished differently, depending on several factors. Some of those factors that determine punishment today were described in Chapter 1, in the discussions of the wedding cake, crime control, and due process models of crime.

Four Goals

Punishment (and its purposes): penalties imposed for committing criminal acts, to accomplish retribution, deterrence, incapacitation, and/or rehabilitation.

Retribution: punishment that fits the crime, that is "equitable" for the offense.

Deterrence: the effect of punishments and other actions to deter people from committing crimes.

Following are the historical reasons and goals for **punishment** of our fellow citizens—what is hoped will be achieved (note that all of these goals and justifications may also be achieved by methods employed in community-based corrections, discussed in Chapter 11):

- *Retribution*: **Retribution** has its roots in Old Testament law where, in Exodus 21:24, the phrase "eye for eye, tooth for tooth, hand for hand, foot for foot" appears for the first time—labeled *lex talionis*, the law of equitable retribution. Death penalty supporters also quote this phrase often as justification for their position. However, neither the phrase "eye for eye" nor the verse itself is a complete sentence, and the death penalty is not mentioned. For many people, "eye for eye" dictates that offenders should be punished in a manner that reflects their crime: It is instinctive for people to want to get even when wronged by another, and it is deeply engrained within ourselves and our society that punishment should be meted out when someone breaks the law.
- *Deterrence*: **Deterrence** probably makes more sense than retribution in terms of betterment of society because it is not grounded on our primal human emotions and instincts. The concept of deterrence stems from the classical school of criminology. Since people will typically avoid unpleasant things, they are much less likely to commit a crime if they know that punishment will result if they are caught. Deterrence has two components: general and specific. By seeing others being punished for their

crimes, the public experiences a *general* deterrent effect because they can see what will befall them should they engage in similar behavior. *Specific* deterrence involves using punishment against specific offenders for their criminal acts in order to discourage them from committing such acts again in the future.

- *Incapacitation*: **Incapacitation**, by its very meaning, is beneficial in that it prevents criminals from victimizing others by placing them in a situation where they are physically unable to commit crimes. The best examples, of course, are incarceration in jails and prisons as well as execution. Bear in mind, however, that one can still commit crimes while in a state of incarceration. The methods of community corrections—probation and parole and other alternatives to incarceration—are forms of incapacitation that also aim to prevent offenders from committing new crimes.

Incapacitation: rendering someone as unable to act or move about, either through incarceration or by court order.

- *Rehabilitation*: Almost since its beginning, the modern criminal justice system has had as a primary goal—indeed a responsibility—to change criminals so that they become law-abiding citizens, in other words, to rehabilitate them. In fact, for most of the 20th century, **rehabilitation** was the system's primary goal in terms of how it was to function and be organized. Since the mid-1960s, however, that ideology has been modified to the point that it is hardly recognizable.[4] The reasons for this ideological change are several, and include changing governmental priorities, other concerns of the public (as revealed in national polls), and institutional and political resistance to change.[5] Furthermore, politicians can point to the historical "nothing works" idea put forth by Robert Martinson, who studied prison programs. Still, a wide array of correctional programs (i.e., vocational, educational, counseling) continues to be offered in prisons and jails, and with marked success. One cost-benefit analysis of 14 correctional treatment programs found that, in all but one of the programs, program benefits outweighed program costs. Such programs can be crucial for assisting offenders to reenter society and not recidivate (commit more crimes).[6]

Rehabilitation: attempts to reform an offender through vocational and educational programming, counseling, and so forth, so that he or she is not a recidivist and does not return to crime and prison.

Sentencing and punishment must accomplish one or more of these goals if the public is to be supportive of them. If increases in prison and jail sentences do not provide effective means of preventing crime, then a more cost-effective strategy must be found that will target the offenders most likely to commit serious crimes at high rates. As a federal report noted:

> It is frequently observed that a small number of offenders commit a disproportionately large number of offenses. If prison resources can be effectively targeted to high-rated offenders, it should be possible to achieve . . . levels of crime control. The key to such a policy rests on an ability to identify high-rate offenders . . . and at relatively early stages in their careers.[7]

Factors Influencing Punishment

As noted earlier, sentencing and punishment involve issues concerning their financial and societal benefit. On average it costs about $35,000 per year to house an inmate in prison,[8] but costs vary widely by jurisdiction. For example, the annual cost of imprisoning a single California inmate is greater than $81,000, exceeding the cost of attending Harvard with room and board by about $14,000.[9] For that amount of money, society expects to be able to accomplish one or more of the four goals described earlier. Regarding the cost-benefit effect of prisons, some researchers argue that prisons should be used to greater advantage, believing it is at least twice as costly to let a prisoner be loose in society than it is to lock him or her up. For example, comparing the cost of incarceration with the human and financial toll of crime, prison expert John DiIulio Jr. believes that prisons are a "real bargain."[10]

In addition to victims making their wishes known early on to the prosecution concerning punishment, as well as victim impact statements and aggravating or mitigating circumstances involved in a crime (discussed briefly in Chapter 1 and later in this chapter), there are other factors that influence sentencing and punishment. First, the U.S. Constitution speaks briefly but forcefully regarding the use of punishment. The Eighth Amendment provides

Juvenile Justice Journal

Goals and Methods of the System

Discussions of the U.S. juvenile justice system in earlier chapters have made clear that juveniles are to be treated differently when they are to be detained, taken to a hearing, and adjudicated; following are more specific forms of treatment that they receive, which may be viewed as goals of the juvenile justice system.

The four primary goals of this system are as follows:[11]

- *Separation from adults:* This is clearly the most important goal of the juvenile justice system. Reformers argued that children and families needed (1) a different form of justice, (2) separate courtrooms, (3) separate detention centers and institutions to avoid corruption of juveniles by adult criminals (the so-called sight and sound separation), and (4) separate sentencing guidelines to avoid the harsh penalties of adult sentencing. Furthermore, a separate group of specially trained professionals—judges, probation officers, and detention staff—are dedicated to working with youths and their families.
- *Youth confidentiality:* Confidentiality of juvenile court proceedings and services reinforces the belief that juveniles will mature beyond a criminal lifestyle if given proper guidance and alternatives. Because of their immaturity, youths lack sound judgment and should not be held fully accountable. Consequently, no criminal record should hinder adult advancement. From a developmental standpoint, confidentiality minimizes stigma and labeling, and helps youths to maintain a positive self-image, thereby reducing the likelihood that they will perceive themselves as criminals.
- *Community-based corrections:* Reformers strongly believed that young people should learn and grow in their own communities. Offering probation as a method for monitoring youth behavior in the community, while providing services that allow youths to grow to adulthood, is seen as the primary dispositional alternative.
- *Individualized justice of minors:* Each case is to be viewed separately. Case workers, social workers, and probation officers develop a casework plan, based on the total social circumstances of the youth and his or her family, to encourage appropriate development and prevent future criminality. Probation staff members are to look into the social situation early in the process and be involved in the decision to file a case. Whenever possible, the case is not filed formally, and an informal outcome is encouraged.

that incarceration will not involve "cruel and unusual punishment," and that fines will not be excessive. Furthermore, the Thirteenth Amendment states that U.S. citizens have a right against involuntary servitude.

Prosecutors can influence the sentencing decision by agreeing to engage in plea negotiation in terms of the number of charges filed or the maximum penalty the judge may impose, or by explaining to the sentencing judge that the offender was particularly cruel in his or her crime or, alternatively, was very cooperative with the police and/or remorseful about the crime. In many states, prosecutors can also make a specific sentencing recommendation to the court that has been agreed upon with the defense.

Defense attorneys probably have less influence than prosecutors over sentencing decisions. Nonetheless, they can seek to obtain the lightest sentence possible, including probation or other alternatives to sentencing, as well as emphasize prior to sentencing such things as the defendant's minor involvement in the crime, the victim's participation, and so on.

The seriousness of the offense is the most important factor in determining the sentencing received for an offense. For instance, a violent crime against a person warrants a harsher penalty than an offense against one's property, and the judge, jury, prosecutor, and defense attorney must take into account the victim's suffering when carrying out their roles in arriving at a proper punishment.

You Be the . . .

Judge

A habitual offender is essentially one who has been convicted of a crime several times (either a misdemeanor or a felony, and typically at least twice). Habitual offender laws usually impose additional punishments on such offenders. Such laws can even address traffic violations. For example, a first-time driving under the influence (DUI) offense is usually a misdemeanor that results in a fine and jail time of less than one year. However, upon being arrested for a second or third DUI, the offender may be charged with a felony (depending on state law). Being classified as a habitual offender can thus result in higher criminal fines, longer jail or prison sentences, and loss of various rights and privileges (e.g., the right to own a firearm or to possess a driver's license).

An example is a Florida law stating that if you receive three (3) or more convictions of serious offenses on separate occasions you will be deemed a habitual offender. Examples of serious traffic offense include voluntary manslaughter while driving, involuntary manslaughter while driving, [and] felony while driving. Other serious traffic offenses include not stopping at an accident with a personal injury or death, or driving with a suspended or revoked license. Three convictions of these offenses and you will be considered a habitual traffic offender.

1. Do you agree with the spirit and intent of habitual offender laws?
2. Are such laws too harsh or too lenient?

Sources: See LegalMatch, "What Is a Habitual Offender," http://www.legalmatch.com/law-library/article/what-is-a-habitual-offender.html; see also Florida Drivers Association, "How to Get a Hardship License for Habitual Offenders," http://www.123driving.com/habitual-offender.shtml.

Where sentencing is concerned, next in importance is the defendant's prior criminal record. The existence of a lengthy criminal record—particularly a record of violence or even habitual crimes against property, such as home invasion—can weigh heavily in terms of sentencing and punishment. Many states have habitual offender laws, which are related to and often viewed as identical to three-strikes laws (discussed in Chapter 1). (See the accompanying "You Be the . . . Judge" box.) These laws vary widely from state to state but typically apply only to felonies and require third-time felons to serve a mandatory 25 years to life. Furthermore, many police departments have repeat offender units dedicated solely to surveilling known offenders.

Punishment Models, Methods, and Reforms

Philosophies of crime and punishment have changed significantly since the late 1700s, when the United States was relatively sparsely populated and predominantly rural. However, with the Industrial Revolution came a new concept of criminal punishment embracing various correctional methods (see timeline in Figure 9.1).[12]

- *The colonial model (1600s–1790s):* During the colonial period most Americans lived under laws that were transferred from England. Puritans rigorously punished violations of religious laws, and banishment from the community, fines, death, and other punishments were the norm. Use of the death penalty was common.
- *The penitentiary model (1790s–1870s):* With the Industrial Revolution—and increasing populations—came a new concept of criminal punishment. Criminal offenders were to be isolated from the bad influences of society and from one another so that, while engaged in productive labor, they might reflect on their past misdeeds and be "penitent" or remorseful for their crimes. As a result, what has been termed "the first American penitentiary, if not the first one in the world," was established in Philadelphia, in 1790, in the Walnut Street Jail. This penitentiary introduced the

institutional pattern of outside cells with a central corridor and the use of solitary confinement as the central method of reforming inmates to "the good life." Inmates were also segregated according to "age, sex, and the type of the offenses charged against them."[13] Auburn Prison, built in New York in 1821, reflected this shift, emphasizing individual cell-block architecture to create an environment to rehabilitate and reform, separate criminals from all contact with corruption, and teach them moral habits, by means of severe discipline. Inmates worked as contract convict labor 10 hours per day, 6 days per week. The Auburn model influenced the emergence of reform schools and workhouses in the 1820s. Then, Eastern State Penitentiary, a huge fortress with thick walls near Philadelphia, was built in 1829, emphasizing complete solitary confinement rather than Auburn's contract labor. New inmates wore hoods when marched to their cells so that they would not see other prisoners. Regimentation included use of the lockstep (marching everywhere in single file), shuffling with the head turned right, practices that continued into the 1930s. No visitors, mail, or newspapers were allowed. The design of this prison became the most influential in U.S. history.

- *The reformatory model (1870s–1890s):* By the middle of the 1800s, reformers became disillusioned with the results of the penitentiary movement, and soon a new generation of reform came to the fore, motivated by humanitarian concerns. This new approach to penology emphasized inmate change and indeterminate sentences. Fixed sentences, lockstep, silence, and isolation were seen as destructive to inmate initiative. This wave of prison reform began with the founding of today's American Correctional Association in 1879 and the building of Elmira Reformatory in 1876. At Elmira, Zebulon Brockway began classification and segregation of prisoners, as well as providing vocational training and rewards for good behavior—including early release for good behavior and parole. Brockway's "New Penology" included the creation of specialized institutions to care for the young, females, and the mentally impaired. The juvenile court system, created in Chicago in 1899, gave wide discretionary powers to judges, and Indiana's Female Prison and Reformatory Institution for Girls and Women was opened in Indianapolis in 1873. Inmates began producing license plates, constructing public highways, and working at prison farms and factories that produced food and items for internal consumption. In 1927, the first federal prison for women opened in Alderson, West Virginia; the minimum-security campus-like prison used residential cottages for inmates. Elsewhere, the camps that housed inmates working on roads became models for minimum-security prisons that emerged in the 1930s. Then, with Congress recognizing the need to build federal penitentiaries, the Three Prisons Act of 1891 authorized the first federal penitentiaries. The old army prison at Fort Leavenworth, Kansas, became the first U.S. penitentiary in 1895; the second opened in 1902 at Atlanta, Georgia; and the third was located at the old territorial prison on McNeil Island in Puget Sound, Washington.
- *The progressive model (1890s–1930s):* The first two decades of the 20th century saw the Progressives—activists seeking to address perceived social problems caused

Figure 9.1 /// A Timeline of Correctional Models

1600s–1790s	The Colonial Model
1790s–1870s	The Penitentiary Model
1870s–1890s	The Reformatory Model
1890s–1930s	The Progressive Model
1930s–1960s	The Medical Model
1960s–1970s	The Community Model
1970s–2000s	The Crime Control Model

by corruption in government, urbanization, industrialization, and immigration—wanting to understand and cure crime; they sought, first, to improve social conditions that appeared to breed crime and, second, to treat criminals so that they would lead crime-free lives. Treatment would be focused on the individual and his or her specific problem. Probation was launched as an alternative to incarceration, allowing offenders to be treated in the community under supervision, and indeterminate sentences came into being.

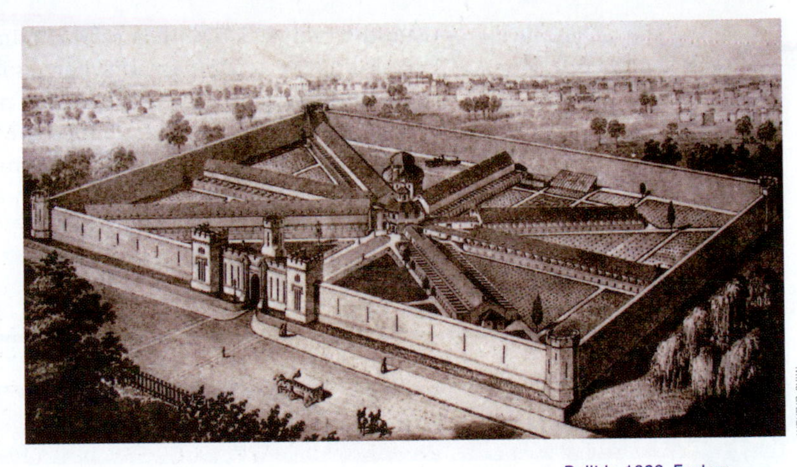

Mike Graham

Built in 1829, Eastern State Penitentiary, a huge fortress near Philadelphia, emphasized complete solitary confinement.

- *The medical model (1930s–1960s):* Grounded in positivist criminology, this model generally included the idea that criminals are mentally ill, and the emphasis of corrections shifted to treatment. Criminals were seen as persons whose social, psychological, or biological deficiencies had caused them to engage in illegal activity and who should receive treatment. Rehabilitation took on national legitimacy and became the primary purpose of incarceration. The Federal Bureau of Prisons (discussed in Chapter 10) was established in 1930 to oversee the 11 federal prisons then in existence. In 1933, Alcatraz was acquired from the U.S. Army for a federal prison. The gangster era was in full swing, and national Prohibition wrought violent crime waves. Alcatraz was the ideal solution—serving the dual purpose of holding public enemies and being a visible icon to warn this new brand of criminal. Under Warden James A. Johnston, Alcatraz's rules of conduct were among the most rigid in the correctional system, and harsh punishments were delivered to inmates who defied prison regulations. More in keeping with the medical model was the appointment in 1937 of James V. Bennett as director of the Federal Bureau of Prisons. In 1941, he built the Federal Correctional Institution at Seagoville, Texas, a prison without walls. Similar prisons became widespread in the 1960s. Different treatment programs were offered to prisoners, and the Federal Prison Industries program, which began in 1934, allowed inmates to be furloughed out of prison for work and other purposes.
- *The community model (1960s–1970s):* Following the inmate riot and hostage taking at New York state's Attica Correctional Facility in 1971, prisons were seen as artificial institutions that interfered with the offender's ability to develop a crime-free lifestyle. Community reintegration was the dominant idea until the 1970s, when it gave way to a new punitive stance in criminal justice.
- *The crime control model (1970s–2000s):* The pendulum swung again in the late 20th century—and continues today—with the public becoming concerned about rapidly rising crime rates and studies of inmate treatment programs challenging their success and worth. Critics attacked the indeterminate sentence and parole, calling for longer sentences for career criminals and violent offenders. Legislators, judges, and officials responded with determinate sentencing laws, three-strikes laws, mandatory sentencing laws

AP Photo/Anonymous

In 1941, James V. Bennett, director of the Federal Bureau of Prisons, oversaw construction of the Federal Correctional Institution at Seagoville, Texas, a prison without walls.

(e.g., doubling one's sentence for a crime committed with a weapon), and so forth. (Regarding the accompanying changes in prisons, in 1961, Texas prison system director George Beto, in opposition to the medical model, had begun emphasizing strict discipline. In 1972, David Ruiz sued the Texas prison system, claiming that the system constituted "cruel and unusual punishment" prohibited by the Eighth Amendment to the U.S. Constitution. His class action suit was settled in 1980, the longest-running prisoners' lawsuit in U.S. history.)

Practitioner's Perspective

Court Administrator

Name: Maxine Cortes
Position: Court Administrator
Location: Carson City, Nevada

What are the primary duties and responsibilities of a practitioner in this position? As a court administrator, I am responsible for facilitating policies, procedures, and operational needs of the courts. I report directly to the judges. So in that capacity, I write grants, handle the budget, am responsible for recruiting facilities, technology projects, and jury systems.

What are some challenges you face in this position? The biggest challenge is that you don't really have control of your day. So every day, you have things that you need to do, meetings that you need to attend. But at a moment's notice, whether it is a phone call, an email, somebody walking through the door, everything can change. Being able to pivot and be professional and positive, and continue to handle any task or issue, being solution-oriented, flexible, and patient—I think that is the biggest challenge that I have every day.

What role does diversity play in this position? Diversity is huge. We serve diverse individuals from all walks of life, all socioeconomic backgrounds, all different ethnicities. The people we work with are also very diverse. Working in the court system, I'll sometimes see people come in, and it looks like they're going to the dentist—they really don't want to be there. To be able to maneuver and help them understand that not everyone thinks alike and sees the world through the same lenses, and being able to take the opportunity with diversity to approach people in different manners that will help them maneuver through the court system, is a major part of the role.

Do you see any common trends in this position? The trends in the court system in the last 10 years have been very exciting. Technology has played a huge role. We used to hire court reporters every day, but we do everything pretty much now through audio-video systems. In the future, I think we will see more technology playing a role in the courts, bringing an opportunity to the public to have that access to the system easily, whether it's through an iPad or a mobile phone, or a PC at home.

What advice would you give to someone either wishing to study, or now studying, criminal justice and wanting to become a practitioner in this position? My advice for students who are looking for any career, whether it be the court system or anywhere else, is to shadow. Find someone who's willing to have them shadow for one to two weeks if they're amenable. Walk in their shoes, learn what they do. I think that is the best way to really determine if you want to spend your life in a career. I also think if you do choose a career in the court system, I highly recommend taking a few classes: leadership, conflict resolution, mediation, budgeting, and communication.

To learn more about Maxine Cortes' experiences as a Court Administrator, watch the Practitioner's Perspective video in SAGE Vantage.

Making Punishment Fit the Crime

Mikael Karlsson/Alamy Stock Photo

Inmates often appear before their parole board to express remorse for their crimes, to explain how they have progressed while in prison, and to offer reasons they should be released and placed on parole.

We might consider punishment from this perspective: A report by Amnesty International discusses capital punishment around the world, including beheadings in Saudi Arabia; hangings in Japan, Iraq, Singapore, and Sudan; firing squads in Afghanistan, Belarus, and Vietnam; stonings in Iran; and "the only country in the Americas that regularly executes: the United States." Amnesty reports that in the 54 countries retaining the death penalty, at least 19,336 people are now under sentence of death, and at least 690 people were executed worldwide in 2018, excluding China (which does not release its figures).[14] This represents the lowest number of executions recorded in the past decade. The death penalty was administered to 25 people in 2018 in the United States (and that number has generally declined since 2000).[15]

The concept of "justice" and acts deserving of punishment vary across nations. Consider that in many other countries people are executed for their political thoughts, apostasy (improper religious beliefs), and "highway robbery"—or that people are flogged 80 times for possessing alcohol, or, as in Singapore, flogged with a rattan cane for vandalism.[16] Although the crime rates in these venues are likely to be relatively low, the question to be asked is this: Are any of these punishments proportional to these crimes?

As you will learn in Chapter 10, for many offenders in the United States, being sentenced to prison is a "step up," partly because there they receive "three hots and a cot" without having

You Be the . . .

Legislator

In June 2019, Alabama enacted a law that requires convicted child sex offenders (those who abused children under the age of 13) to undergo "chemical castration" as a condition of parole. Although one representative initially advocated a surgical approach to castration, the approved method involves administering drugs to reduce testosterone levels. The law is highly controversial, with some critics arguing that it violates basic human rights and others pointing to research findings that suggest a high libido is not linked to child sexual abuse. Think about this form of punishment from a lawmaker's standpoint:

1. Do you support this approach to managing sex offender recidivism? Would you alter your opinion if you were told that chemical castration was effective in only a very small number of cases? What if you were told it was not effective at all?
2. How do politics play a role in proposing new laws for heinous crimes?
3. Which, if any, of the four goals for punishment outlined at the beginning of the chapter does this punishment achieve?
4. Does it matter if lowering testosterone levels, even if only for a limited time, produces long-term health consequences?
5. Does this punishment violate the Eighth Amendment?

Source: James Hamblin, "Alabama Moves to State-Ordered Castration: A New Law for Child Sex Offenders Harkens Back to a Time When Much Less Was Known About Human Sexuality," *The Atlantic*, June 11, 2019, https://www.theatlantic.com/health/archive/2019/06/alabama-chemical-castration/591226/.

to support a family, and they might even be surrounded by their family, friends, and affiliate gang members. If that sad commentary on American life is true, then one is left to wonder how our society can allow that to happen—or, perhaps, why we devote the time, effort, and money—at least $71 billion annually for state and local corrections expenditures alone[17]—to basically "warehouse" 1.5 million people.[18] Some observers might say we should abandon the warehousing approach and make every effort to try to identify those individuals for whom the prison experience can be beneficial. Then, while those individuals are a "captive audience" in prison, we should make all manner of rehabilitative educational and vocational programming available to them. But others might argue that those inmates for whom prison is a "step up" should be put to work at hard labor, to make prison life less attractive and thus discourage them from repeat offending. But which way should the pendulum swing?

The accompanying "You Be the . . . Legislator" box poses some thought-provoking questions regarding punishment. As you read them, consider the four goals of punishment discussed earlier, as well as your own philosophy concerning punishment.

TYPES OF SENTENCES TO BE SERVED

How offenders serve their sentences—in a determinate or indeterminate (defined in Chapter 1), as well as a concurrent or consecutive, fashion—is a crucial distinction, particularly in terms of how long an offender must remain in prison and where parole is concerned. Next we distinguish between these types.

Determinate and Indeterminate Sentences

Determinate sentencing is either legislatively determined or judicially determined. In states using a determinate sentencing structure, convicted offenders are sentenced for a fixed term, such as 10 years. There is no opportunity for a paroling authority to make adjustments in time served when making release decisions. Offenders are released at the expiration of their term, minus any good-time credits. Under a legislatively determined structure, the legislature fixes by law the penalty for specific offenses or offense categories. In a judicially determined system, the judge has broad discretion to choose a sanction, but once imposed, it is not subject to change.

Conversely, in an indeterminate sentencing format, the convicted individual will be sentenced for a set range of time, such as 5–10 years, so that his or her conduct inside the prison system, amenability to rehabilitative efforts (e.g., educational and vocational programs, counseling), apparent remorsefulness, and so forth can be taken into account in deciding a release date. The legislature sets a broad range of time, expressed as minimum and maximum sentences, for a particular offense or category of offenses, and the responsibility for determining the actual term of incarceration is divided between the judge and the parole board. The judge's sentence is also made in terms of a minimum and a maximum term.

The authority of a parole board to grant discretionary release to a prisoner before the expiration date of the maximum term varies from state to state. The parole board determines the actual release date, typically using a formula for determining earliest parole eligibility, which may occur after a percentage of the minimum, after a percentage of the maximum, or after the entire minimum has been served, depending on the state.[19] Those persons supporting the rehabilitative ideal for offenders will obviously be more in favor of indeterminate sentencing, which allows for the length of sentence to be adjusted in response to the offender's positive responses to treatment and programs.[20]

Concurrent and Consecutive Sentences

Assume that a man is convicted for committing three separate offenses as part of a night's crime spree: He unlawfully entered a couple's home (burglary), stole several valuable items (a felony theft), and violently assaulted the husband in making his escape. He is sentenced to 15 years, 3 years, and 20 years, respectively, for each of these crimes. Depending on the sentencing court's decision, the offender will serve his three sentences for those crimes either concurrently or consecutively.

A *concurrent* sentence means he will serve all three sentences *at the same time*, with each sentence running along the same timeline, along parallel tracks as follows:

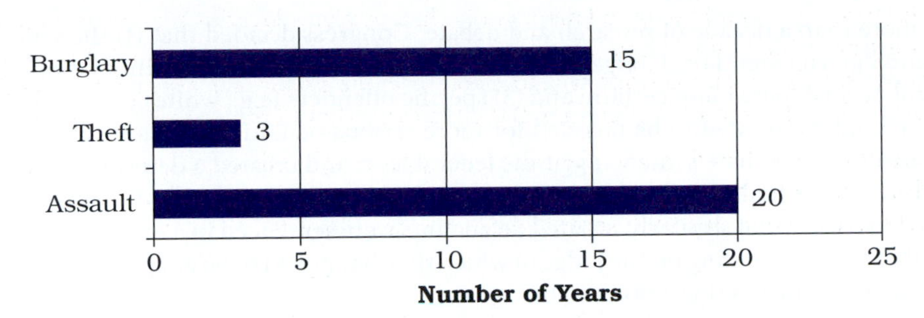

Because these sentences are running at the same time, the maximum amount of time the man will do is 20 years—his sentences of 3 and 15 years will have already run during the 20-year sentence.

If, instead, the offender is to serve the three sentences *consecutively*, each sentence will be served separately—in other words, when he finishes serving the sentence for the first crime, he immediately begins serving the sentence for the second crime, and so on. This approach is often referred to as "stacking" the sentences—one after the other—to maximize the sentence length, as follows:

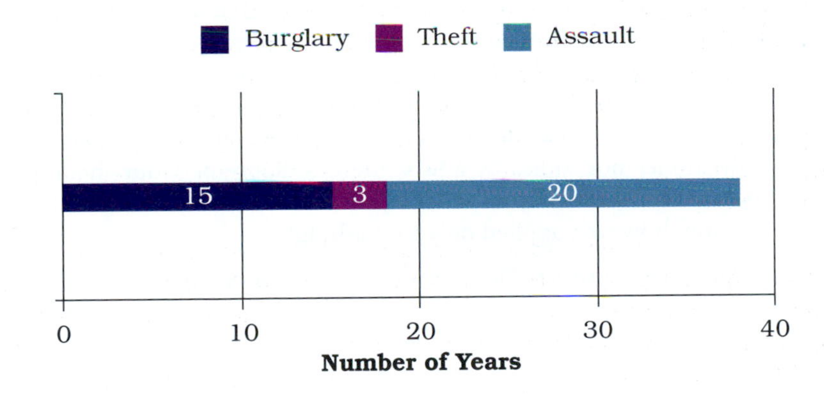

In this sentencing approach, the offender would serve a total of approximately 38 years (15 + 3 + 20), subject to reductions for good behavior and parole eligibility.

SENTENCING GUIDELINES

The growing complexity and importance of the **sentencing guidelines** now found in criminal justice pose a bit of a dilemma for this introductory course textbook: On the one hand, they are too complicated to discuss comprehensively or in great detail; on the other hand, they are far too important to ignore. Therefore, this section strives to achieve an appropriate balance by looking at the guidelines in summary form.

Sentencing guidelines: an instrument developed by the federal government that uses a grid system to chart seriousness of the offense, criminal history, and so forth and thus allows the court to arrive at a more consistent sentence for everyone.

Background: Legislation and Court Decisions

In the mid- to late 1970s and early 1980s, many people were becoming discontented with the indeterminate sentencing process; they witnessed inmates often being released after serving only a fraction of their sentences (some jurisdictions even allowed 30 days of "good-time" reduction of a sentence for each 30 days served). That, coupled with renewed concern about the rising crime rate throughout the nation, resulted in wide experimentation with sentencing systems by many states and the creation of sentencing guidelines at the federal level.

Federal Sentencing Guidelines

After more than a decade of research and debate, Congress decided that (1) the sentencing discretion given federal trial judges needed to be structured; (2) the administration of punishment needed to be more certain; and (3) specific offenders (e.g., white collar and violent, repeat offenders) needed to be targeted for more serious penalties.[21] As a result, Congress abolished indeterminate sentencing at the federal level and created a determinate sentencing structure through the federal sentencing guidelines. The Sentencing Reform Act[22] was enacted to ensure that similarly situated defendants were sentenced in a more uniform fashion rather than depending on the judge to which they happened to be assigned.[23] The federal sentencing system was thus reformed by

- Dropping rehabilitation as one of the goals of punishment
- Creating the U.S. Sentencing Commission and charging it with establishing sentencing guidelines
- Making all federal sentences determinate
- Authorizing appellate review of sentences[24]

A long line of legal challenges ensued involving these sentencing guidelines. The first such challenge came from the state of Washington, where the petitioner had pleaded guilty to kidnapping. Under Washington's law, the maximum penalty for that offense is 10 years. A separate range-of-sentence provision limited the maximum allowable sentence to 53 months, but it authorized an upward departure (i.e., longer sentences) for "exceptional" judge-determined factors. The trial judge increased the sentence to 90 months because the crime was committed with deliberate cruelty. Because the facts supporting the enhanced penalty were neither admitted by the petitioner nor found by a jury, the U.S. Supreme Court held in *Blakely v. Washington* (2004) that the sentence violated the petitioner's Sixth Amendment right to trial by jury.[25] This decision, however, applied only to Washington.

Then, six months later, the Supreme Court decided *United States v. Booker,* this time addressing sentencing guidelines nationally.[26] The defendant, Booker, was found guilty by a jury of possessing at least 50 grams of crack cocaine (he actually had 92.5 grams). Under those facts, the guidelines required a possible 210- to 262-month sentence. Although the jury never heard any such evidence, the judge, finding by a preponderance of the evidence that Booker possessed the much larger amount of cocaine, rendered a sentence that was almost 10 years longer than what the guidelines prescribed. By a vote of 5–4, the U.S. Supreme Court found that the U.S. sentencing guidelines violated the Sixth Amendment by allowing judicial, rather than jury, fact-finding to form the basis for the sentencing; in other words, letting in these judge-made facts is unconstitutional. The guidelines also allowed judges to make such determinations with a lesser standard of proof than the jury's "beyond a reasonable doubt" and to rely on hearsay evidence that would not be admissible at trial.[27] The Court did not discard the guidelines entirely. The guidelines, the Court said, are to be merely advisory and not mandatory. Thus, the guidelines are a resource a judge can look at but may choose to ignore. Although courts still must "consider" the guidelines, they need not follow them. In addition, sentences for federal crimes will become subject to appellate review for "unreasonableness," allowing appeals courts to clamp down on particular sentences that seem far too harsh.

In late 2007, the U.S. Supreme Court went further and explained what it meant in 2005 by "advisory" and "reasonableness," deciding two cases that together restored federal judges to their traditional central role in criminal sentencing. The Court found that district court judges do not have to justify their deviations from the federal sentencing guidelines and that

Table 9.1 /// State of Washington Sentencing Grid

SERIOUSNESS LEVEL	OFFENDER SCORE									
	0	1	2	3	4	5	6	7	8	9+
LEVEL XVI	LIFE SENTENCE WITHOUT PAROLE/DEATH PENALTY									
LEVEL XV	180–240	187.5–249.75	195.75–260.25	203.25–270.75	210.75–280.5	218.25–291	234–312	253.5–337.5	277.5–369.75	308.25–411
LEVEL XIV	92.25–165	100.5–175.5	108–183	115.5–190.5	123.75–198.75	131.25–206.25	146.25–221.25	162–237	192.75–267.75	223.5–297.75
LEVEL XIII	92.25–123	100.5–133.5	108–144	115.5–153.75	123.75–164.25	131.25–174.75	146.25–195	162–216	192.75–256.5	223.5–297.75
LEVEL XII	69.75–92.25	76.5–102	83.25–110.25	90–120	96.75–128.25	103.5–138	121.5–162	133.5–177	156.75–207.75	180–238.5
LEVEL XI	58.5–76.5	64.5–85.5	71.25–93.75	76.5–102	83.25–110.25	90–118.5	109.5–145.5	119.25–158.25	138.75–183.75	157.5–210
LEVEL X	38.25–51	42.75–56.25	46.5–61.5	50.25–66.75	54–72	57.75–76.5	73.5–97.5	81–108	96.75–128.25	111.75–148.5
LEVEL IX	23.25–30.75	27–36	30.75–40.5	34.5–45.75	38.25–51	42.75–56.25	57.75–76.5	65.25–87	81–108	96.75–128.25
LEVEL VIII	15.75–20.25	19.5–25.5	23.25–30.75	27–36	30.75–40.5	34.54–5.75	50.25–66.75	57.75–76.5	65.25–87	81–108
LEVEL VII	11.25–15	15.75–20.25	19.5–25.5	23.25–30.75	27–36	30.75–40.5	42.75–56.25	50.25–66.75	57.75–76.5	65.25–87
LEVEL VI	9–10.5	11.25–15	15.75–20.25	19.5–25.5	23.25–30.75	27–36	34.5–45.75	42.75–56.25	50.25–66.75	57.75–76.5
LEVEL V	4.5–9	9–10.5	9.75–12.75	11.25–15	16.5–21.75	24.75–32.25	30.75–40.5	38.25–51	46.5–61.5	54–72
LEVEL IV	2.25–6.75	4.59	9–10.5	9.75–12.75	11.25–15	16.5–21.75	24.75–32.25	32.25–42.75	39.75–52.5	47.25–63
LEVEL III	0.75–2.25	2.25–6	3–9	6.75–9	9–12	12.75–16.5	16.5–21.75	24.75–32.25	32.25–42.75	38.25–51
LEVEL II	0–67.5 days	1.5–4.5	2.25–6.75	3–9	9–10.5	10.5–13.5	12.75–16.5	16.5–21.75	24.75–32.25	32.25–42.75
LEVEL I	0–45 days	0–67.5 days	1.5–3.75	1.5–4.5	2.25–6	3–9	9–10.5	10.5–13.5	12.75–16.5	16.5–21.75

Source: Reprinted with permission, copyright © 2012 State of Washington/John C. Steiger, PhD.

they have broad discretion to disagree with the guidelines and to impose what they believe are reasonable sentences—even if the guidelines call for different sentences. Both cases—*Gall v. United States*[28] and *Kimbrough v. United States*[29]—were decided by the same 7–2 margin and chided federal appeals courts for failing to give district judges sufficient leeway. Further, in 2017, by a vote of 7–0, the U.S. Supreme Court struck down an argument that the federal sentencing guidelines are too vague, again noting that these are not strict guidelines, that they merely provide advice to judges who use their own discretion when determining appropriate sentences.[30]

State-Level Sentencing Guidelines

Several states have enacted sentencing guidelines. As an example, Table 9.1 shows the sentencing grid used by the state of Washington, which has developed a sophisticated and objective sentencing tool called *The Adult Sentencing Guidelines Manual*.[31] This manual provides comprehensive information on adult felony sentencing as set forth under state law, identifying the seriousness level of the offense and "scoring" the offender's criminal history. The seriousness of crimes listed on the vertical axis ranges from Level I (which includes offenses such as simple theft, malicious mischief, attempting to elude a pursuing police vehicle, and possessing stolen property) to Level XVI (aggravated murder, which includes first-degree murder with one or more of a number of aggravating circumstances). The offender score listed on the horizontal axis ranges from 0 to 9+, and it is determined based on five factors: (1) the number of prior convictions; (2) the relationship between any prior offense(s) and the current offense of conviction; (3) the presence of other current convictions; (4) the offender's custody status at the time the crime was committed; and (5) the length of time between offenses.

Victim impact statements: information provided prior to sentencing by the victims of a crime (or, in cases of murder, the surviving family members) about the impact the crime had on their lives; allowed by the U.S. Supreme Court.

Boston Globe/Getty Images

Victim impact statements are increasingly common in the sentencing phase of criminal cases, where the focus has traditionally been on the offender rather than the victims and their loved ones. These statements help judges and juries to more fully understand the impact of a crime and what punishment is appropriate.

VICTIM IMPACT STATEMENTS

Victim impact statements are written or oral information provided in court—most commonly at sentencing—and at offenders' parole hearings concerning the impact of the crime on the victim and the victim's family. These statements generally inform the court of the crime's financial, emotional, psychological, and/or physical impact on their lives. They provide a means for the court to refocus its attention on the human cost of the crime, as well as for the victim to participate in the criminal justice process. The right to make an impact statement is generally available not only to the victim, but also to survivors of homicide victims, the parent or guardian of a minor victim, and a person representing an incompetent or incapacitated victim. Victims can present oral or written impact statements, with some states allowing victims to submit video- or audio-taped statements. Some states allow child victims to present drawings to explain how the crime affected them.[32]

Victim impact statements were upheld in 1991 by the U.S. Supreme Court, in *Payne v. Tennessee*.[33] Payne was convicted of two counts of murder for stabbing to death a mother and her 2-year-old daughter and also wounding her 3-year-old son. During the sentencing hearing, the boy's grandmother described how the killings affected the surviving grandson. On appeal, the Supreme Court held that the victim has a right to be heard, and quoted from a 1934 opinion by Justice Benjamin Cardozo: "Justice, though due to the accused, is due to the accuser also."[34] In 2016, the U.S. Supreme Court reiterated that victim impact statements concerning how the crime has impacted the individual are permissible, but that statements containing victims' opinions concerning the crime, defendant, or appropriate sentence violate the Eighth Amendment.[35]

Marsy's Law: Given that Marsy's Law focuses on victim's rights, you might be surprised to learn that the American Civil Liberties Union's political director is against the national campaign to adopt these laws. Jeanne Hruska argues that the law is "poorly drafted and a threat to existing constitutional rights."[36] She notes that our constitutional rights (at both the national and state levels) protect individuals against government abuse, and that defendants' rights apply only when the state might harm an accused—not a victim. In her view, the law "pits victims' rights against defendants' rights" and strengthens the power of the state. Consider how victim impact statements might strengthen the state's power and increase the chance of an overly harsh sentence. Do you think this is a significant concern, or do the benefits of victim impact statements outweigh potential costs?

CAPITAL PUNISHMENT

Because of its finality, debate concerning the death penalty has always been emotionally charged. Strong arguments have been put forth by those who favor and those who oppose **capital punishment**.

Capital punishment: a sentence of death, or carrying out same via execution of the offender.

Arguments For and Against

For a politician or political entity to try to fashion a state or national policy on the death penalty that would definitively appeal to most Americans would be nearly impossible. The U.S. Supreme Court has halted death sentences and then approved them, and several states have done likewise (largely because many convicted murderers on death row have been found to be innocent). Some studies indicate that the death penalty works to prevent crimes of murder, and other studies show the opposite. Americans themselves seem fickle in their views of whether or not killers should be put to death.

Figure 9.2 shows the percentages of Americans who favor and oppose the death penalty. The percentages of people in favor of the death penalty rose steadily for the most part from about 1967 to 1995; from that point, however, there has been an overall decline in such support, dropping from about 80 percent to less than 60 percent. Meanwhile, the percentage of people saying they are in opposition to the death penalty is about the same today as it was in 1937, with 41 percent of the population reporting that they are not in favor, despite a significant drop to about 16 percent in the mid-1990s.

Figure 9.2 /// Percentage of Americans Favoring and Opposing the Death Penalty

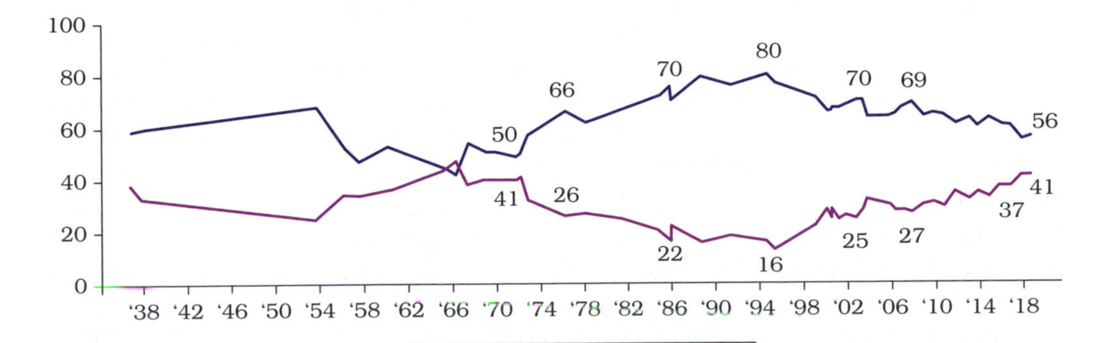

Source: Copyright © 2018 Gallup, Inc. All rights reserved. The content is used with permission; however, Gallup retains all rights of republication.

Arguments in Favor

People who support the death penalty often believe that it deters other people from committing murder, while others base their stand on theological grounds—the commandment "Thou shalt not kill." Others favor capital punishment because of the retribution it provides to family members and friends of the victim; "getting even" is a proper punishment in the eyes of many people, rather than some "rehabilitative" ideal. It is society's nod to *lex talionis*—"eye for an eye": "If you kill one of us, we will kill you." One thing the death penalty most certainly accomplishes, however, is the prevention of future murders by that particular offender. As James Q. Wilson noted, "Whatever else may be said about the death penalty, it is certain that it incapacitates."[37]

Does the existence of a death penalty deter individuals from committing murder? That is an important question, and one that is at the tip of the spear in this debate. The answer, unfortunately, is elusive, and it depends largely on which set of studies one looks at and gives credence to. Beginning in 2003, a number of studies were published that reportedly demonstrated the death penalty saves lives by acting as a deterrent to murder. One study that garnered a large amount of attention, conducted by Mocan and Gittings, analyzed 6,143 death sentences imposed in the United States between 1977 and 1997. Results indicated that each execution resulted in five fewer murders, and each commutation of a death sentence to a long or life prison term resulted in five additional homicides. Further, each additional removal from death row when one's sentence was vacated resulted in one additional homicide.[38] Another study, by professors at Emory University, using a panel data set of more than 3,000 counties from 1977 to 1996, found that each execution resulted in 18 fewer murders and that the implementation of state moratoria was associated with an increased incidence of murders.[39] Then, two studies by a Federal Communications Commission economist also supported the deterrent effect of capital punishment, finding that each additional execution, on average, resulted in 14 fewer murders[40] and that executions conducted by electrocution were the most effective at providing deterrence.[41] However, the findings of these studies were challenged by other criminologists.

Arguments in Opposition

Arguments against the death penalty include that it does not have any deterrent value, that it is discriminatory against minorities, that retribution is unfitting for a civilized society, and that it can (and does) claim the lives of innocent people.

At least two criminologists have identified serious flaws in the Mocan-Gittings study, which found that each execution resulted in five fewer murders. Richard Berk, using Mocan and Gittings's original data set, removed the Texas data, ran the model exactly as the original authors did for the other 49 states, and found that the deterrent effect disappeared.[42]

A second reexamination of the Mocan-Gittings study was conducted by Jeffrey Fagan, who, by modifying their measure of deterrence, also found that all the deterrent effects disappeared. Rather than prove that Mocan and Gittings erred in their assumptions, Fagan showed that small changes in their assumptions could produce wild fluctuations in their deterrence estimates.[43] Regarding the contention that the death penalty is discriminatory, many people would agree with a finding by the U.S. Government Accountability Office that there is "a pattern of evidence indicating racial disparities in the charging, sentencing, and imposition of the death penalty."[44] Amnesty International argues, furthermore, that "from initial charging decisions to plea bargaining to jury sentencing, African-Americans are treated more harshly when they are defendants, and their lives are accorded less value when they are victims. All-white or virtually all-white juries are still commonplace in many localities."[45] Following are other findings regarding the death penalty discrimination thesis:

- A report sponsored by the American Bar Association concluded that one-third of African American death row inmates in Philadelphia would have received sentences of life imprisonment if they had not been African American.

- A study of death sentences in Connecticut conducted by Yale University School of Law revealed that African American defendants receive the death penalty at three times the rate of white defendants in cases where the victims are white. In addition, killers of white victims are treated more severely than people who kill minorities, when it comes to deciding what charges to bring.
- A study released by the University of Maryland concluded that race and geography are major factors in death penalty decisions. Specifically, prosecutors are more likely to seek a death sentence when the race of the victim is white and are less likely to seek a death sentence when the victim is African American.[46]

Finally, of course, is the argument that innocent people can be—and have been—executed for crimes they never committed, which is discussed in the "Wrongful Convictions" section.

Key Supreme Court Decisions

In *Furman v. Georgia* (1972), by vote of 5–4, the U.S. Supreme Court, for the first time, struck down the death penalty under the cruel and unusual punishment clause of the Eighth Amendment. The decision involved not only Furman, convicted for murder, but also two other men (in Georgia and Texas) convicted for rape; juries at the trials of all three men had imposed the death penalty without any specific guides or limits on their discretion. The justices for the majority found this lack of guidelines or limits on jury discretion to be unconstitutional, as it resulted in a random pattern among those receiving the death penalty; one justice also felt that death was disproportionately applied to the poor and socially disadvantaged, and felt those groups were denied equal protection under the law.[47]

Four years after *Furman*, the Supreme Court rendered another major death penalty decision. A Georgia jury had found Troy Gregg guilty of armed robbery and murder and sentenced him to death. Gregg challenged his death sentence, claiming that it was, *per se*, a "cruel and unusual" punishment that violated the Eighth and Fourteenth Amendments. In a 7–2 decision, the Court held that a punishment of death did not violate the Constitution. Where a defendant has been convicted of deliberately killing another person, the careful and judicious use of the death penalty may be appropriate if employed carefully. The Court noted that Georgia's death penalty statute required a bifurcated proceeding (where the trial and sentencing are conducted separately), as well as specific jury findings as to the severity of the crime and the nature of the defendant and a comparison of aggravating and mitigating circumstances.[48] Another significant decision was rendered by the Supreme Court in 2005, in *Roper v. Simmons*, holding that the Eighth and Fourteenth Amendments forbid the execution of offenders who were under the age of 18 when their crimes were committed.[49] In addition, the Supreme Court has held that

- The Eighth Amendment prohibits the execution of persons who are mentally incompetent at the time of their execution.[50]
- The death penalty cannot be applied to adults who rape children (and, more broadly, that a state cannot impose the death penalty for a crime that did not result in the death of the victim, except for crimes committed against the state—i.e., espionage, treason).[51]
- To show their legal counsel was ineffective, defendants in capital cases must prove that the attorney's performance was less than reasonable (e.g., counsel did not present mitigating circumstances), and that there is a reasonable likelihood that this substandard performance changed the outcome of the trial.[52]
- People who are opposed to the death penalty cannot be automatically excluded from serving on juries in capital cases; however, people whose opposition is so strong as to "prevent or substantially impair the performance of their duties" (*Witherspoon v. Illinois*, 1968) may be removed from the jury pool during *voir dire* (preliminary examination).[53]

Methods of Execution

Here we discuss the methods of execution in use today in the United States. Figure 9.3 shows methods of execution, by state.

Although death by lethal injection has been widely thought to be much more humane than other methods, several challenges to this method of execution have been raised in the U.S. Supreme Court in recent years (and botched executions have ignited the out-of-court debate). Such challenges typically claim that the drugs used in the executions cause extreme and unnecessary pain, while masking the pain being experienced by the inmate, and thus violate the Eighth Amendment's ban on cruel and unusual punishment. In *Baze v. Rees* (2008), the U.S. Supreme Court held that the three-drug "cocktail" used by 35 states and the federal government did not violate the Eighth Amendment.[54] Chief Justice John Roberts observed that a method of execution would violate the Constitution only if it was "deliberately designed to inflict pain." Two years earlier, Clarence Hill, an inmate on Florida's death row, challenged the use of lethal injection, *per se*, as causing unnecessary pain contrary

Figure 9.3 /// Method of Execution, by State, 2019

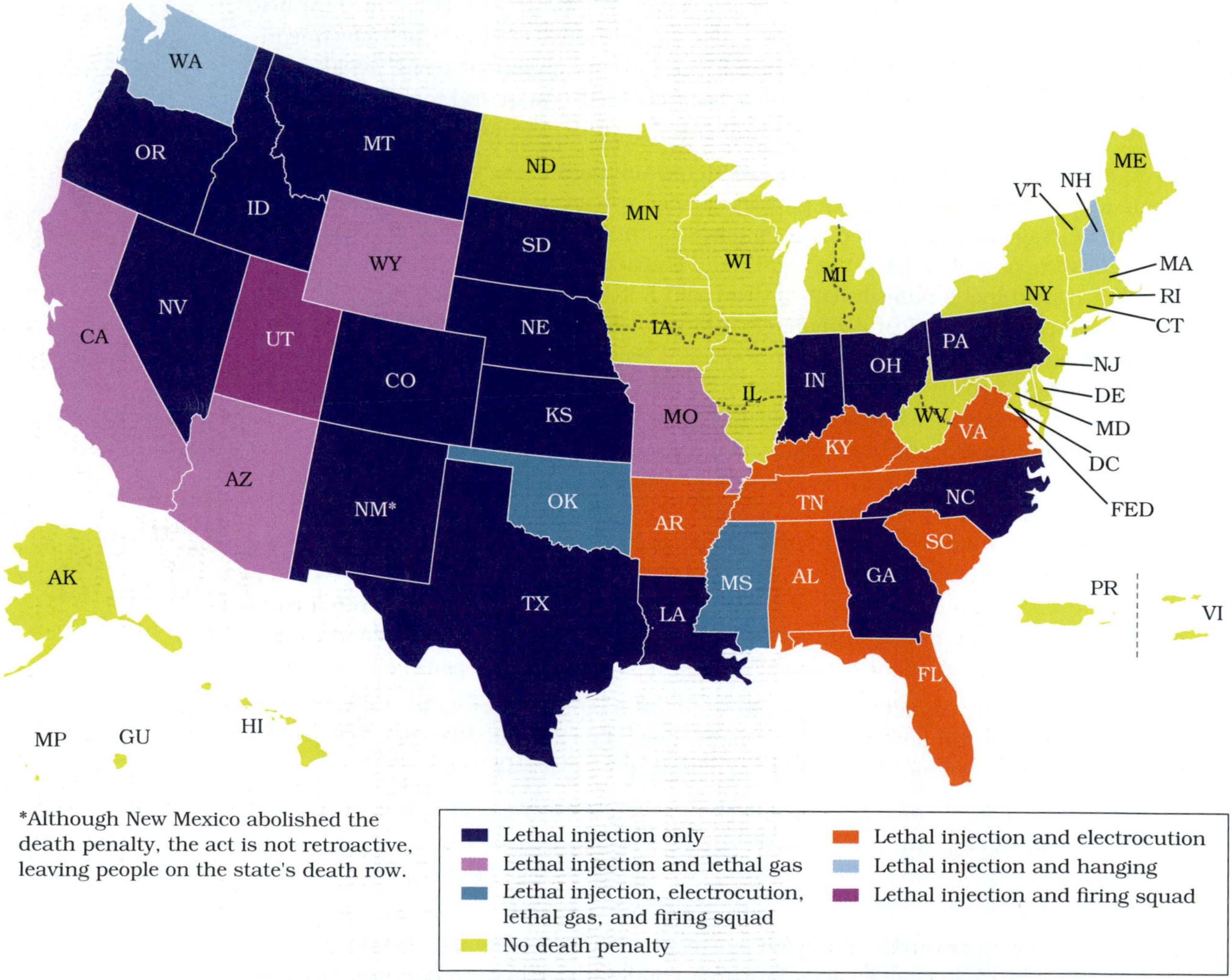

*Although New Mexico abolished the death penalty, the act is not retroactive, leaving people on the state's death row.

- Lethal injection only
- Lethal injection and lethal gas
- Lethal injection, electrocution, lethal gas, and firing squad
- No death penalty
- Lethal injection and electrocution
- Lethal injection and hanging
- Lethal injection and firing squad

Source: Death Penalty Information Center, Methods of Execution, http://www.deathpenal tyinfo.org/executions/methods-of-execution.

to contemporary standards of decency; however, by a vote of 5–4, the U.S. Supreme Court denied a stay of execution (Justice Antonin Scalia noted that "lethal injection is much less painful than hanging"), and Hill was executed in September 2006.[55]

Wrongful Convictions: Rethinking the Death Penalty

Being convicted of a crime is no guarantee that one is actually guilty as charged. It means simply that a jury or judge was persuaded by the state's case against the defendant or, in the case of plea bargaining, the defendant was persuaded to plead guilty to avoid trial. Today, we know that many innocent people have nonetheless been convicted and, in some cases, sentenced to death, until their names were cleared—they were exonerated—by DNA evidence. As of May 2019, the Death Penalty Information Center reports that 165 death row inmates in the United States have been exonerated since 1973, including at least 20 people previously facing death whose innocence was proved through DNA evidence.[56]

These **exonerations** and problems with executions have not gone unnoticed. Since 2000, there has been an average of five death row exonerations a year.[57] Further, 40 states have not carried out an execution in the past five years, and 8 states have abolished the death penalty since 2004.[58] The U.S. Supreme Court held in 2002 that the execution of the intellectually disabled was unconstitutional[59] and that the finding of an aggravating factor justifying a death sentence must be made by a jury, not merely by the sentencing judge.[60] Then, in 2005, the Court limited the death penalty even further, ruling that executing offenders who were juveniles at the time of their crimes is cruel and unusual punishment.[61] In 2019, the U.S. Supreme Court agreed to hear a death penalty case involving mitigating factors in the case of Arizona prisoner James Erin McKinney.[62] The outcome of this case may have significant ramifications for the application of the death penalty across the United States.

Exoneration: to absolve someone of criminal blame, or find someone not guilty.

Recent Trends in Federal Executions

In July 2019, U.S. Attorney General William Barr ordered the Bureau of Prisons to schedule executions for five federal death row inmates. All five of these inmates had been convicted of crimes involving the murder of children.[63] Barr's Justice Department directive followed a 16-year period in which the federal government had not carried out capital punishment sentences for those awaiting execution.

The U.S. federal government has executed only three inmates since 1988: two in 2001 and one in 2003.[64] All three were executed using lethal injection. The five inmates subject to Barr's order for capital punishment will likely be executed using a new protocol that requires the use of a single drug, pentobarbital, instead of the three-drug cocktail used in previous federal executions. In ordering the executions, Barr said, "Congress has expressly authorized the death penalty through legislation adopted by the people's representatives in both houses of Congress and signed by the President . . . . Under Administrations of both parties, the Department of Justice has sought the death penalty against the worst criminals, including these five murderers, each of whom was convicted by a jury of his peers after a full and fair proceeding. The Justice Department upholds the rule of law—and we owe it to the victims and their families to carry forward the sentence imposed by our justice system."[65]

AGGRAVATING AND MITIGATING CIRCUMSTANCES

In passing sentence, judges (and, in death penalty cases, the jury) will look at factors other than the crime itself. They consider the manner in which the crime was committed in order to determine whether or not they should increase or decrease the severity of the punishment. In other words, the judge or jury might "look behind" the crime to see if, for example, the victim was tortured prior to being killed or if the defendant played a minor role in the offense. These are termed aggravating and mitigating circumstances or factors.

Aggravating circumstances can include use of a weapon, an especially heinous or cruel crime, commission of murder for hire, or the offender's being a peace officer engaged in official duties.[66] *Mitigating circumstances* include little or no prior criminal history, the offender's acting under duress or under the influence of mental illness or extreme emotional disturbance, or the offender's being young.[67] Under the Supreme Court's decision in *Gregg v. Georgia*, discussed earlier, the judge will also instruct the jury members that they may not impose the death penalty unless they first determine the existence of one or more statutory aggravating circumstances beyond a reasonable doubt, as well as determine that the aggravating circumstances outweigh the mitigating circumstances beyond a reasonable doubt.

CRIMINAL APPEALS

In an appeal, one who has been convicted for a crime attempts to show that the trial court made a legal error that affected the decision in the case, or that he or she had ineffective counsel or that due process was violated in some other way. The defendant now asks a higher (appellate) court to review the transcript of the case for such errors—and possibly to have the conviction overturned or be granted a retrial. In addition, the defendant may contest the trial court's sentencing decision without challenging the underlying conviction. Then, if the appellate court grants the appeal, it may reverse the lower court's decision in whole or in part. However, if the appellate court denies the appeal, the lower court's decision stands.[68]

Article I, Section 9, of the Constitution speaks only briefly and indirectly to criminal appeals, saying that the privilege of the *writ of habeas corpus* (discussed in Chapter 9 in the case of *Kibbe v. Henderson*) shall not be suspended, "unless when in Cases of Rebellion or Invasion the public Safety may require it." The name of this writ, often referred to as "the great writ," is Latin for "you have the body." *Habeas corpus* is the inmate's means of asking a court to grant a hearing to determine whether or not he or she is being held illegally.

AP Photo/Gerald Herbert

Henry James served nearly 30 years in Louisiana's infamous Angola prison for a sexual assault he did not commit. He was convicted in 1982 largely on the basis of an incorrect eyewitness identification—the cause of approximately 75 percent of all wrongful convictions. James was exonerated through DNA testing and finally released from Angola in 2011.

Convicted persons who are indigent (poor) are entitled to free legal counsel for their initial appeal[69] and a free copy of their trial transcript.[70] (However, the Supreme Court has held that prison inmates are not entitled to free legal counsel for subsequent "discretionary" appeals.[71]) Furthermore, as is noted in Chapter 10, prison inmates also have a right to access to law libraries, and many law schools have defender clinics in which law students represent state and federal prisoners in appellate and postconviction litigation in state and federal courts. This is in addition to prison "writ-writers"—inmates who over time develop considerable expertise in constitutional law and means of filing writs and petitions.

Not only are inmates able to challenge their criminal conviction on a variety of grounds (the evidence introduced against them; the police arrest, search, and seizure of their person and property; the judge's instructions to the jury; and so on), but after being incarcerated they may attempt to obtain what are called postconviction remedies—making what are termed "collateral attacks." These lawsuits are civil in nature (unlike the original appeal of their conviction) and challenge, for example, the conditions of their confinement. Such cases often list the prison warden as the respondent and can involve the inmate's filing a *writ of habeas corpus*, typically in a federal court having jurisdiction over his or her place of incarceration.

/// IN A NUTSHELL

- Historically, we have punished people for the following reasons: retribution, deterrence, incapacitation, and rehabilitation. The U.S. Constitution speaks only briefly but forcefully regarding the use of punishment. The Eighth Amendment provides that incarceration will not involve "cruel and unusual punishment" and that fines will not be excessive.
- Prosecutors can influence the sentencing decision by agreeing to engage in plea negotiations concerning the number of charges filed or the maximum penalty the judge may impose, by explaining that the accused was very cooperative with the police and/or remorseful for the crime, and so on. Defense attorneys can seek to obtain the lightest possible sentence, or other alternatives to sentencing, or they can emphasize such things as the defendant's minor involvement in the crime.
- The seriousness of the offense is the most important factor in determining the sentencing received for an offense, followed by the defendant's prior criminal record.
- Philosophies of crime and punishment have changed significantly since the late 1700s; today people are calling for longer sentences for career criminals and violent offenders. Legislators, judges, and officials have responded with determinate sentencing laws, three-strikes laws, mandatory sentencing laws (e.g., doubling one's sentence for a crime committed with a weapon), and so forth.
- Under *determinate* sentencing, convicted offenders are sentenced for a fixed term, such as 10 years, with no opportunity for a paroling authority to make adjustments in time served. Conversely, in an *indeterminate* sentencing format, the convicted individual will be sentenced for a set range of time, such as 5–10 years, allowing for the length of sentence to be adjusted in response to the offender's positive responses to treatment and programs.
- Becoming discontented with the sentencing process and the rising crime rate throughout the nation, Congress abolished indeterminate sentencing at the federal level and created a determinate sentencing structure through the federal Sentencing Reform Act. The U.S. Supreme Court found that the guidelines violated the Constitution by allowing judicial, rather than jury, fact-finding to form the basis for the sentencing. The Court said the guidelines are to be merely advisory and not mandatory. Thus, the guidelines are a resource a judge can look at but may choose to ignore.
- Victim impact statements are written or oral information provided in court—most commonly at sentencing—and at offenders' parole hearings, concerning the impact of the crime on the victim and the victim's family. These statements generally inform the court of the crime's financial, emotional, psychological, and/or physical impact on their lives.
- Supporters of the death penalty often believe that it deters other people from committing murder, while others base their stand on theological grounds. Others favor capital punishment because of the retribution it provides to family members and friends of the victim. Arguments against the death penalty include that it does not have any deterrent value, that it is discriminatory against minorities, that retribution is unfitting for a civilized society, and that it can (and does) claim the lives of innocent people. There is some, but not unanimous, research in support of the deterrence argument.
- In 1972, the U.S. Supreme Court struck down all death penalty laws as being cruel and unusual punishment, due to the manner in which the sanction was being administered. The Court later approved the death sentence in concept. Today, lethal injection is the method of execution authorized in a majority of the states.
- Since 1973, a total of 165 persons on death row in the United States have been exonerated. According to the Death Penalty Information Center, DNA evidence led to at least 20 of those death row exonerations.
- Convicted, indigent persons are entitled to free legal counsel for their initial appeal as well as a free copy of their trial transcript. However, the Supreme Court later held that prison inmates are not entitled to free legal counsel for subsequent "discretionary" appeals.

/// KEY TERMS & CONCEPTS

Review key terms with eFlashcards at **edge.sagepub.com/peakbrief**.

Capital punishment 211
Deterrence 198
Exoneration 215
Incapacitation 199
Punishment (and its purposes) 198
Rehabilitation 199
Retribution 198
Sentencing guidelines 207
Victim impact statements 210

/// REVIEW QUESTIONS

Test your understanding of chapter content. Take the practice quiz at **edge.sagepub.com/peakbrief**.

1. How would you describe the four goals of punishment? Which one of them do you believe works the best? Which goal or function is now predominant in our society?
2. What are the factors that influence the degree—and harshness—of the punishment that a convicted person will receive?
3. How would you delineate the different philosophies regarding crime and punishment that evolved from the colonial era to today? How did prison construction change in accordance with those changes in punishment models?
4. What forms of punishment used around the world would you point to that are clearly excessive in terms of the offenses committed? Explain your answer.
5. What are the differences between, and purposes of, both determinate and indeterminate sentences? Which is likely used when the crime control model or due process model is more predominant in a community?
6. How would you explain the rationale for, and operation of, the federal sentencing guidelines?
7. What is the legal basis for victim impact statements? How do such statements work, and for what purpose?
8. What are the fundamental arguments for and against capital punishment? What did the Supreme Court say about capital punishment in *Furman* and *Gregg?*
9. What are the prevailing methods of execution in use today?
10. What changes have been brought by DNA with regard to the death penalty?
11. What are examples of both aggravating and mitigating circumstances, and how do they apply to sentencing decisions? To the death penalty?
12. What rights are possessed by a convicted person regarding access to legal counsel, trial transcripts, and law libraries?

/// LEARN BY DOING

1. Your state's governor is considering a moratorium on all executions because of DNA and death row exonerations. Knowing you are a criminal justice student, you are asked by a state senator to prepare a pro-con paper concerning the benefits and issues involved with doing so, and of DNA in general. How would you respond? Include in your response an assessment of the deterrent effects of capital punishment laws.
2. Assume your criminal justice professor has assigned you to read a journal article by Landy and Aronson titled "The Influence of the Character of the Criminal and His Victim on the Decisions of Simulated Jurors," published in the *Journal of Experimental Social Psychology* 5 (1969), pages 141–152. You find the article at https://www4.uwsp.edu/psych/s/389/landy69.pdf. Read Experiment II, including the instructions and case study (involving an incident with both an attractive victim and an unattractive victim) as given to university sophomores, as well as the experiment's results and discussion, on pages 146–151. Summarize and explain the above in written form.
3. Your criminal justice class is to debate the following: "RESOLVED: Deterrence is lost for the general public when an inmate remains on death row a dozen or more years." Plan how you would respond on both the pro and con sides of the debate.

/// STUDY SITE

SAGE edge™

Get the tools you need to sharpen your study skills. SAGE edge offers a robust online environment featuring an impressive array of free tools and resources.

Access practice quizzes, eFlashcards, video, and multimedia at **edge.sagepub.com/peakbrief**

CORRECTIONS

PART 4

This part includes two chapters and examines many aspects of correctional organizations and operations.

Chapter 10 examines federal and state prisons and local jails in terms of their evolution and organization, inmate population trends (including mass incarceration) and classification, and some technologies.

Chapter 11 reviews community corrections and alternatives to incarceration: probation, parole, and several other diversionary approaches. Included are discussions of the origins of probation and parole, functions of probation and parole officers, and several intermediate sanctions (e.g., house arrest, electronic monitoring).

©istockphoto.com/josefkubes

CHAPTER 10

PRISONS AND JAILS

Structure and Function

Men who had been in prison as much as five years still knew next to nothing on the subject. It probably took a decade behind bars for any real perception on the matter to permeate your psychology and your flesh.

—Jack Henry Abbott, *In the Belly of the Beast: Letters From Prison*[1]

I can think of nothing, not even war, that has brought so much misery to the human race as prisons.

—Clarence Darrow, 1936

LEARNING OBJECTIVES

As a result of reading this chapter, you will be able to

1. Describe correctional populations, employment, expenditures, and institutional types
2. Review some of the factors that contribute to recent declines in prison and jail populations
3. Explain the basic structure and function of state prisons
4. Explain the basic structure and function of the federal prison system
5. Describe how supermax prisons function, how they differ from other prisons, and findings concerning their effects on inmates and constitutionality
6. Define and review the operation of private, for-profit prisons
7. Review how local jails are organized and constructed, including the new generation jail
8. Describe some of the technologies now in use in corrections, with particular emphasis on methods for halting drone and cell phone use by inmates

ASSESS YOUR AWARENESS

Test your basic knowledge of prisons and jails by responding to the following eight true-false items; check your answers after reading this chapter.

1. Factors influencing the recent (i.e., in the past decade, roughly) decline in prison and jail populations include reductions in violent crimes, greater use of alternatives to incarceration, and the early release of thousands of nonviolent offenders.
2. A person who is convicted of committing a murder will likely be forced to serve a lengthy sentence in a local jail.
3. The general mission of correctional institutions is to securely hold criminals while providing them with opportunities to become productive and law-abiding citizens.
4. There are basically two custody levels of prisons: maximum and minimum.
5. Supermax prisons are so named because they offer the maximum amount of freedom and programming permitted by the courts.
6. Today, very few prison inmates, and no jail inmates, are involved in productive work or treatment programs.
7. Owing to the latest technologies used by prison and jail personnel, inmate use of drones and cell phones has been eradicated.
8. The detention of juveniles is typically a last resort, but there are many types of secure and nonsecure approaches available.

<< Answers can be found on page 293.

There has probably never been a more pervasive, universal push for prison and jail reform than at the present time. The United Nations (UN), the U.S. Congress, and even a U.S. president's son-in-law (Jared Kushner) are involved in a growing movement to reform the manner in which our correctional institutions function. Because of the concerns that people and organizations have about the effects of incarceration on human rights, society in general (and persons living in poverty, in particular), the personal health of inmates, and governmental budgets, there is certainly ample room for concerted action. The UN Office on Drugs and Crime has delineated four specific "areas of work":

- Pretrial detention (e.g., Is it overused? Are detainees being abused?)
- Prison management (Is it in line with the rule of law with respect to individuals' human rights? Is imprisonment being used to prepare individuals for life outside prison following release?)

- Alternative measures and sanctions (Can more community-based sanctions be used?)
- Social reintegration (Are noncustodial sanctions—instead of isolation from society—and purposeful activities and programs in prisons, being considered and employed?)

Of course, the provision of adequate health care for detainees cuts across all of these four domains.[2]

As you read this chapter, consider if or where prison reform might or should be undertaken, as well as where more human and financial resources might be needed to bring about those reforms.

INTRODUCTION

What are the differences in definitions, missions, structures, and functions of prisons and jails and in the duties of persons working in them? How are decisions made concerning the type of institution to which one is sent? Are more people incarcerated when the economy takes a downturn? These are valid questions to ask, particularly given the expense associated with institutionalizing offenders rather than having them remain in the community.

In his classic 1961 book, *Asylums*, Erving Goffman described life inside what he termed "total institutions," or those places "organized to protect the community against what are felt to be intentional dangers to it: jail, penitentiaries, POW camps, and concentration camps."[3] Goffman said that total institutions share the following features:

- All aspects of life are conducted in the same place and under the same single authority.
- Each phase of the member's daily activity is carried on in the immediate company of a large batch of others.
- All phases of the day's activities are tightly scheduled.
- The various enforced activities are brought together in a single rational plan . . . to fulfil the official aims of the institution.[4]

Goffman appears to have captured the essence of our correctional organizations, which are typically viewed as the end result of one's movement through the criminal justice system (as described and illustrated in Chapter 1). However, it might well be argued that the correctional process begins at the point of one's *arrest*, when the individual is incarcerated in jail awaiting trial and official attempts are initiated to identify and change his or her criminal tendencies.

David R. Frazier Photolibrary, Inc./Alamy

Inmates watch television in the recreational area of a county jail. Jails differ from prisons in that they are typically used as a short-term, temporary holding facility for persons recently arrested and awaiting trial.

In any case, corrections comprises agencies and programs that are responsible for carrying out the sentences and punishment that the courts have administered to those who have been accused, tried, and convicted for their criminal acts. Keep in mind that most of the work of corrections is accomplished not by locking people away but, rather, in the community, where offenders serve terms of probation or parole (see Chapter 11).

Unfortunately, most of what the public "knows" about prisons and jails is probably based on Hollywood's stylized (and often very unrealistic) portrayal in movies such as *The Shawshank Redemption, The Green Mile, Escape From Alcatraz, Cool Hand Luke, The Longest Yard,* and *The Great Escape*. These and other such portrayals of prison and jail life typically show the administrators and their staff being cruel, bigoted, corrupt, and morally base.

One truth, however, is that, for many members of the prison population, incarceration is anything but punishing and instead offers them an overall improvement in lifestyle and conditions. This chapter addresses that observation.

Although correctional populations began to decline in 2011 and continue to decline, corrections remains a boom industry, with about 6.6 million Americans under some form of correctional supervision.[5] This chapter first considers some reasons for the decline in correctional populations, and then the focus shifts to correctional agencies as organizations, including some demographic and cost information, their mission, and purposes of inmate classification systems. After examining local jails, we look next at the state prison organization, supermax prisons, and the federal prison system. Finally, we consider some of the good and bad uses of technologies as they concern correctional institutions, the latter of which includes inmates' use of drones and cell phones.

CORRECTIONAL FACILITIES AS ORGANIZATIONS

Like police and courts organizations, the correctional component of the U.S. criminal justice system also represents organizations—agencies composed of elements that are made up of collective functions and that contribute to their overall mission.

Distinguishing Jails and Prisons

The words *jail* and *prison* are often used interchangeably—and erroneously, as there are major differences between the two. The major differences between whether someone is sentenced to jail or prison concern the nature of the crime and the length of the sentence to be served. A **jail** is a short-term, often temporary holding facility for persons recently arrested and awaiting trial, and who often are unable to pay bond or bail for their release; or they might be serving short sentences for misdemeanors, generally one year's duration or less.

Jail: a facility that holds persons who have been arrested for crimes and are awaiting trial, persons who have been convicted for misdemeanors and are serving a sentence (up to a year in jail), federal offenders, and others.

Prison: a state or federal facility housing long-term offenders, typically felons, for a period greater than one year.

Conversely, **prisons** are designed for longer confinement. The majority of convicted felons serve their sentences in a prison. People convicted of committing federal crimes are typically sentenced to federal prisons, and those who break state laws go to state prisons.

Another important distinction concerns who administers the facility. Jails are generally run by a county sheriff's office, whereas prisons are operated by state or federal governments.[6]

Spencer Weiner/Los Angeles Times/Getty Images

Being under court order to reduce its crowded prisons, the state of California has done so significantly since May 2011.

Inmates, Employment, and Expenditures

As shown in Figure 10.1, a little more than 1.5 million Americans are now held in state and federal prisons.[7] After increasing annually until 2009, that population has been declining overall. In addition, about 740,000 are housed each day in local jails.[8] Furthermore, prisons hold about 2,750 persons on death rows in the jurisdictions allowing capital punishment (30 states and the federal government).[9]

Federal, state, and local governments combined spend more than $87 billion per year on correctional institutions (about 90 percent of which is spent at the state and local levels).[10]

Figure 10.1 /// Total Population Under the Supervision of U.S. Adult Correctional Systems, 2006–2016

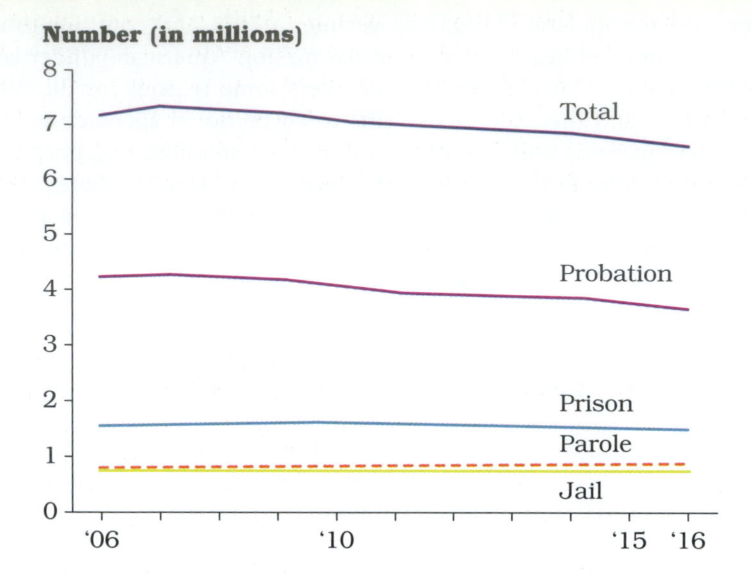

Note: Estimates may not be comparable to previously published Bureau of Justice Statistics reports because of updated information or rounding

Sources: Bureau of Justice Statistics, Annual Probation Survey, Annual Parole Survey, Annual Survey of Jails, and National Prisoner Statistics program, 2006–2016; see also Fig. 1 at p. 1 of Bureau of Justice Statistics, *Correctional Populations in the United States,* 2016, https://www.bjs.gov/content/pub/pdf/cpus16.pdf.

The Bureau of Labor Statistics estimates that about 468,000 individuals in the United States are employed as correctional officers.[11]

Why the Decline in Correctional Populations?

What factors have led to the recent reductions in prison and jail populations? Several reasons are now evident: First, in 2015 the federal prison system began releasing thousands of nonviolent drug offenders (and in federal prisons, a little more than half of all inmates are drug offenders); and second, a number of states are now seeking to save high costs of incarceration by enacting prosecution and sentencing policies and changes in laws to reduce prison populations. As an example, California's Proposition 47, approved by voters in 2014, retroactively reduced some drug and property crimes from felonies to misdemeanors. In addition, some states have offered expanded substance abuse treatment programs, established specialty courts, and spent more money on reentry programs aimed at reducing recidivism.[12]

Furthermore, incarceration rates have fallen slightly each year since 2008 due to sharp declines in violent and property crime rates, the former by about 50 percent since 1993 and the latter by as much as 70 percent.[13]

Another aspect of incarceration affecting local jail—and eventually prison—populations that is advancing nationally is that of money bail reform. Here again, California was at the forefront by passing legislation in August 2018 to eliminate money bail; several other states and local jurisdictions have followed and undertaken some form of bail reforms. Indeed, an impressive coalition of organizations (including the American Bar Association)[14] supports reform of the money bail system that often requires people who have been arrested, but not convicted, to spend weeks, months, or years in jail awaiting trial simply because they cannot afford to post bail.[15] This reform will have an impact on prison populations, as studies show that for felony cases, the conviction rate increases from 50 percent to 85 percent for those defendants jailed pretrial who cannot make bail.[16] Indeed, it has been argued that what is termed mass incarceration (discussed in the next section) cannot end without reforming the money-bail system.[17] With respect to community-based supervision of offenders (as opposed to incarceration), about 4.6 million adults are now supervised in the community, with about 3.7 million of these adults being on probation and the remainder on parole.[18] In short, expanded use of several of the alternative-to-incarceration approaches to be discussed in Chapter 11 (e.g., probation and parole, house arrest, electronic monitoring, shock incarceration, and day reporting centers) are having a major impact.

Mass Incarceration in America: Time to Rethink and Reform?

Each year, a new crop of youths in their upper teens constitutes the majority of those arrested for serious crimes. As these offenders are arrested and removed from the crime scene, a new crop replaces them: "The justice system is eating its young. It imprisons them, paroles them, and rearrests them with no rehabilitation in between," according to prison researcher Dale Secrest.[19] Still, large-scale, long-term imprisonment unquestionably keeps truly serious offenders behind bars, preventing them from committing more crimes.

Although corrections populations have been declining for several years, many observers believe much work and analysis need to be done in terms of who is being incarcerated. David Garland, a sociologist at New York University, coined the term ***mass incarceration*** in 2001, to characterize the extraordinarily high incarceration rates (now at more than two million people) in the United States.[20] Together, the incarcerated population would constitute the fourth largest city in the nation (and if offenders on probation and parole were included, the population would be second only to New York City).

Mass incarceration: a term generally referring to what is perceived as the United States's disproportionately high rates of imprisonment of young African American men; some believe it deters crime and incapacitates offenders, while others say that it weakens poor families and keeps them socially marginalized.

Michelle Alexander, Ohio State University law professor and civil rights advocate, wrote in *The New Jim Crow* (which spent 70 weeks on the *New York Times* best-seller list) that mass incarceration "emerged as a stunningly comprehensive and well-disguised system of racialized social control."[21] Alexander asserts that no other country imprisons as many of its racial or ethnic minorities. Alexander emphasizes that "nothing has contributed more to the systematic mass incarceration of people of color . . . than the War on Drugs." (*Note:* A prize-winning documentary film, *The House I Live In*, also explores the effects of U.S. drug policy on the nation's poor and minority communities.) Alexander predicts that, given current trends, one in three young African Americans will serve time in prison, and in some cities, more than half of all such men are currently under some form of correctional control.[22] In addition to the negative effects of one's coming out of prison with a felony record and typically few job skills, the additional effects of mass incarceration on such individuals include that for the rest of their lives, they can be denied the right to vote (nearly half of the states deprive voting rights during incarceration and for a period of time after), automatically excluded from juries, and legally discriminated against in employment, housing, and access to education and public benefits.[23] Furthermore, mass incarceration is said to cripple families (causing home instability and food insecurity) and communities, perpetuate poverty, and institutionalize a form of racial control.[24]

The Vera Institute of Justice has examined incarceration rates in *The New Dynamics of Mass Incarceration* and finds that the tendency to use state prison population figures (which, as noted earlier, have declined) as a sole indication of sentencing reform actually belies a more complex picture: In some states, the incarcerated population has shifted away from prisons and into local jails, so there has not been a reversal of mass incarceration. The report finds many jail populations are on the rise and that jails hold about 18 times as many people as state and federal prisons. Furthermore, while about a dozen states have increased their prison populations, they have also increased their jail populations; at least eight states hold more people in jails than in prisons, and four states continue to lock up people in both prisons and jails at all-time high levels. In sum, Vera's view is that an urgent need exists to rethink our approach to ending mass incarceration.[25]

©AP Photo/Nick de la Torre

Many people believe that, rather than using punishment alone, it is also important to try to rehabilitate offenders—that is, providing treatment programs to help them to recover from a criminal lifestyle or personality disorders, from substance abuse or addiction, and so on, in order to become productive citizens.

General Mission and Features

Correctional organizations are complex, hybrid organizations that utilize two distinct yet related management subsystems to achieve their goals: One is concerned primarily with managing correctional employees, and the other is concerned primarily with delivering correctional services to a designated offender population. The correctional organization, therefore, employs one group of people—correctional personnel—to work with and control another group: offenders.

The mission of correctional agencies has changed little over time. It is as follows: to protect the citizens from crime by safely and securely handling criminal offenders while providing offenders some opportunities for self-improvement and increasing the chance that they will become productive and law-abiding citizens.[26] An interesting feature of the correctional organization is that *every* correctional employee who exercises legal authority over offenders is a supervisor, even if the person is the lowest-ranking member in the agency or institution. Another feature of the correctional organization is that—as with the police—everything a correctional supervisor does may have civil or criminal ramifications, both for himself or herself and for the agency or institution. Therefore, the legal and ethical responsibility for the correctional (and police) supervisor is greater than it is for supervisors in other types of organizations.

Finally, it is probably fair to say there are two different philosophies concerning what a correctional organization should be: (1) a custodial organization, which emphasizes the caretaker functions of controlling and observing inmates, and (2) a treatment organization, which emphasizes rehabilitation of inmates. These different philosophies contain potential conflict for correctional personnel.

Punishment for Some, a "Step Up" for Others

Most Americans probably assume that sending offenders to prisons and jails—depriving them of their freedom of movement and many amenities while living under an oppressive set of rules—serves a useful purpose, will bring them to the "good life," and instills in them a desire to obey the laws and avoid returning to prison after their release. In fact, experts have said that such punishments will work only under the following two conditions: (1) if they injure the offender's "social standing by the punishment," and (2) if they make "the individual feel a danger of being excluded from the group."[27]

Unfortunately, however, this view overlooks two very important facts—facts that are perhaps a sad commentary on the kinds of lives being led by many people in the United States:

- Most serious offenders neither accept nor abide by those two conditions.
- Most incarcerated people today come from communities where conditions fall far below the living standards that most Americans would accept.[28] As stated by prison expert Joan Petersilia, the grim fact and national shame is that for many people who go to prison, the conditions inside are not all that different from (and might even be better than) the conditions outside.[29] For some members of our society, going to prison or jail—and obtaining "three hots and a cot" (three meals and a bed)—may actually represent an *increased* standard of living. Obviously, for those individuals, the threat of imprisonment no longer represents a horrible punishment and therefore has lost much of its deterrent power. When such people go to prison, they seldom feel isolated but are likely to find friends, if not family, already there.[30]

Practitioner's Perspective

Jail Administrator

Name: Mitch Lucas
Position: Jail Administrator
Location: Charleston, South Carolina

What is your career story? I came to the Charleston County Sheriff's Office to be the public information officer (PIO). In my role as PIO, I had to deal with the folks at the jail, and I was absolutely astonished at how motivated and dedicated people could be in a very important role like that, but nobody outside the jail knew what they did. Most people have no idea what goes on in a jail—they only know what they see in the media. And the people at the jail were just so incredible when it came to helping me in my job and doing theirs. I told the sheriff that if I ever had the opportunity, I'd like to work at the jail in some capacity. A few years later, he made me the jail administrator. I served as the jail administrator for seven and a half years, absolutely the best job I ever had, mainly because of the people—both inside the jail and outside in the field—with whom I worked.

What are some challenges and misconceptions you face in this position? The biggest paradigm shift I had to make when I became the jail administrator was to realize that the jail is the common denominator of the criminal justice system. Everyone who is arrested, goes to court, whoever is indicted and ends up going to prison, all pass through the jail. The jail's main purpose is to maintain inmate behavior, and keep the community safe by keeping the jail running smoothly, and keeping the people that are in the jail safe. That was a bit of a shift for me. Because you have to sort of remove yourself from whether they're guilty or not guilty, and just deal with the situation at hand, and that's providing an environment that is safe for everyone.

There are two valleys of misconception. One is from law enforcement. Law enforcement officers believe that people work in jails because they can't be cops. I felt the same way at one point, but when I got to meet jail workers and work with them, I realized that's not the case. The other misconception comes from the media. All other public safety aspects—law enforcement, firefighters, dispatchers—all of them are portrayed as heroes in most of the stories. But the corrections officer, whether it's at a jail or in prison, is typically portrayed as cruel and vicious.

What characteristics and skills are most helpful to succeed in this position? There are two really important skills to be successful in law enforcement corrections. First are communication skills. The person should be able to communicate in a variety of ways, and they must be able to listen, analyze information, and act on it. The other one is writing. Probably the number one problem in all areas for criminal justice is poor report writing. You have to be able to write effectively, and I encourage anybody that talks to me about getting into law enforcement corrections to take as many English and public speaking courses as they can.

Do you see any common trends in this position? We're seeing more and more that education plays a greater role because of everything with the Freedom of Information Act. Everything that's done in the jail is no longer in a vacuum, at both the state and federal level. If a reporter or family member asks to see a video of a particular inmate or particular situation, they can get it. So the field is no longer just what we used to call a lockup, where you put someone in a cell, lock the door, and walk away.

To learn more about Mitch Lucas' experiences as a Jail Administrator, watch the Practitioner's Perspective video in SAGE Vantage.

Furthermore, it appears that prison life is not perceived as being as difficult as it once was. Inmates' actions speak loudly in this respect: About two-thirds of today's parolees are rearrested within five years; evidently they believe the "benefits" of committing a new crime outweigh the costs of being in prison.[31] Finally, the stigma of having a prison record is not the same as in the past, because so many of the offenders' peers and family members also have done time. Imprisonment also confers status in some neighborhoods. To many people, serving a prison term is a badge of courage. It also is their source of food, clothing, and shelter.[32] Any discussion of jails and prisons should be prefaced with these facts concerning the inmates' world.

Classification of Inmates: A Cornerstone of Corrections

Classification (of inmates): an inmate security and treatment plan based on one's security, social, vocational, psychological, and educational needs while incarcerated.

If the prison or jail experience is to carry any benefit, classifying inmates into the proper levels of housing, programming, and other aspects of their incarceration must be accomplished so as to have an influence on their behavior, treatment, and progress while in custody—as well as for the general safety of inmates and staff. Correctional staff must make **classification** decisions in at least two areas: the inmate's level of *physical restraint* or "security level," and the inmate's level of supervision or *custody grade*. These two concepts are not well understood and are often confused, but they significantly impact a prisoner's housing and program assignments,[33] as well as an institution's overall security level.

The most recent development in classification is unit management, in which a large prison population is subdivided into several mini-institutions analogous to a city and its neighborhoods. Each unit has specified decision-making authority and is run by a staff of six, whose offices are on the living unit; this enables classification decisions to be made by personnel who are in daily contact with their inmates and know them fairly well.[34] Robert Levinson delineated four categories into which correctional officials classify new inmates:[35]

- *Security* needs are classified in terms of the number and types of architectural barriers that must be placed between the inmates and the outside world to ensure they will not escape and can be controlled. Most correctional systems have four security levels: supermax (highest), maximum (high), medium (low), and minimum (lowest).
- *Custody* assignments determine the level of supervision and types of privileges an inmate will have. A basic consideration is whether or not an inmate will be allowed to go outside the facility's secure perimeter, so some systems have adopted a fourfold array of custody grades—two inside the fence (one more restrictive than the other) and two outside the fence (one more closely supervised than the other).
- *Housing* needs were historically determined by an "assign to the next empty bed" system, which could place the new, weak inmate in the same cell with the most hardened inmate. A more sophisticated approach is known as internal classification, in which inmates are assigned to live with prisoners who are similar to themselves. This approach can involve the grouping of inmates into three broad categories: heavy—victimizers; light—victims; and moderate—neither intimidated by the first group nor abusers of the second.
- *Program* classification involves using interview and testing data to determine where the newly arrived inmate should be placed in work, training, and treatment programs. These programs are designed to help the prisoner make a successful return to society.

Justin Sullivan/Getty Images News/Getty Images

Physical fitness and sports activities are popular among inmates, and jails and prisons at all levels typically offer a variety of such opportunities.

In the past, most prison systems used a highly subjective system of classifying inmates that involved a review of records pertaining to the inmate's prior social and criminal history, test scores, school and work performance, and staff impressions developed from interviews. Today, however, administrators employ a much-preferred objective system that is more rational, efficient, and equitable. Factors used in making classification decisions are measurable and valid, and are applied to all inmates in the same way. Criteria most often used are escape history, any detainers (requests filed with the institution holding an inmate asking that the jurisdiction hold the prisoner for that agency, or notify the agency when release of the prisoner is about to occur), prior commitments, criminal history, prior institutional adjustment, history of violence, and length of sentence.[36]

STATE PRISONS AS ORGANIZATIONS

As noted earlier in this chapter, the mission of most prisons is to provide a safe and secure environment for staff and inmates, as well as programs for offenders that can assist them after release.[37] This section describes how **state prisons** are organized to accomplish this mission. First is a look at the larger picture—the typical organization of the central office within the state government that oversees *all* prisons within its jurisdiction—and then a look at the characteristic organization of an individual prison.

State prison: a correctional facility that houses convicted felons.

Prison organizational structures (see Figure 10.2) have changed considerably over time. Until the beginning of the 20th century, prisons were administered by state boards of charities, boards composed of citizens, boards of inspectors, state prison commissions, or individual prison keepers. Most prisons were individual provinces; wardens, who were given absolute control over their domain, were appointed by governors through a system of political patronage. Individuals were attracted to the position of **warden** because it carried many

Warden: the chief administrator of a federal penitentiary or state prison.

Figure 10.2 /// Organizational Structure for a Medium-Security Prison

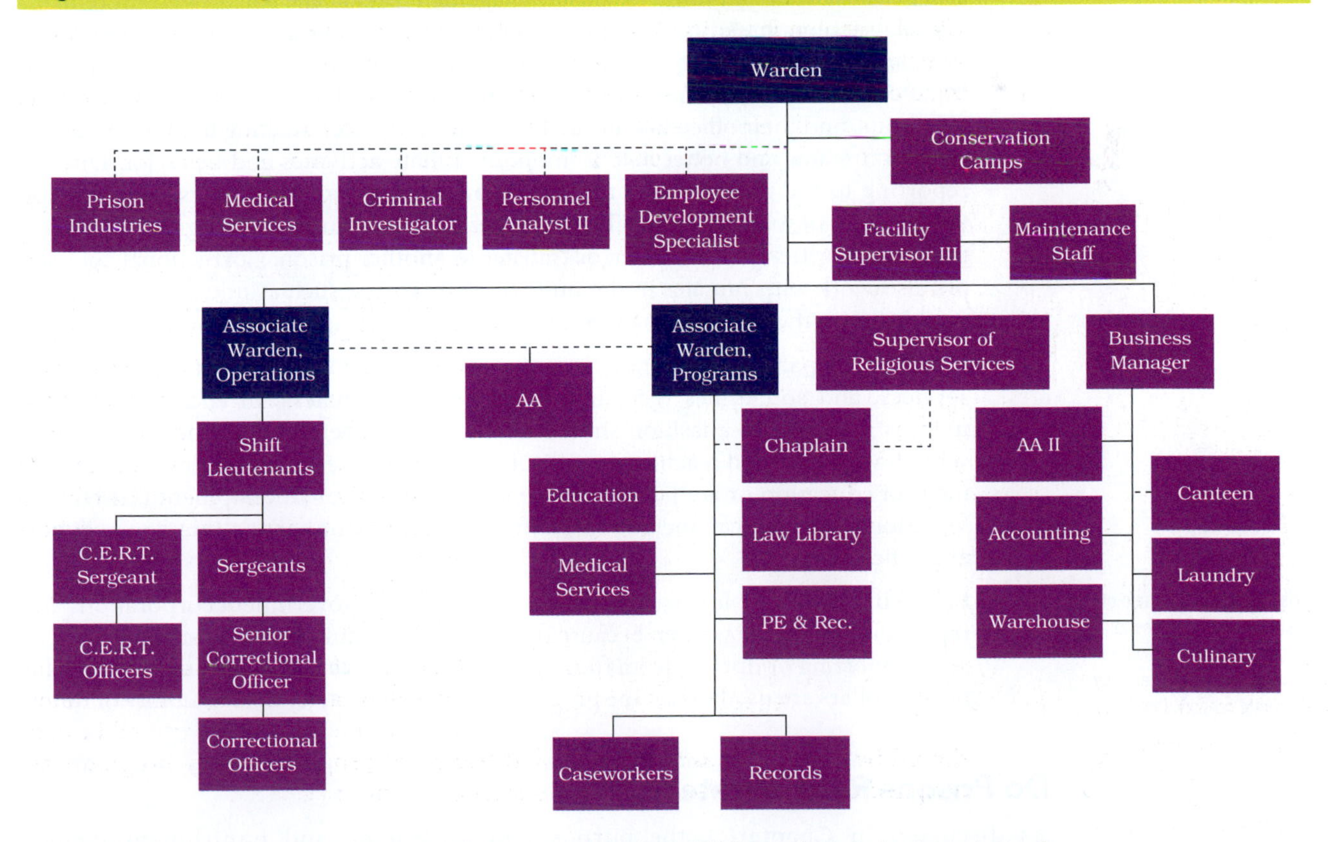

Note: AA = administrative aide; C.E.R.T. = correctional emergency response team; PE & Rec. = physical education and recreation.

fringe benefits, such as a lavish residence, unlimited inmate servants, food and supplies from institutional farms and warehouses, furnishings, and a personal automobile. Now most wardens or superintendents are civil service employees who have earned their position through seniority and merit.[38] We discuss the warden's position more later in this chapter.

Reporting to the warden are deputy or associate wardens, each of whom supervises a department within the prison. The deputy warden for operations will typically oversee correctional security, unit management, the inmate disciplinary committee, and recreation. The deputy warden for special services will generally be responsible for the library, mental health services, drug and alcohol recovery services, education, prison job assignments, religious services, and prison industries. The deputy warden for administration will manage the business office, prison maintenance, laundry, food service, medical services, prison farms, and the issuance of clothing.[39]

We now discuss correctional security, unit management, education, and prison industries in greater detail:

- The correctional security department is usually the largest department in a prison, with 50 to 70 percent of all staff. It supervises all of the security activities within a prison, including any special housing units, inmate transportation, and the inmate disciplinary process. Security staff wears military-style uniforms; a captain typically runs each eight-hour shift, lieutenants often are responsible for an area of the prison, and sergeants oversee the rank-and-file correctional staff.
- The unit management concept was originated by the federal prison system in the 1970s and now is used in nearly every state to control prisons by providing a "small, self-contained, inmate living and staff office area that operates semi-autonomously within the larger institution."[40] The purpose of unit management is twofold: to decentralize the administration of the prison and to enhance communication among staff and between staff and inmates. Unit management breaks the prison into more manageable sections based on housing assignments; assignment of staff to a particular unit; and staff authority to make decisions, manage the unit, and deal directly with inmates. Units usually comprise 200–300 inmates; staff members are assigned to specific units, and their offices are located in the housing area, making them more accessible to inmates and better able to monitor inmate activities and behavior. Directly reporting to the unit manager are "case managers," or social workers, who develop the program of work and rehabilitation for each inmate and write progress reports for parole authorities, classification, or transfer to another prison. Correctional counselors also work with inmates in the units on daily issues, such as finding a prison job, working with their prison finances, and creating a visiting and telephone list.[41]
- Education departments operate the academic teaching, vocational training, library services, and sometimes recreation programs for inmates. An education department is managed in a fashion similar to that of a conventional elementary or high school, with certified teachers for all subjects that are required by the state department of education or are part of the General Educational Development (GED) test. Vocational training can include carpentry, landscaping or horticulture, food service, and office skills.

Prison industries: use of prison and jail inmates to produce goods or provide services for a public agency or private corporation.

- **Prison industries** are legislatively chartered as separate government corporations and report directly to the warden because there is often a requirement that the industry be self-supporting or operate from funds generated from the sale of products. Generally, no tax dollars are used to run the programs, and there is strict accountability of funds.

Do Prisons Rehabilitate?

As discussed in Chapter 9, the purposes of sentencing and punishment include rehabilitation—the goal of changing the offender so that he or she becomes a law-abiding citizen. Can and do prisons really rehabilitate? The data on recidivism (from the Latin word *recidivus,* which means "recurring"; an offender's committing new crimes) will address the question. Recidivism is one of the most important concepts in criminal justice, as it concerns a person's relapse into criminal behavior; recidivism is generally measured by one's criminal acts resulting in rearrest, reconviction, or return to prison *within a three-year period* after he or she was released from incarceration or placement on supervision for a previous criminal conviction.[42]

A study of state prisoners by the Bureau of Justice Statistics looked at recidivism beyond the aforementioned three-year threshold, with a nine-year follow-up of state prisoners; the study provides a dismal view of rearrest after release. In sum, five in six (83 percent) of released state prisoners committed a new offense for which they were rearrested. An estimated 68 percent were rearrested within three years; over the nine-year period, 82 percent of those persons arrested during the first three years were arrested during years four through nine, and almost half (47 percent) of those who did not have an arrest within three years were arrested during years four through nine.[43] Clearly, the longer one has been released from prison, the greater the likelihood of reoffending. More needs to be done to prepare inmates for reentry, providing educational opportunities, job training and counseling, and outreach and support services post-release.[44]

FEDERAL PRISONS

The Federal Bureau of Prisons (BOP) was established in 1930 to provide care for federal inmates. Today, the BOP has more than 36,000 employees and consists of 144 institutions of many types (including prisons, camps, medical centers, penitentiaries, transfer centers, and detention centers) and nearly 180,000 federal offenders.[45] Approximately 85 percent of these inmates are confined in BOP-operated facilities, while the remainder are confined in privately managed or community-based facilities and local jails.[46]

Prison Types and General Information

The BOP operates institutions at five different security levels in order to confine offenders in an appropriate manner.[47] The levels (and associated numbers of institutions) are discussed in the sections that follow; the federal supermax facility is discussed later in this chapter. Security levels are based on such features as the presence of external patrols, towers, security barriers, or detection devices; the type of housing within the institution; internal security features; and the staff-to-inmate ratio.

Minimum Security (7)

Minimum-security institutions, also known as federal prison camps, have dormitory housing, a relatively low staff-to-inmate ratio, and limited or no perimeter fencing. These institutions are work and program oriented, and many are located adjacent to larger institutions or on military bases, where inmates help serve the labor needs of the larger institution or base.

Low Security (42)

Low-security federal correctional institutions have double-fenced perimeters, mostly dormitory or cubicle housing, and strong work and program components. The staff-to-inmate ratio in these institutions is higher than in minimum-security facilities.

Medium Security (47)

Medium-security correctional institutions and penitentiaries designated to house medium-security inmates have strengthened perimeters (often double fences with electronic detection systems), mostly cell-type housing, a wide variety of work and treatment programs, an even higher staff-to-inmate ratio than low-security federal correctional institutions, and even greater internal controls.

High Security (17)

High-security institutions, also known as U.S. penitentiaries, have highly secured perimeters (featuring walls or reinforced fences), multiple- and single-occupant cell housing, the highest staff-to-inmate ratio, and close control of inmate movement.

Administrative Security (20)

These are institutions with special missions, such as the detention of pretrial offenders; the treatment of inmates with serious or chronic medical problems; or the containment of extremely dangerous, violent, or escape-prone inmates (such as in the supermax prison, in Florence, Colorado).

In addition, a number of BOP institutions have a small, minimum-security camp adjacent to the main facility. These camps, often referred to as satellite camps, provide inmate labor to the main institution and to off-site work programs.

A Career With BOP

The General Schedule that is used for BOP position classification and pay has 15 grades—GS-1 (lowest) to GS-15 (highest). Agencies establish (classify) the grade of each position based on the level of difficulty, responsibility, and qualifications required. For example, individuals with a high school diploma and no additional experience typically qualify for GS-2 positions; those with a bachelor's degree, for GS-5 positions; and those with a master's degree, for GS-9 positions. Individuals who wish to investigate career opportunities with the BOP should visit the following websites:

1. Pay classification system: https://www.opm.gov/policy-data-oversight/pay-leave/pay-systems/general-schedule/
2. The hiring process: https://www.bop.gov/jobs/hiring_process.jsp
3. The correctional officer position: https://www.bop.gov/jobs/positions/index.jsp?p=Correctional%20Officer
4. Job benefits, which are exceptional: https://www.bop.gov/jobs/life_at_the_bop.jsp

ROBYN BECK/AFP/Getty Images

Prison housing units vary, but they usually consist of 200–300 inmates and place staff members in the housing area, to be more accessible to inmates and better able to monitor their activities and behavior.

Community Corrections in the Federal System

Parole was abolished in the federal prison system in 1987, when the federal sentencing guidelines went into effect (discussed in Chapter 9). Since then, inmates can have no more than 54 days a year subtracted from a federal sentence, commencing after the person has served 12 months; this policy was upheld in 2010 by the U.S. Supreme Court.[48] Nonetheless, the BOP is actively supporting and using community-corrections techniques.

SUPERMAX PRISONS

Super-maximum—or "supermax"—prisons represent the most secure form of incarceration now in existence in the United States and abroad (although they are given different names in other countries). They began to proliferate in the United States in the mid-1980s as a means of holding extremely disruptive or violent inmates. Such prisons have not been without controversy, however, and this section includes a brief history and description, some research findings on effects on inmates, and some constitutional questions that have been raised.

©Lizzie Himmel/Sygma/Getty Images

The "Administrative Maximum" prison, or ADX, located in Florence, Colorado, is the only federal supermax in the nation.

Origin

Supermax prisons exist in both state and federal prison systems and effectively originated in 1983 in Marion, Illinois, when two correctional officers were murdered by inmates on the same day and the warden put the prison in "permanent lockdown." Thus began 23-hour-a-day cell isolation and no communal yard time for inmates, who were also not permitted to work, attend educational programs, or eat in a cafeteria.[49]

According to Amnesty International, today at least 40 states operate supermax prisons in the United States and house about 25,000 inmates.[50] To understand what supermax prisons are and how they operate, one can look at the federal "Administrative Maximum" prison, or ADX, located in Florence, Colorado, 90 miles south of Denver.

ADX is the only federal supermax facility in the country (the others are state prisons). It is home to a "who's who" of criminals: Boston Marathon bomber Dzhokhar Tsarnaev, "Unabomber" Ted Kaczynski; "Shoe Bomber" Richard Reid; Ramzi Yousef, who plotted the 1993 World Trade Center attack; Oklahoma City bomber Terry Nichols; and Olympic Park bomber Eric Rudolph—and, possibly soon, drug kingpin "El Chapo," discussed in Chapter 12. Ninety-five percent of the prisoners at ADX, known as the "Alcatraz of the Rockies," are the most violent, disruptive, and escape-prone inmates from other federal prisons.[51]

Supermax prison: a penal institution that, for security purposes, affords inmates very few, if any, amenities and a great amount of isolation.

Two Key Components: Operation and Design

Upon viewing its external aspect for the first time, one immediately sees that this is not the usual prison: Large cables are strung above the basketball courts and track; they are helicopter deterrents.[52] Supermax inmates rarely leave their cells; in most such prisons, an hour a day of out-of-cell time is the norm. Inmates eat all of their meals alone in the cells, and typically no group or social activity of any kind is permitted. They are also typically denied access to vocational or educational training programs. Inmates can exist for many years separated from the natural world around them.[53]

Effects on Inmates

Given the high degree of isolation and lack of activities in supermax prisons, a major concern voiced by critics of these facilities is their "social pathology" and potential effect on inmates' mental health. Although there is very little research to date concerning the effects of supermax confinement,[54] some authors point to previous isolation research that shows greater levels of deprivation lead to psychological, emotional, and physical problems. For example, studies show that as inmates face greater restrictions and social deprivations, their levels of social withdrawal increase; limiting human contact, autonomy, goods, or services is detrimental to inmates' health and rehabilitative prognoses, and tends to result in depression, hostility, severe anger, sleep disturbances, and anxiety. Women living in a high-security unit have been found to experience claustrophobia, chronic rage reactions, depression, hallucinatory symptoms, withdrawal, and apathy.[55]

Some researchers argue that supermax facilities are not effective management tools for controlling violence and disturbances within prisons, nor are they effective in reducing violence or disturbances within the general population; they conclude that supermax prisons should not be used for their current purpose.[56]

Constitutionality

Because of the relatively recent origin of supermax prisons, their constitutionality has been tested in only a few cases. The first, *Madrid v. Gomez* (1995), addressed conditions of confinement in California's Pelican Bay Security Housing Unit.[57] The judge observed that its image was "hauntingly similar to that of caged felines pacing in a zoo"; however, the judge concluded that he lacked any constitutional basis to close the prison. In the most recent case, *Jones-El v. Berge* (2004), a federal district court in Wisconsin concluded that "extremely isolating conditions . . . cause SHU [Security Housing Unit] syndrome in relatively healthy prisoners . . . . Supermax is not appropriate for seriously mentally ill inmates."[58] The judge ordered several prisoners to be removed from the supermax facility.

PRIVATE PRISONS

Privatization: the operation of existing prison facilities, or the building and operation of new prisons, by for-profit companies.

Perhaps one of the most rapidly growing segments of corrections has been the outsourcing or **privatization** of prisons, a term that includes either the operation of existing prison facilities by a private company or the building and operation of a new prison by a for-profit company. At present about 8.5 percent of state and federal prison inmates (128,000) and 73 percent of persons (34,900) held in immigration detention centers are in privately operated facilities.[59] The two largest private prison companies are CoreCivic (formerly the Corrections Corporation of America, or CCA) and the Geo Group, which together have annual revenues of more than $3.5 billion.[60]

Historically, strong arguments have been made, both pro and con, regarding the privatization of prisons. Proponents of the concept argue that private prisons provide an effective, cost-saving alternative for governments seeking to address significant capacity needs while taking pressure off of corrections budgets. Conversely, opponents maintain that there is no guarantee that standards will be upheld, no one will maintain security if employees go on strike, there will be different inmate disciplinary procedures, the company will be able to refuse certain inmates or could go bankrupt, and the company can increase its fees to the state.[61]

A legitimate question here is "which is better—public or private prisons?" The totality of both types of prisons has not often been studied, and it is known that custody level, size, crowding, age, and architecture can all have a strong influence on measures of quality. One comprehensive study, however, examined 1,129 institutions—105 of which were private. Only one significant difference was found: State prisons were much more likely to be under a court order for the conditions of confinement than private prisons. Given the negative effects of crowding, this is a significant finding.[62]

Prior to the 2016 presidential election, the U.S. Department of Justice announced it would discontinue the use of private prisons in the federal system, affecting 13 federal prisons. However, in 2017 the new Trump administration reversed that policy and now houses about 34,000 inmates in such facilities, including halfway houses.[63] President Trump is known to support privatization, and has claimed that private prisons work better than government-operated prisons.[64]

JAILS AS ORGANIZATIONS

Across the United States, approximately 3,283 jails are administered locally,[65] holding an average daily population of 727,000. As with state and federal prisons, the number of confined inmates in local jails has been declining since 2008.[66]

The primary purposes of local jails are (1) to hold accused law violators who cannot post bond so as to ensure their appearance at trial, and (2) to hold those persons convicted of lesser offenses until they complete their court-ordered sentences.

The incarceration of persons charged with crimes and awaiting trial, as well as some persons serving sentences for lesser offenses, distinguishes local jails from state and federal prisons. Whereas prisons house persons convicted of more serious offenses—usually felons with sentences of a year or more—jails generally hold persons charged with misdemeanors for up to a year.[67] Jail organization and hierarchical levels are determined by several factors: size, budget, level of crowding, local views toward punishment and treatment, and even the levels of training and education of the jail administrator. An organizational structure for a jail serving a population of about 250,000 is suggested in Figure 10.3.

The New Generation/Direct Supervision Jail

Federal courts have at times abandoned their traditional hands-off doctrine toward prison and jail administration, largely in response to the deplorable conditions and inappropriate treatment of inmates. The courts became more willing to hear inmate allegations of constitutional violations ranging from inadequate heating, lighting, and ventilation to the censorship of mail.[68]

In response to lawsuits and to improve conditions, many local jurisdictions explored new ideas and designed new jail facilities. The first **new generation/direct supervision jail** opened in the 1970s in Contra Costa County, California. This facility quickly became a success, was deemed cost-effective to build and safer for inmates and staff, and carried several advantages: Officers "live" with the inmates and are encouraged to mingle with them and to provide them privileges and activities (thus increasing good behavior and reducing idleness) while being better able to control inmate movement. As a result, there is a low level of tension in the unit, as fights are broken up quickly, weapons are not involved, and sexual assaults are almost nonexistent. Bathroom and shower areas are monitored closely, and noise levels are low due to the architecture and close supervision.[69] To the extent possible, symbols of incarceration are removed in these new jails, which have no bars in the living units; windows are generally provided in every prisoner's room; and padded carpets, movable furniture, and colorful wall coverings are used to reduce the facility's institutional atmosphere. Inmates are to be divided into small groups of approximately 40–50 for housing purposes. All of these facility features were designed to reduce the "trauma" of incarceration.[70] Figure 10.4 provides three views of how new generation/direct supervision jails are configured.

New generation/direct supervision jail: a jail that, by its architecture and design, eliminates many of the traditional features of a jail, allowing staff members greater interaction and control.

Jail Treatment Programs

Rehabilitation and reintegration are considered secondary goals of local jails; however, many of them provide a wide array of programs for inmates, such as the following types as noted by the National Institute of Justice (NIJ):[71]

- Adult basic education (including computer and other desirable classes)
- Alcoholics Anonymous
- Behavior change programs
- Consumer education
- Criminal justice system programs
- Educational enrichment
- English as a Second Language
- Faith-based living skills
- General educational development (GED)
- Health education

Figure 10.3 /// Organization Structure for Jail Serving County of 250,000 Population

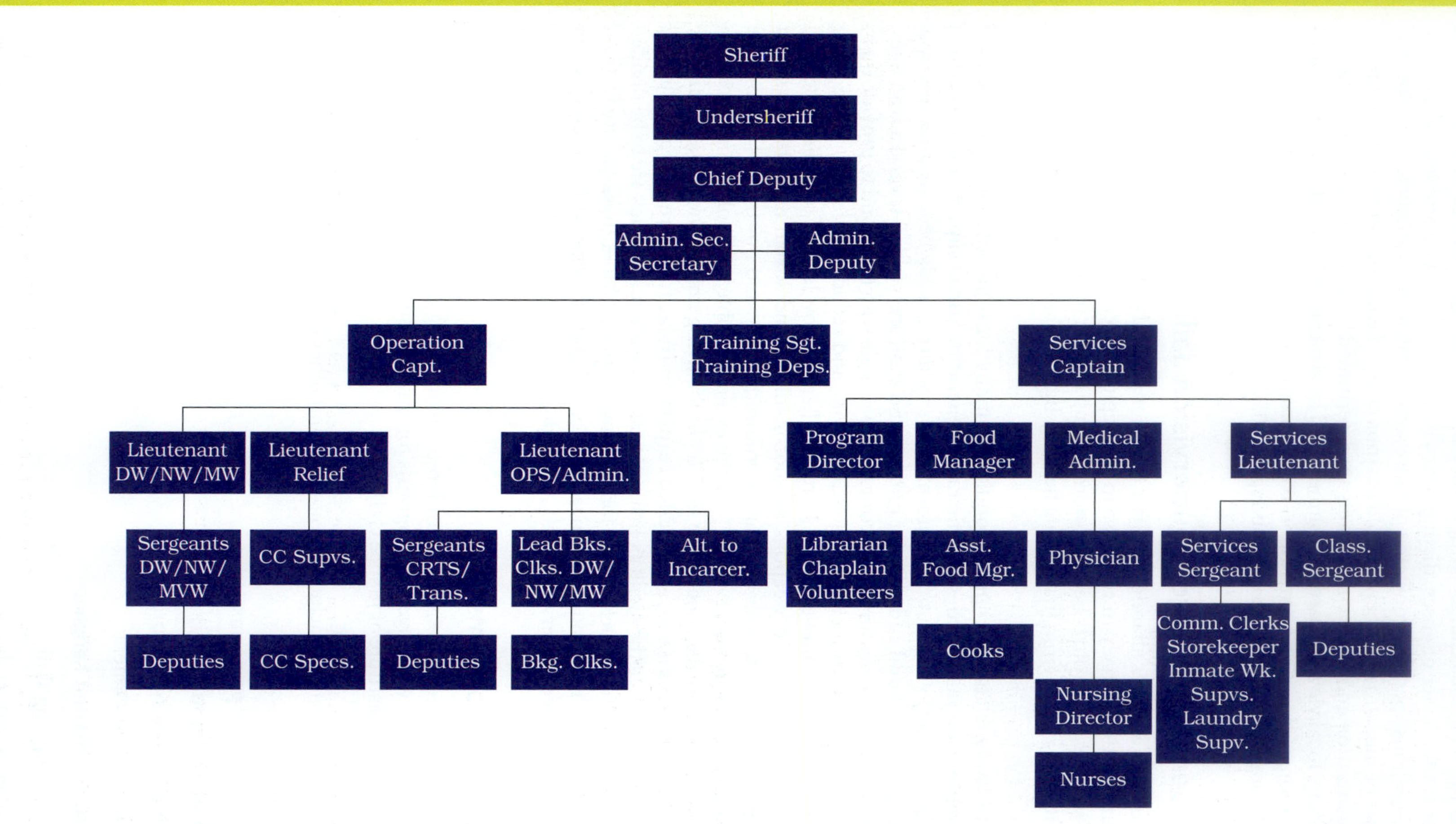

Note: DW = day watch; NW = night watch; MW = mid watch; CC = conservation camps; CRTS/Trans = court transportation; OPS/Admin = operation/administration; Comm. Clerks = commissary clerks.

Figure 10.4 /// Direct Supervision Jails

Jails can have a combination of design styles—Jails that are not predominantly direct supervision in design or management can have an addition or section of inmate housing that uses direct supervision.

Podular/direct supervision jails—Inmates' cells are arranged around a common area, usually called a "dayroom." An officer is stationed in the pod with the inmates. The officer moves about the pod and interacts with the inmates to manage their behavior. There is no secure control booth for the supervising officer, and there are no physical barriers between the officer and the inmates. The officer may have a desk or table for paperwork, but it is in the open dayroom area.

David R. Frazier Photolibrary, Inc./Alamy

Linear/intermittent supervision jails—Includes jails with cells arranged along the sides of a cell block. Officers come into the housing unit on scheduled rounds or as needed to interact with the inmates.

©Andrew Aitchison/Corbis Historical/Getty Images

Podular/remote supervision jails—Includes jails that have a podular design with cells around a dayroom, but no officer is permanently stationed inside the pod. Indirect supervision is provided through remote monitoring at a console.

MediaNews Group/Orange County Register via Getty Images/MediaNews Group/Getty Images

Source: U.S. Department of Justice, National Institute of Corrections, *Direct Supervision Jails: 2006 Yearbook,* p. vii, http://nicic.org/Downloads/PDF/Library/021968.pdf.

- Mental and physical health programs
- Narcotics Anonymous
- Religious involvement
- Substance abuse education

According to the NIJ, which uses a seven-step review and evidence-rating process,[72] about 27 percent of such programs do not accomplish their goals, while about two-thirds are "promising," and 6 percent are "effective."[73]

Washoe County Sheriff's Office

Many local jails have inmate work programs, such as graffiti removal, landscaping, snow removal for elderly citizens, and many others. Here, inmates clean up a municipal river's banks.

Making Jails Productive Through Labor

The 1984 Justice Assistance Act removed some of the long-standing restrictions on interstate commerce of prisoner-made goods. By 1987, private-sector work programs were under way in 14 state correctional institutions and two county jails.[74] Today, many inmates in U.S. jails are involved in productive work. Some simply work to earn privileges, and others earn wages applied to their custodial costs and compensation to crime victims. Some hone new job skills, improving their chances for success following release. At one end of the continuum is the "trusty" (an inmate requiring a low security level) who mows the grass in front of the jail and thereby earns privileges; at the other end are jail inmates working for private industry for real dollars.[75] Some jails have work programs involving training in dog grooming, auto detailing, food service, book mending, mailing service, painting, printing, carpet installation, and upholstering. In addition, some agencies have work programs that use inmates (who pass a review of their criminal history and disciplinary actions) for such activities as graffiti removal, landscaping, setup and teardown of special events, and snow removal for the elderly and/or disabled.[76]

TECHNOLOGIES IN CORRECTIONAL FACILITIES: GOING AFTER CONTRABAND

As with police and court systems, technologies are used in jails and prisons to provide a greater degree of safety for both officers and inmates and, thus, to improve efficiency and effectiveness of correctional practices. They employ video monitoring, GPS devices worn on wrists, biometric entry points that scan an inmate's iris or fingerprints, and remote medical tools so that inmates receive virtual checkups from doctors (discussed later in this section).

However, inmates have uses for technologies as well, including drones, for facilitating their unwavering attempts to bring contraband items into the facility and thus enhance their quality of life. Contraband—including drugs, cell phones, weapons, cigarettes, pornographic materials, and hacksaw blades—has been the bane of prisons and jails since the beginning of time. To obtain these items, inmates use the U.S. mail, visitors, staff members (correctional staff as well as outside persons, such as building contractors), people to throw items over the walls (such as tennis balls containing drugs), and drones. Some jail and prison officials try to control contraband by using a video visitation system, thereby reducing the amount of contact inmates have with family and friends, the need for inmate searches, and reductions in staff time. Short of that, however, several means are available for getting contraband and other such items inside the walls.

Next we focus on the twofold problem of drones and cell phones, and what prison officials are doing to try to stop these items from posing major security issues.

A Perilous Combination—Drones and Cell Phones

Today's drones have become cheap and thus more widespread; they can move quickly, hover, and be evasive when necessary. As noted, they can be used to deliver all manner of contraband items into correctional institutions, two highly sought-after items being drugs and cell phones. Regarding the former, drugs not only get inmates high, but they also create power struggles and disorder, thus threatening the lives of staff and inmates alike. Fentanyl, a synthetic opioid (see Chapter 12), can easily result in overdoses and pose a hazard to staff when exposed to the drug.

Regarding the latter, cell phones, their being smuggled into prisons and jails can cause all manner of problems, from directing gang activities and violent crimes inside and outside the facility to distributing child pornography and intimidating witnesses outside. This is not an insignificant issue, as the use of metal detectors, scanners, and security cameras have all failed to stop cell phones from being smuggled in. As an example, in a recent year California seized more than 13,000 cell phones in its prisons.[77]

What can be done? Regarding drones, federal laws and state statutes have been enacted in the United States to prevent drones from hovering over prison and jail airspace. In addition, private companies have developed technology that is essentially drone radar, to provide advance warning of an incoming drone, the direction it is coming from, and ultimately the identity of the package recipient and the location of the drone pilot.[78]

To combat the threat caused by cell phones, federal laws have stood in the way of "jamming" illicit calls by prisoners, as doing so can threaten the use of phones outside the prison and jail walls. However, the U.S. Department of Justice, the federal Bureau of Prisons, and researchers are testing technology that would disrupt wireless service for the thousands of cell phones inside the walls without posing such a threat to phones on the outside. Furthermore, judges in South Carolina and California have ordered wireless carriers to disable hundreds of contraband cell phones used by inmates.[79]

The Body Scanner

Another tool being used against contraband and adding another layer to the screening process is the body scanner, which functions the same as those used at airports. Penal facilities have long used metal detectors, but they cannot detect illegal drugs and nonmetallic objects. For about $125,000, body scanners can spare corrections officers the need to conduct traditional, unclothed searches of inmates requiring several minutes' time (these can be accomplished in 10 seconds). The devices pay for themselves by reducing the need for officer overtime and workplace injuries. Some institutions have even purchased scanners using funds from the prison canteen (a store where inmates buy snacks, hygiene items, and so on),

You Be the . . .

Warden

You are a prison warden in a city where three drone drops have occurred over the prison yard over the past month. This has led to an escalation in the institution's drug problems, particularly with synthetic opioids like fentanyl. You managed to capture one drone pilot, and suspect that the others, like this one, are not local but instead are experts who are hired by well-off inmates to fly in the illicit drugs and other items, which are then sold at exorbitant prices. While you are working on a solution, you learn from inside sources that more drops are planned in the near future.

You also hear of two possible means for intercepting the drones. One is the use of eagles, which police in The Netherlands have employed for grabbing and taking down rogue drones. Their sharp talons and keen eyesight make it possible for them to pinpoint and pluck a drone from the sky in a simple maneuver. The other means is the use of a laser weapons system that effectively blows drones out of the sky.

1. Assuming cost is not an issue, should your prison and others use one or both of these tools?
2. Can you foresee serious problems in the deployment of either method? If so, what are they?

See the Notes section at the end of the book for assistance in arriving at a decision.[80]

Juvenile Justice Journal

Use of Detention

Although the detention of juveniles is typically used as a last resort, juvenile courts may hold delinquents in a secure facility if believed to be in the best interest of the community or the child.[81] Several options are available to juvenile judges who feel that some sort of custodial living arrangement in a residential or institutional facility is warranted for a particular youth. Generally, these facilities fall into two categories: nonsecure and secure.

Nonsecure confinement facilities afford the youth some movement within the community and include foster home placement, shelter care facilities, group homes, halfway houses, and camps or ranches. Foster homes provide youths with a substitute family in cases where their own homes lack consistent adult supervision or the youths are unruly.[82] The typical stay is short, averaging about 26 months, and the majority of foster children are returned to their parents.[83] Shelter care facilities provide youths with a home-like environment on a short-term basis; house parents provide the children care and safety for a few hours up to a few weeks. Youths in shelter care are nonviolent and are awaiting placement in a foster home or group home.[84] Group homes are an intermediate alternative available to the judge; community-based, they have counselors or residents who act as parental figures for youths in groups of 10 to 20. They are able to offer residential community placement, treatment, and supervision.[85] Halfway houses are for juveniles who are in transition and for youths who have often spent a period of time in a secure residential facility because of delinquent acts; they are often operated by private agencies rather than a unit of government.[86]

Secure confinement is generally used as a last resort and includes placing youths in shock probation, boot camps, and industrial schools or other institutions. In shock probation (or, more accurately, parole), a judge sends the offender to an institution providing secure placement; after a period of usually 60–90 days, the youth is brought back for a hearing and is placed on probation. The youth does not know he or she will soon be released from custody, so the idea is to "shock" the youth into displaying good behavior after getting a taste of institutionalization.[87] Boot camps are highly structured, paramilitary, short-term correctional programs that last from 90 to 180 days and strive to impart physical fitness and teach youths strict discipline. However, no evaluation of juvenile boot camps has found this type of program to reduce recidivism.[88] Finally, industrial or training schools emphasize vocational training, coping skills, education, and substance abuse and mental health counseling. As with boot camps, evaluations have not shown them to be effective in reducing recidivism; in fact, they are generally believed to increase reoffending.[89]

Is There a "School-to-Prison Pipeline"?

Some authors believe our schools are plagued by a **school-to-prison pipeline**—often perceived as grounded in racial discrimination.[90] It is argued that students who are expelled from school for disruptive behavior, and are then compelled to live in the homes and neighborhoods where their problems began, will become stigmatized, more hardened and embittered, and often more engaged in criminality.

According to Rocque and Paternoster, African American youth fare worse in schools than whites in several areas: They show less interest in school activities, have lower grades, are more likely to be held back and to be in special education, and have higher rates of incarceration.[91] Added to those factors is their lower socioeconomic status and their environment, in which one's peers may dismiss and ridicule academic success, as well as the increased likelihood that they will be singled out for punishment in school—and a school-to-prison pipeline trajectory for minorities is present.[92]

What can be done about this pipeline or trajectory? First, according to some educators, realize that suspensions do not really work—they merely get the individual out of the classroom. Next is the need to examine existing discipline structures to ensure that they help, rather than hurt, students (such as the seven teenagers who were arrested and charged with disorderly conduct for an end-of-the-year water balloon fight).[93]

which are normally used to buy items that benefit inmates; they justify this use by saying the device benefits inmates by sparing them the time and embarrassment of an unclothed search.

A county sheriff's office in Madison (Somerset County), Maine, averaged about eight incidents per year of smuggling of illicit materials into the jail's housing areas, but with the use of the body scanner, the jail has had no contraband instances, the deterrent effect was total, and the device paid for itself in personnel time savings.[94]

School-to-prison pipeline: the notion that certain policies and practices push schoolchildren, particularly those who are most at risk, out of classrooms and into the juvenile and adult criminal justice and prison systems.

Offender Programming and Management

Technology is changing the methods of offender management through the use of web-based systems that provide educational programs to prisoners, treat prisoners who are addicted to drugs or are sex offenders, and provide vocational training. Prison administrators now keep accurate records of inmates' purchases for items in the prison store, payments to victims and their families, and other reasons for which money flows in and out of prisoners' bank accounts.[95] Automated systems also control access gates and doors, individual cell doors, and the climate in cells and other areas of the prison. Correctional agencies have also used computers to conduct presentence investigations, supervise offenders in the community, and train correctional personnel. With computer assistance, jail administrators receive daily reports on court schedules, inmate rosters, time served, statistical reports, maintenance costs, and other data.

"Virtual Visits" by Inmates and the Public

Prison and jail inmates make frequent visits to hospitals and courtrooms, which creates public safety concerns. In Ohio alone, 40,000 inmate trips to and from medical facilities were eliminated because videoconferencing technology—virtual visits—made medical consultations available from within the institution. An onsite prison physician or nurse assists with the physical part of the examination, taking cues from the offsite specialist via video and relaying information such as electrocardiogram and blood pressure data.[96]

/// IN A NUTSHELL

- Erving Goffman described the characteristics of "total institutions." His definition seems to capture the essence of correctional organizations. Unfortunately, most of what the public "knows" about prisons and jails is probably obtained through Hollywood's fictional versions.
- Sending people to prisons and jails is intended to serve a useful purpose. However, for many inmates, the conditions inside are not all that different from (and might even be better than) the conditions outside. Indeed, for some members of society, going to prison or jail may represent an increased standard of living.
- Several factors contribute to the recent decline in prison and jail populations; essentially, these are federal and state efforts to early-release nonviolent offenders. In addition, violent crimes are decreasing, and there is greater use of alternatives to incarceration.
- Mass incarceration is said to be a form of racialized social control, primarily affecting people of color, resulting in the imprisonment of large numbers of minorities for nonviolent drug offenses, and taking a huge toll on such individuals (in terms of rights lost) as well as minority families. Some researchers point to national increases in jail populations, which would override the belief that reductions in prison populations indicate that sentencing reform is occurring.
- Correctional organizations are complex, hybrid organizations that use two related management subsystems to achieve their goals: One is managing correctional employees, and the other is concerned primarily with delivering correctional services to a designated offender population.
- The mission of correctional agencies has changed little over time. It is to protect the citizens from crime by safely and securely handling criminal offenders while providing offenders some

opportunities for self-improvement and increasing the chance that they will become productive and law-abiding citizens.

- The correctional security department is typically the largest department in a prison, with 50 to 70 percent of all staff. It supervises all of the security activities within a prison.
- The unit management concept was originated by the federal prison system in the 1970s, to control prisons by providing a "small, self-contained, inmate living and staff office area that operates semi-autonomously within the larger institution." Unit management breaks the prison into more manageable sections based on housing assignments.
- Education departments operate the academic teaching, vocational training, library services, and sometimes recreation programs for inmates.
- Prison industries are separate government corporations that provide meaningful, productive employment that helps reduce inmate idleness and supplies companies with a readily available and dependable source of labor.
- Correctional staff must make classification decisions in at least two areas: the inmate's level of physical restraint or "security level," and the inmate's level of supervision or custody grade.
- The Federal Bureau of Prisons operates institutions at five different security levels in order to confine offenders in an appropriate manner. Security levels are based on such features as the presence of external patrols, towers, security barriers, or detection devices; the type of housing within the institution; internal security features; and the staff-to-inmate ratio. Each facility is designated as minimum, low, medium, high, or administrative security.
- Supermax prison operations are quite different from traditional prisons. Inmates rarely leave their cells; they eat all of their meals alone in their cells, and typically no group or social activity of any kind is permitted; and they are typically denied access to vocational or educational training programs. Although little research has examined the effects of supermax confinement, some authors point to previous isolation research that shows greater levels of deprivation lead to psychological, emotional, and physical problems.
- A rapidly growing segment of corrections is the privatization of prisons, or operation by a for-profit company; proponents argue that they are an effective, cost-saving alternative; opponents say there is generally less accountability and that a private prison company might go bankrupt or increase its fees to the state.
- The primary purpose of jails is to hold accused law violators who cannot post bond to ensure their appearance at trial, and to hold those persons convicted of lesser offenses until they complete their court-ordered sentences.
- The term *new generation jail* refers to a style of architecture and inmate management that is totally new and unique to local detention facilities. There is a greater level of personal safety for both staff and inmates, greater staff satisfaction, more orderly and relaxed inmate housing areas, and a better maintained physical plant; these facilities are also cost-effective to construct and to operate.
- Today, many inmates in U.S. jails are involved in productive work and treatment programs. Some simply work to earn privileges, and others earn wages applied to their custodial costs and compensation to crime victims. Some hone new job skills, improving their chances for success following release.
- Technological developments have improved the operations and safety of correctional institutions, but inmate use of drones and cell phones have created serious security concerns.

/// KEY TERMS & CONCEPTS

Review key terms with eFlashcards at **edge.sagepub.com/peakbrief**.

Classification (of inmates) 228
Jail 223
Mass incarceration 225
New generation/direct supervision jail 235
Prison 223
Prison industries 230
Privatization 234
School-to-prison pipeline 240
State prison 229
Supermax prison 233
Warden 229

/// REVIEW QUESTIONS

Test your understanding of chapter content. Take the practice quiz at **edge.sagepub.com/peakbrief**.

1. What factors have been recognized as decreasing prison and jail populations?
2. What is meant by mass incarceration, and what are its causes?

3. What is the general mission of a correctional organization?
4. How is it that, for many members of our society, being incarcerated is not "punishment" but rather an improved lifestyle?
5. What is the basic purpose underlying prison inmate classification?
6. What are the basic elements of the federal prison system?
7. How would you describe supermax prisons and how they differ from conventional prisons?
8. How do private prisons differ in structure and function from traditional, government-operated institutions?
9. How do jails differ in structure and function from prisons, what programs are provided for inmate rehabilitation, and how does the new generation jail differ from the traditional model?
10. Describe some of the technologies now in use in corrections, with particular emphasis on methods for halting drone and cell phone use by inmates.

/// LEARN BY DOING

1. Assume you are on a high school recruiting trip for your university's criminal justice department, and a member of a large student group mentions that he wishes to major in criminal justice. He is motivated to do so by an uncle's criminal experiences, for which he is "serving a three- to five-year *supermax* sentence *in a private city jail* for *robbing* people's homes while they were away at work." Because he is overheard by the group in obviously misspeaking about several terms and concepts, you wish to tactfully correct his miscues. What will you say?
2. You are assigned as part of a "Current Correctional Practices" class project to explain the differences between supermax and traditional prisons as well as traditional and new generation jails. What will be your response?
3. Your criminal justice professor assigns the class to prepare a paper on the major differences between state prisons and local jails, including their structure and function. How will you delineate the differences between them?

/// STUDY SITE

SAGE edge™

Get the tools you need to sharpen your study skills. SAGE edge offers a robust online environment featuring an impressive array of free tools and resources.

Access practice quizzes, eFlashcards, video, and multimedia at **edge.sagepub.com/peakbrief**

AP Photo/Suchat Pederson

CHAPTER 11

CORRECTIONS IN THE COMMUNITY

Probation, Parole, and Other Alternatives to Incarceration

The mood and temper of the public in regard to the treatment of crime and criminals is one of the most unfailing tests of the civilization of any country.

—Winston Churchill

LEARNING OBJECTIVES

As a result of reading this chapter, you will be able to

1. Describe what is meant by community corrections
2. Explain why the criminal justice system uses alternatives to incarceration
3. Describe the definitions and origins of probation and parole, as well as the differences between them
4. Identify the eligibility and rights accorded to people serving terms of probation or parole
5. Explain the functions of probation and parole officers—and the impact of high caseloads
6. List and explain the purposes and functions of several intermediate sanctions and alternatives to incarceration
7. Provide an overview of the risk-need-responsivity (RNR) model for addressing reoffending
8. Explain the rationale that underlies the use of restorative justice

ASSESS YOUR AWARENESS

Test your knowledge of probation and parole by responding to the following seven true-false items; check your answers after reading this chapter.

1. Probation began with the voluntary work of a simple Boston shoe cobbler.
2. Probation, because it is more costly than prison, is used sparingly in the United States.
3. An offender placed on parole does not serve any time in prison or jail and serves the entirety of his or her sentence in the community.
4. A person may have his or her probation revoked and then be sent to prison for behaviors such as using alcohol, violating curfew, and associating with other known criminals.
5. Persons whose probation or parole status might be revoked, and who thus might be sent to prison, enjoy no legal rights or benefits.
6. In addition to probation and parole, other alternatives to prison include house arrest, electronic monitoring, and boot camps.
7. Restorative justice places emphasis on giving victims a voice and placing greater accountability on the offender.

<< Answers can be found on page 293.

Carlos Eduardo Arevalo Carranza, 24, was arrested in March 2019, accused of stalking and then killing 59-year-old Bambi Larson of San Jose, California. When Larson did not show up for work, coworkers notified her son. He found his mother dead in her bedroom with multiple lacerations and stab wounds. Video surveillance in the area was not clear enough to provide a positive suspect identification, but police were able to trace the suspect's path and find a shirt containing both the victim's and the suspect's DNA. Police eventually arrested Carranza in connection with the murder, finding Larson's phone and e-reader in Carranza's possession at the time of arrest. The Santa Clara County District Attorney's Office charged Carranza with first-degree murder and two special circumstances of burglary and mayhem.

Carranza is a self-identified gang member and a transient, with an extensive criminal history. He had been previously charged with (and in several cases convicted of) a wide variety of crimes, including burglary, drug possession, entering and occupying a property, resisting arrest, and battery.

Although the media has given much attention to Carranza's illegal immigration status, it is also important to note that he was on probation at the time that he allegedly murdered Larson. When a person commits a

Carlos Eduardo Arevalo Carranza

crime while serving a sentence in the community (e.g., while under probation or parole supervision), we are naturally critical of the decisions made by the judge or parole board members involved in the case. If you had been the judge in any of Carranza's previous cases, would you have allowed this individual to serve his sentence on probation in the community? How much money should we spend to punish offenders, and what are our alternatives? As you read this chapter, think about how justice can be achieved (or not) when offenders are sentenced to corrections in the community.

INTRODUCTION

"I would like to use my college degree to help people who have gotten into trouble. Should I work in probation or parole? Should I work with adults or juveniles? And what kind of work would I be doing?" These questions have a ring of familiarity to most if not all criminal justice professors, as many students today seek to make their contributions to society by working with and trying to reform criminal offenders.

The word *probation* is probably familiar to most college and university students. If they don't "make the grades," they can be placed on academic probation; if their athletic programs fail to adhere to the rules of the National Collegiate Athletic Association (NCAA), the institution and its athletic program(s) can be placed on probationary status. The common thread here is that each group or individual is served warning that it had better change its behavior, or more severe penalties will follow. And so it is with criminal offenders—most of whom are sentenced to a *community-based* form of punishment, with probation being a commonly used form.

Community corrections: probation, parole, and a variety of other measures that offer convicted offenders an alternative(s) to incarceration.

A majority of persons under correctional supervision are on probation or parole—together termed **community corrections**. Probation and parole alone represent a significant component of the U.S. criminal justice system, with about 4.5 million adults now being supervised in the community; about 3.7 million of these adults are on probation in lieu of incarceration, and the remainder are on early release from incarceration but under conditions as per terms of their parole.[1]

This chapter examines probation, parole, and a variety of other measures that constitute a broad array of alternatives to incarceration. It begins by looking at both philosophical and economic arguments for having alternatives to incarcerations, followed by discussions of the origins and contemporary aspects of probation and parole. Included are the rights that probationers and parolees have when the state wishes to remove their freedom and send them to prison, as well as the functions of probation and parole officers—including the impact of high caseloads and the debate concerning whether or not these officers should be armed. The chapter then discusses intermediate sanctions. These alternatives to incarceration may be lesser known but are still quite beneficial to offenders and help to decrease prison and jail populations; they include intensive supervision (of probation), house arrest, electronic monitoring, shock incarceration/boot camps, day reporting centers, halfway houses, and furloughs. Following a review of the risk-need-responsivity (RNR) model, which is used to assess what programs to use with offenders to stop the cycle of reoffending and reincarceration, is a discussion of a relatively new movement in criminal justice: restorative justice.

Probation: an alternative to incarceration in which the convict remains out of jail or prison and in the community and thus on the job, with family, and so on, while subject to conditions and supervision of the probation authority.

WHY ALTERNATIVES TO INCARCERATION?

The leading alternative to incarceration is **probation**, which was discussed briefly in Chapter 1 and is defined as the court's allowing a convicted person to remain at liberty in the community, while being subject to certain conditions and restrictions on his or her activities.

The United States is not soft on crime, but there are several valid reasons for using **alternatives to incarceration**:

Alternatives to incarceration: a sentence imposed by a judge other than incarceration, such as probation, parole, shock probation, or house arrest.

- It allows the offender to remain in the community, which has a greater rehabilitative effect than incarceration, thereby reducing recidivism (repeat offending).
- It allows the offender to take greater advantage of treatment or counseling options.
- It allows the offender to avoid the "pains" of imprisonment.
- It is far less expensive.
- It permits ongoing ties with family, employment, and other social networks.

However, to be effective, a real alternative to incarceration needs to have three elements: It must incapacitate offenders enough so that it is possible to interfere with their lives and activities to make committing a new offense extremely difficult; it must be unpleasant enough to deter offenders from wanting to commit new crimes; and it has to provide real and credible protection for the community.[2]

More than a half century ago, the President's Commission on Law Enforcement and Administration of Justice (1967) endorsed community-based corrections—the use of probation and parole—as a humane, logical, and effective approach for working with and changing criminal offenders. According to the commission, that includes

> building or rebuilding solid ties between the offender and the community, obtaining employment and education, securing in the large sense a place for the offender in the routine functioning of society. This requires . . . efforts directed towards changing the individual offender (and) mobilization and change of the community and its institutions.[3]

The demand for prison space has created a reaction throughout corrections.[4] With the cost of prison construction now exceeding a quarter of a million dollars per cell in maximum-security institutions, cost-saving alternatives are becoming more attractive, if not essential.

The realities of prison construction and overcrowding have led to a search for intermediate punishments.[5] This in turn has brought about the emergence of a new generation of programs, making community-based corrections, according to Barry Nidorf, a "strong, full partner in the fight against crime and a leader in confronting the crowding crisis."[6] Economic reality dictates that cost-effective measures be developed, and this is motivating the development of intermediate sanctions.[7]

ORIGINS OF PROBATION AND PAROLE

The concepts of probation and parole have long and interesting histories.

Probation Begins: The Humble Shoe Cobbler

Although contemporary probation has roots dating to biblical times,[8] its history in the United States dates to the 19th century. "Judicial reprieve" was used in English courts to serve as a temporary suspension of sentence to allow the defendant to appeal to the Crown for a pardon. In the United States, the suspended sentence was used as early as 1830 in Boston and became widespread in American courts, even though there was no statutory provision for it. By the mid-19th century, though, many courts were using a judicial reprieve to suspend sentences.[9] This posed a legal question: Could judges suspend sentences wholesale, after trials that were scrupulously fair, simply to give the defendant a second chance?[10]

In 1916, the U.S. Supreme Court, in a decision affecting only the federal courts, held that judges did not have the discretionary authority to suspend sentences. However, the Court

Inmates on parade in Elmira's yard.

ruled that Congress could authorize the temporary or indefinite suspension of sentences; this led to the development of probation statutes.[11]

John Augustus, a Boston shoe cobbler, is credited as being the "father of probation." In 1841, Augustus appeared in court on behalf of a drunkard; as Augustus later explained, "I was in court one morning . . . in which the man was charged with being a common drunkard. He told me that if he could be saved from the House of Correction, he never again would taste intoxicating liquors; I bailed him, by permission of the court."[12] During his first year of service as an unpaid, volunteer probation officer, Augustus assisted 10 drunkards. By Augustus's own account, he eventually bailed "eleven hundred persons, both male and female."

Augustus performed several tasks that are reminiscent of modern probation. He investigated each case—inquiring into the offender's character, age, and influences—and kept careful records of each person's progress. His probation work soon caused him to fall into financial difficulties, however, requiring his friends' monetary assistance. Augustus died in 1859.

By 1869, the Massachusetts legislature had required that a state agent be present if court actions might result in the placement of a child in a reformatory (the forerunner of today's caseworkers). Then, in 1878, Massachusetts passed the first probation statute, mandating an official state probation system with salaried probation officers. Other states quickly followed suit:

- By 1900, Vermont, Rhode Island, New Jersey, New York, Minnesota, and Illinois passed probation laws.
- By 1910, 32 more states passed legislation establishing juvenile probation.
- By 1930, juvenile probation was legislated in every state except Wyoming.[13]

Parole Origins: Alexander Maconochie

The word *parole* stems from the French *parol*, or "word of honor," which was a means of releasing prisoners of war who promised not to resume arms in a current conflict.[14] One writer cited the year 1840 as one in which "one of the most remarkable experiments in the history of penology was initiated."[15]

In that year, Alexander Maconochie became superintendent of the British penal colony on Norfolk Island, about 930 miles northeast of Sydney, Australia. He began a philosophy of punishment based on reforming offenders: The convict was to be punished for the past while being trained for the future. Maconochie advocated open-ended ("indeterminate") sentences. His system worked, although it was harshly ridiculed by some Australians as "coddling criminals."

Returning to England in 1844, Maconochie began writing and speaking of his experiment. One of those impressed by Maconochie was Walter Crofton, who in 1854 became director of the renowned Irish system of penal management. Crofton implemented, among many other things, a "ticket of leave" system, allowing inmates to be conditionally released from prison, to be supervised by the police. Crofton recommended a similar system for the United States.

In 1876, when Zebulon Brockway was appointed superintendent of the Elmira Reformatory in New York, he drafted a statute providing for indeterminate sentences. Continued good behavior by inmates resulted in early release—America's first parole system. Paroled inmates remained under the jurisdiction of reformatory authorities for an additional 6 months, during which the parolee was required to report on the first day of every month to his appointed guardian and provide an account of his conduct and situation (a decade after Elmira began operations, New York opened three women's reformatories). This system was copied by other states; it was expanded further by the Great Depression, which abolished economic exploitation of convict labor.[16] Once introduced in the United States, parole spread fairly rapidly. In doing so, it survived an early series of constitutional challenges.[17] A 1939 survey reported that, by 1922, parole existed in 44 states, the federal system, and Hawaii.[18] Mississippi adopted a parole law in 1944, becoming the last state to do so.

A few reasons have been offered for the relatively rapid spread of parole legislation:

- There was general dissatisfaction with the determinate sentencing provisions of the time, and parole was seen as a response to some of the criticisms: Parole would promote reformation of prisoners by providing an incentive to change; at the same time, it would serve as a means of equalizing disparate judicial sentences.[19]
- Release before sentence expiration was already an aspect of most prison systems—through "good-time" deductions, which began in New York in 1817, and through gubernatorial clemency, which was used far more extensively than today.
- Parole was believed useful for enforcing prison discipline and for controlling prison population levels.[20]

PROBATION AND PAROLE TODAY

Probation: Eligibility and Legal Rights

As noted earlier, today about 3.7 million adults are on probation in lieu of being incarcerated. As seen in Table 11.1, more than half (59 percent) of those persons were convicted for committing a felony; furthermore, about one-fourth of them committed a drug offense, while another 26 percent committed a property offense, and 20 percent committed a violent offense.[21]

Table 11.1 /// Characteristics of Adults on Probation, 2000, 2015, and 2016

Characteristic	2000	2015	2016
Sex	100%	100%	100%
Male	78	75	75
Female	22	25	25
Race/Hispanic origin*	100%	100%	100%
White	54	55	55
Black/African American	31	30	28
Hispanic/Latino	13	13	14
American Indian/Alaska Native	1	1	1
Asian/Native Hawaiian/other Pacific Islander	1	1	1
Two or more races	. . .	—	—
Status of supervision	100%	100%	100%
Active	76	76	75
Residential/other treatment program	. . .	1	1

(Continued)

Table 11.1 /// (Continued)

Characteristic	2000	2015	2016
Financial conditions remaining	. . .	2	2
Inactive	9	4	4
Absconder	9	7	7
Supervised out of jurisdiction	3	2	2
Warrant status	. . .	5	5
Other	3	4	4
Type of offense	100%	100%	100%
Felony	52	57	59
Misdemeanor	46	41	40
Other infractions	2	2	2
Most serious offense	100%	100%	100%
Violent	. . .	20	20
Domestic violence	. . .	4	4
Sex offense	. . .	4	4
Other violent offense	. . .	13	13
Property	. . .	28	26
Drug	24	25	24
Public order	24	15	17
DWI/DUI	18	13	14
Other traffic offense	6	2	2
Other**	52	12	13

Note: Detail may not sum to total due to rounding. Estimates are based on most recent data and may differ from previously published statistics. See *Methodology.* Characteristics are based on probationers with a known type of status.

—Less than 0.05%.

. . . Not available.

*Excludes persons of Hispanic or Latino origin, unless specified.

**Includes public order offenses.

Source: Bureau of Justice Statistics, *Probation and Parole in the United States,* April 2018, https://www.bjs.gov/content/pub/pdf/ppus16.pdf, p. 17.

Judges consider a number of factors when evaluating the eligibility of an offender for probation:

- The nature and seriousness of the current offense
- Whether a weapon was used and the degree of physical or emotional injury, if any, to the victim
- Whether the victim was an active or a passive participant in the crime

- The length and seriousness of the offender's prior record
- The offender's previous success or failure on probation
- The offender's prior incarcerations and success or failure on parole[22]

Although a decision is made for the offender to remain free and avoid incarceration, the probationer must still abide by certain conditions that the court and probation officers will put in place to govern his or her behavior. If the probationer does not comply with those conditions, there are two possible types of violations that might be committed—**technical violations** or **substantive violations**:

Technical violation: in probation and parole, when one violates certain conditions that must be obeyed to remain out of prison, such as violating curfew, using drugs or alcohol, or not maintaining a job.

Substantive violation: an allegation that one was arrested for a new criminal offense while serving probation.

- *Technical violations*: These may include failing to pay court costs or fines, missing a probation meeting, using alcohol, violating curfew, associating with other known criminals, failing to submit to a mandatory drug test, failing a drug test, failing to complete community service, or engaging in out-of-state travel or changing address without permission.
- *Substantive violations*: These occur when the probationer commits a new criminal offense.

Following are several factors that the judge and prosecutor may take into account when considering a probation violation:

- The seriousness and nature of the probation violation
- The history of previous probation violations
- New criminal activity surrounding the probation violation
- Aggravating and mitigating circumstances of the probation violation
- The probation officer's and/or probation department's view of the probation violation
- The probation violation with respect to the probation term (whether it occurred at the beginning, middle, or end of the probationary term)

About 18 percent of probationers exit probation supervision each year for some untoward reason, for example, they were incarcerated for a new crime, violated a condition of supervision, or absconded.[23] The violation of probation is a serious offense, can be a felony, and can result in a judge's revoking probation. As soon as a probation violation occurs, an arrest may soon follow, and the defendant may be ordered to appear in court for a probation violation hearing. If a judge revokes probation, state laws often allow the judge to impose the maximum penalty for the charge.

Certain rights have been afforded probationers who are about to undergo a probation **revocation** hearing. First, in *Mempa v. Rhay* (1967), the U.S. Supreme Court held that such hearings are a "critical stage" where substantive rights could be lost, and that the Sixth Amendment therefore required the presence of counsel to help in "marshaling facts."[24] Later, in 1973, the Supreme Court decided *Gagnon v. Scarpelli*, in which a probationer had his probation revoked without a hearing; the court determined that probationers had certain due process rights, including the following:[25]

Revocation: the court's revoking probation or parole status for the purpose of returning an offender to prison (usually for not following the conditions of probation or parole, or for committing a new offense).

- Notice of the alleged violation
- A preliminary hearing to determine probable cause
- The right to present evidence
- The right to confront adverse witnesses
- A written report of the hearing
- A final revocation hearing

You Be the . . .

Judge

After 33 days of testimony from 60 witnesses and 400 pieces of evidence, a jury acquitted 25-year-old Casey Anthony in July 2011 on charges of first-degree murder in the June 2008 death of her 2-year-old daughter, Caylee. As you read in Chapter 7, after lying many times to detectives and appearing culpable under the evidence, Anthony was charged with killing Caylee by covering her mouth with duct tape (her body was found six months later in woods near Anthony's home). The defense maintained that Caylee had accidentally drowned while in her parents' care and that Anthony's father attempted to cover up the death.

After acquittal for murder (the actual cause of Caylee's death was never established, nor was Anthony's DNA found on her daughter's skeletal remains, among other evidentiary shortcomings), the only convictions that remained were for Anthony's lying to detectives about her actions and what happened to the child. The task then facing the judge was to determine whether and how to sentence Anthony.

In the end, Anthony was given one year's probation for a 2010 check fraud conviction, which she served in an undisclosed Florida location. The judge who presided over the trial later stated that he believed Anthony did in fact kill Caylee, by accidentally administering too much chloroform in an effort to sedate her (during the trial, evidence showed that Anthony researched the use of chloroform, which was once used as an inhaled anesthetic during surgery).

1. Had you been the judge in this case, would you have granted probation to Anthony?
2. Why did this case, relative to most, receive and continue to receive such extraordinary public attention?
3. Do you believe Anthony would pose a danger to other children she might bear or with whom she might otherwise be associated in the future?
4. What elements might have contributed to Anthony's acquittal?

Sources: For more background and legal analysis of the case, see David Lat, "Trials and Error: The Casey Anthony Case," *Above the Law,* June 1, 2012, http://abovethelaw.com/2012/06/trials-and-error-the-casey-anthony-case/?rf=1.

Parole: Eligibility and Legal Rights

Table 11.2 shows that about one-third of all parolees were incarcerated for a drug offense, one-third were incarcerated for a violent offense, and 21 percent had committed a property offense.[26]

Similar to probation decision making, judges will take into account the following general criteria when considering whether or not an inmate should be granted parole:

- The nature and seriousness of the current offense, including aggravating and mitigating circumstances
- Statements made in court concerning the sentence
- The length and seriousness of the offender's prior record
- The inmate's attitude toward the offense, family members, the victim, and authority in general
- The attitude of the victim or victim's family regarding the inmate's release
- The inmate's insight into causes of past criminal conduct
- The inmate's adjustment to previous probation, parole, or incarceration
- The inmate's participation in institutional programs
- The adequacy of the inmate's parole plan, including residence and employment[27]

Like probationers, the parolee must still abide by certain conditions that the court and parole officers will put in place. As with probation, if the parolee does not

Table 11.2 /// Characteristics of Adults on Parole, 2000, 2015, and 2016

Characteristic	2000	2015	2016
Sex	100%	100%	100%
Male	88	87	87
Female	12	13	13
Race/Hispanic origin*	100%	100%	100%
White	38	44	45
Black/African American	40	38	38
Hispanic/Latino	21	16	15
American Indian/Alaska Native	1	1	1
Asian/Native Hawaiian/other Pacific Islander	—	1	1
Two or more races	. . .	—	—
Status of supervision	100%	100%	100%
Active	83	83	82
Inactive	4	5	5
Absconder	7	6	7
Supervised out of state	5	4	4
Financial conditions remaining	. . .	—	—
Other	1	3	2
Maximum sentence to incarceration	100%	100%	100%
Less than one year	3	6	6
One year or more	97	94	94
Most serious offense	100%	100%	100%
Violent	. . .	32	30
. . . Sex offense	. . .	8	8
. . . Other violent offense	. . .	24	22
Property	. . .	21	21
Drug	. . .	31	31
Weapon	. . .	4	4
Other**	. . .	13	13

Note: Detail may not sum to total due to rounding. Estimates are based on most recent data and may differ from previously published statistics. See *Methodology.* Characteristics are based on probationers with a known type of status.

—Less than 0.05%.

. . . Not available.

*Excludes persons of Hispanic or Latino origin, unless specified.

**Includes public order offenses.

Source: Bureau of Justice Statistics, *Probation and Parole in the United States*, April 2018, https://www.bjs.gov/content/pub/pdf/ppus16.pdf, p. 24.

comply with conditions of parole, there are two possible types of violations that might be committed:

- *Technical violations*: These may include failing to report regularly to the parole officer (PO), not keeping the PO advised of changes in address, not providing notice of change in employment, not reporting any new arrest, associating with other felons, traveling in violation of time or distance constraints, possessing weapons, using alcohol or drugs, wearing electronic monitoring device improperly, or contacting a partner who was victimized in a domestic violence case.
- *Substantive violations*: These occur when the parolee commits a new criminal offense.

If either or a combination of these violations occurs, the parole officer may initiate charges of parole violation. The violation of parole conditions is viewed as a serious offense and can be a felony and result in a judge's revocation order. As soon as a parole violation occurs, an arrest may soon follow, and the defendant may be ordered to appear in court for a violation hearing (described more below). Over 25 percent of adult parolees exit that status each year due to some negative event; for example, they received a new sentence, parole was revoked because they committed a new crime, they violated a condition of supervision, or they absconded.[28]

Until 1972, decisions to revoke one's parole and return him or her to prison could be made arbitrarily by individual parole officers. However, the U.S. Supreme Court, in *Morrissey v. Brewer,* held in that year that parolees who faced the possibility of losing their freedom possessed certain rights under the Fourteenth Amendment:[29]

- Written notice of the alleged violation(s)
- A preliminary hearing to establish whether there is probable cause that the parolee violated the conditions of parole
- Disclosure of the evidence against the parolee
- The opportunity to be heard in person and to present witnesses and documentary evidence
- The right to confront and cross-examine adverse witnesses
- A neutral and detached body to hear the evidence
- A written statement by the fact finders concerning the evidence relied upon for any revocation decision

If the hearing officer (not necessarily a judge) finds that probable cause exists to believe that a violation has occurred, the hearing will move to the adjustment phase, which is where the information is introduced about why the individual should or should not be continued on supervision.

LUCY NICHOLSON/REUTERS/Newscom

One of the typical conditions placed on a probationer is that he or she submit to mandatory drug tests.

One of the most important groups of individuals that parole board members come in contact with is victims of crime. At one time, victims were not typically involved with the parole process in the United States, but that has changed. Through strong victim advocacy, crime victims have begun to be recognized as key stakeholders in the criminal justice process. Today, paroling authorities typically provide victims with information concerning any activity in their offender's case, provide opportunities for input to the board—in person and/or in writing—and take into account the needs of and dangers to victims as part of their decision-making procedures. A number of states now appoint victims of crime or victim advocates as members of their paroling authorities.[30] Figure 11.1 shows the parole decision-making guidelines employed by the Pennsylvania Board of Probation and Parole.

Figure 11.1 /// Pennsylvania Board of Probation and Parole, Parole Decision-Making Guidelines

PENNSYLVANIA BOARD OF PROBATION AND PAROLE

PAROLE DECISION MAKING GUIDELINES

Name ____________________

Parole No. __________ **SID No.** __________ **Institution No.** __________

Date of Interview __________ **Institution** __________

Interview Type ___ **Minimum** ___ **Review** ___ **Reparole Review** ___ **Parole Application**

Violence Indicator

1. Instant Offense

Violent	☐	+3
Non-Violent	☐	+1

(1) Murder, Voluntary Manslaughter, Aggravated Assault, Robbery, Arson, Burglary (Residential), Assault by Prisoner, Assault by Life Prisoner, Kidnapping, Extortion Accompanied by Threats of Violence, all Sex Crimes, and criminal attempt, criminal conspiracy, and/or criminal solicitation to commit any of the above-noted offenses.

Risk/Needs Assessment

2. Level of Service Inventory - Revised

Raw Score: _______

High Risk	☐	+3
Medium Risk	☐	+2
Low Risk	☐	+1

Sex Offender Risk Assessment (Static 99)

Raw Score: _______

High Risk	☐	+3
Medium Risk	☐	+2
Low Risk	☐	+1

(All offenders considered for parole shall be assessed using the Level of Service Inventory - Revised ("LSI-R"). Offenders convicted of a sex offense shall be assessed using the LSI-R as well as the Sex Offender Risk Assessment Instrument. ***The higher level of risk shall be used for all sex offenders.***

Institution Adjustment

3. Institutional Programming

Unacceptable Program Compliance	☐	+3
Reasonable Efforts (2)	☐	+2
Currently Involved	☐	+1
Completion of Required Programs (3)	☐	+0

(2) No access or on waiting list.
(3) Includes offenders who are currently involved and will complete prior to release

Source: Peggy B. Burke, *A Handbook for New Parole Board Members* (Washington, D.C.: U.S. Department of Justice, National Institute of Corrections, and Association of Paroling Authorities International, April 2003), pp. 41–42, http://www.apaintl.org/documents/CEPPParoleHandbook.pdf.

Note: The Level of Service Inventory—Revised (LSI-R) mentioned in the figure includes both criminal history items and measures of offender needs, such as substance abuse, employment, and special-needs accommodations. The tool is administered during a standardized, one-hour interview, and it evaluates indicators of program success (e.g., residence, family ties, employment) and predictors of program failure (e.g., prior convictions, prior failures to appear, prior violations of sentence). It includes 54 items, and responses are totaled to give a risk-needs score. *Source:* Joan Petersilia, *When Prisoners Come Home* (New York: Oxford University Press, 2003), pp. 72–73.

You Be the . . .

Parole Board

Bernie Madoff's fraudulent Ponzi scheme bilked investors out of tens of billions of dollars, for which he was sentenced to 150 years in prison. Indeed, the former NASDAQ chairman victimized nearly 5,000 clients involving more than 20,000 individual investors, pension funds, corporations, universities, and celebrities; to date, several billion dollars in payouts to victims have begun. Meanwhile, as a federal prison inmate in Durham, North Carolina, Madoff, now nearly 79 years of age, is said to have an easy job selling commissary items to inmates. Reportedly, other inmates accord him great respect due to the scale of his crimes and for his refusal to implicate any of his employees in the scheme (while also providing other inmates stock tips and financial advice). He is also said to be experiencing heart problems and physical pain, having no family visitors (two sons have died since his arrest), and avoiding any responsibility or remorse for his crimes. Instead, he is said to be devoting a large amount of his time to attempting to rehabilitate his public image and shift blame to his investors—whom he argues expected unrealistic returns on their investments, which then led to his being set up for a fraud conviction. These claims have caused a former FBI profiler to deem Madoff a classic sociopath, someone who functions without a conscience and has an antisocial personality.

There appear to be two "faces" of Bernie Madoff. On the one hand, he appears to be an ailing, elderly inmate who is estranged from his family, had no prior record of violence, and is serving a lengthy sentence. On the other hand, he perpetrated the largest financial fraud in history, is apparently unwilling to accept responsibility or express remorse for his crimes, and caused devastating losses to tens of thousands of Americans. Given all that,

1. Would you vote to grant Madoff parole at the soonest possible time?
2. If not, are there rehabilitative or victim restitution programs or activities in which Madoff might participate that would help to support his being granted parole in the future?

Sources: Brian Ross, Rhonda Schwartz, and Megan Christie, "Bernie Madoff 'Doing Fine' in Prison Despite Heart Issues, Few Visitors," ABC News, February 2, 2016, http://abcnews.go.com/US/bernie-madoff-fine-prison-heart-issues-visitors/story?id=36552841.

Do Probation and Parole Work?

Do probation and parole work by reducing arrests? An ambitious study by California's Council of State Governments (CSG) Justice Center attempted to learn the answer, examining more than 2.5 million adult arrest, probation, and parole supervision records from 11 agencies in four cities—Los Angeles, Redlands, Sacramento, and San Francisco—over a 42-month period.[31] The CSG wanted to determine (1) the extent to which people on probation and parole contribute to crime, as measured by arrests, and (2) the types of crimes these people are most likely to commit.[32]

The findings were a bit surprising. First, people under probation or parole supervision accounted for 22 percent of total arrests—which means that nearly 8 of 10 arrestees were not supervised, certainly a respectable finding. However, one in three arrests for *drug* crimes involved someone who was on probation or parole—in fact, people under supervision were more likely to be arrested on drug offenses than for violent, property, or other types of crimes.[33] During the study period, the number of total arrests declined by 18 percent, while the number of arrests of people who were under supervision declined by *40 percent*—61 percent for persons on parole and 26 percent for individuals under probation supervision.[34]

Figure 11.2 /// Do Probation and Parole Work?

RESEARCH FINDINGS

Approximately one in five arrests involved an individual under probation or parole supervision; the majority of total arrests involved people who were not under supervision.

A key objective of this study was to determine to what extent people under correctional supervision drove arrest activity. To make that determination, researchers matched arrest data with parole and probation supervision data.

Supervision Status Among All Adult Arrestees:

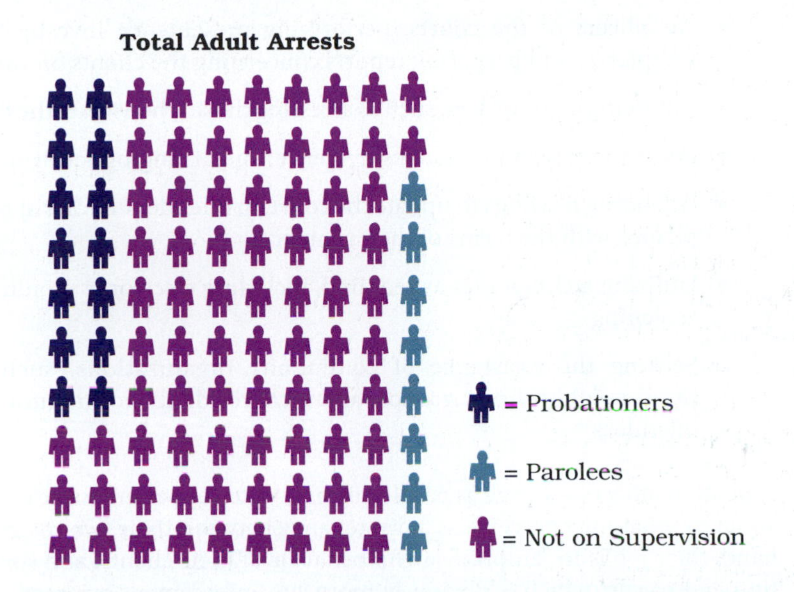

Designation	Adult arrests	% of total
Total	476,054	100%
Parolees	40,476	8.5%
Probationers	66,251	13.9%
Not Supervised	369,327	77.6%

Source: Council of State Governments Justice Center, *The Impact of Probation and Parole Populations on Arrests in Four California Cities*, January 2013, http://www.cdcr.ca.gov/Reports/docs/External-Reports/CAL-CHIEFS-REPORT.pdf, p. 14.

It seems, therefore, that supervision generally works, but more work remains to be done in terms of keeping parolees and probationers from reoffending for drug abuse. Figure 11.2 displays some of the findings.

FUNCTIONS OF PROBATION AND PAROLE OFFICERS

Both probation and parole officers perform similar functions; in fact, in some states, the jobs of parole and probation officers are combined. Both must possess important skills, such as good interpersonal communication, decision-making, and writing skills. They operate independently, with less supervision than most prison staff experience. Both are trained in the techniques for supervising offenders and then assigned a caseload. They may be on call 24 hours a day to supervise and assist offenders at any time.[35]

Probation and "Front-End" Duties

Probation officer: one who supervises the activities of persons on probation.

Probation officers supervise offenders at the *front* end of the sentencing continuum—those offenders with a suspended prison sentence—monitoring their behavior in the community and their compliance with the conditions of probation. Probation officers usually work with either adults or juveniles exclusively. Only in small, usually rural, jurisdictions do probation officers counsel both adults and juveniles.

Probation officers have a number of duties, including

- Reporting to the court any violations of probation
- As officers of the courts, performing presentence investigations (as discussed in Chapter 7) and preparing reports concerning the clients on their caseload
- Enforcing court orders, such as arresting those who violate the terms of their probation
- Performing searches, seizing evidence, and arranging for drug testing
- Attending hearings to update the court on offenders' efforts at rehabilitation and compliance with the terms of their sentences
- Utilizing technologies as required, including electronic monitoring devices and drug screening
- Seeking the assistance of community organizations, such as religious institutions, neighborhood groups, and local residents, to monitor the behavior of many offenders[36]

Probation officers—as well as parole officers—often experience role conflict, which is brought about by what they perceive as a discrepancy between their two main functions: On the one hand, their job is to "enforce" lawful behavior of their clients, and sometimes revoke probation and parole, which is obviously more law enforcement oriented in nature; on the other hand, they must also be empathetic and understanding, and provide guidance and counseling to their clients, which is more of a social worker role.[37] These seemingly contradictory roles can contribute to job stress.

Parole and "Back-End" Duties

Parole officer: one who supervises those who are on parole.

Parole officers perform *back*-end duties of the sentencing continuum, supervising offenders who have been released from prison. Parole officers are most often employed by the state department of corrections; the state criminal justice department; or a youth authority/juvenile corrections, county, or federal justice department. Like probation officers, parole officers supervise offenders through personal contact with the offenders and their families; this can be quite dangerous (and lead to role conflict, discussed previously), as parole officers work with paroled convicts, their friends, and their family. Like probationers, some parolees are required to wear an electronic device so that probation officers can monitor their location and movements. Parole officers' duties include

- Helping parolees adjust back into society, as well as avert any actions that would jeopardize their parole status
- Developing a plan for the parolee before he or she is released from prison
- Planning employment, housing, health care, education, drug screening, and other activities that help the parolee's rehabilitation and functioning in a community environment
- Arranging for offenders to get substance abuse rehabilitation or job training
- Attending parole hearings and making recommendations based on their interviews with and surveillance of parolees[38]

The Burden of Large Caseloads

Caseload refers to the average number of cases supervised by a probation or parole officer in a given period. Each case represents an offender on probation or parole supervised by an individual officer. As John Conrad observed,

Caseload: the number of cases awaiting disposition by a court, or the number of active cases or clients maintained by a probation or parole officer.

> There is much that a good probation/parole officer can do for the people on his or her caseload. A parole officer who makes it clear that, "fellow, if you don't watch your step I'm gonna run your ass right back to the joint," is not in a position to be helpful as a counselor or facilitator. With the best intentions, a[n] officer struggling with the standard unwieldy caseload of 100 or more will deal with emergencies only, and sometime will not be able to do that very well.[39]

As noted earlier, about 4.5 million persons are on either probation or parole in the United States. Is there a precise number of offenders who can be supervised effectively by an officer? The answer is no, because the number of offenders an officer can supervise effectively is a function of the type of offenders being supervised by certain officers; all offenders and officers are unique and bring different knowledge, skills, capacities, and competencies. The American Probation and Parole Association (APPA) asks the question, rhetorically,

> How many patients can a surgeon operate on in a given day? How many cars can a mechanic fix each week? How many haircuts can a barber complete in a month? It does not take an expert in any of these fields to realize the answer normally is that *it depends*.[40]

Of course, caseload size can affect the quality of supervision that an officer is able to provide—and can also bring the glare of the media when something goes horribly wrong. For example, a newspaper in Detroit published an exposé titled "Felons on Probation Often Go Unwatched." The article reported that the county had roughly 30,000 probationers and approximately 250 officers to supervise them—an average of nearly 120 offenders per officer. In one case, an officer was fired after a probationer was arrested for attempted murder and engaging in a shootout with police. The probationer was a fugitive, missing several office visits, but he was never reported as an absconder, nor was he listed as a fugitive at the time of the shooting. According to the article, the probation officer "was so overworked that she failed to get an arrest warrant for [the probationer] when he became a fugitive for missing his monthly probation office appointment. [The officer] still hadn't done so by March 28 when he was arrested."[41]

MediaNews Group/Orange County Register via Getty Images/MediaNews Group/Getty Images

Probation officers monitor offenders' behavior in order to check their compliance with the conditions of their probation. These probation officers are performing a compliance sweep at a probationer's home.

The APPA points out that caseload sizes have long been too large, and for several identifiable reasons:

> For at least the past four decades it has been well-known to professional insiders that probation and parole officer workloads exceed realistic potential for accomplishing the numerous tasks required to supervise offenders. The point here is that many departments are increasing caseloads to well over 200 offenders per officer, making it virtually impossible for offenders to receive adequate attention and interaction from officers to have any substantial rehabilitative effect. Compounding these issues is the current trend of concentrating on sex offenders, the infusion of electronic monitoring technologies, and increasing the use of probation for higher-risk offenders as well as widening the justice net to low risk offenders.[42]

To Arm or Not to Arm

Whether probation and parole officers should be armed has also been debated. Traditionalists believe that carrying a firearm contributes to an atmosphere of distrust between the client and the officer; they argue that if officers carry weapons, they are perceived differently than as counselors or advisors, whose purpose is to guide offenders into treatment and self-help programs. Conversely, some people view a firearm for these officers as essential for protecting them from the risks associated with confronting violent, serious, or high-risk offenders.[43] Officers must visit homes and places of employment in the neighborhoods in which offenders live; some of these areas are not safe. Equally or more dangerous is when officers must revoke parole or probation, which could result in the offender's imprisonment.

There is no national or standard policy regarding weapons, and officers themselves are not in agreement about being armed. Some states classify probation and parole officers as peace officers and grant them the authority to carry a firearm both on and off duty.[44] In sum, it would seem that the prudent decision concerning arming should be based on the need, officer safety, and local laws and policies.

OTHER ALTERNATIVES: INTERMEDIATE SANCTIONS

Intermediate sanctions: forms of punishment that are between freedom and prison, such as home confinement and day reporting.

In addition to the diversionary (or problem-solving) drug, mental health, and other specialty courts discussed in Chapter 7 that are reducing jail and prison populations, there are corrections programs—**intermediate sanctions**—that are less restrictive than total confinement but more restrictive than probation and that reduce those populations as well. In fact, a survey by the federal Bureau of Justice Statistics found that in addition to the 745,200 persons who are confined in U.S. jails, approximately 55,900 are supervised outside the jail facility under electronic monitoring, home detention, day reporting, community service, treatment programs, and other pretrial and work programs.[45] These and other forms of supervision are discussed in the following sections.

Intensive Supervision Probation and Parole

Intensive supervision probation and parole: post-release supervision that usually includes much closer and stricter supervision, more contact with offenders, more frequent drug tests, and other such measures.

Since their beginning, **intensive supervision probation and parole** have been based on the premise that increased client contact would enhance rehabilitation while affording greater client control. Current programs are largely a means of easing the burden of prison overcrowding.[46]

Intensive supervision programs can be classified into two types: those stressing diversion and those stressing enhancement. A *diversion program* is commonly known as a "front door" program because its goal is to limit the number of generally low-risk offenders who enter prison. An *enhancement program* generally selects already-sentenced probationers and parolees and subjects them to closer supervision in the community than they receive under regular probation or parole.[47] As of 1990, jurisdictions in all 50 states had instituted *intensive supervision probation* (ISP). Persons placed on ISP are supposed to be those offenders who, in the absence of intensive supervision, would have been sentenced to imprisonment. Although ISP is invariably more costly than regular supervision, the costs "are compared not with the costs of normal supervision but rather with the costs of incarceration."[48]

ISP is demanding for probationers and parolees and does not represent freedom; in fact, it may stress and isolate repeat offenders more than imprisonment does. Given the option of

Practitioner's Perspective

Probation Officer

Name: Jessica Johnston
Position: Probation Officer
Location: Denver, Colorado

What is your career story? I was always interested in the field of justice, but I never really had a strong concept of how that would look as a career. So I studied it in school. I got a bachelor's degree in sociology and psychology, and then went on to pursue a master's degree in criminology. It was then that I realized I really wanted to have direct service experience. So probation was just a really natural fit given the academic studies that I had pursued. I originally started in a municipal probation office, where I handled municipal violations, city ordinance violations, and other things like that. I ultimately moved into a state and county probation office. I think probation is this wonderful opportunity for intervention in people's lives, and it offers them a chance to stay connected in the community and really make the necessary changes in the least invasive way possible.

What misconceptions do you often hear about this position? I've encountered a couple common misconceptions about probation officers. The first one is that we're the tough guy, that we're really throwing the book at folks and all the emphasis is on accountability. But what you find when you start working in the field is that you wear many hats. Yes, you are a source of accountability for probationers, but you're also a resource for them. You help connect them to services they really need in order to make a meaningful change in their lives. That gets lost on a lot of people who don't work in the field, that the role is much more of a relationship where you build a rapport, and you really try to help people to take control of their circumstances. Another big misconception is that a lot of people who enter this field, including myself, really want to work with the highest risk people, the worst of the worst. But when you start out, you often start with the lower risk caseloads, so you can get your feet wet and learn the ropes. I actually came to enjoy working with low- and moderate-risk clients far more, because what I found was that on probation, most people are actually ordinary people that have made a mistake or have an unresolved problem or issue. It's really an exciting experience to be part of an intervention that helps them self-correct or get them the help they need.

What role does diversity play in this position? The front line of probation is dominated by women, which is different from how it was historically. I do think that management positions continue to be held by more males. That's slowly changing, but I still see that divide between men in management and women as front-line workers. In terms of the clientele, it's really contextualized by the jurisdiction you're in. There are a lot of rural places in Colorado that just don't have the diversity that urban Denver does, so we've had a lot of conversations about what our client population looks like and whether that's reflective of what it should look like. We're moving in the right direction, but we still see a lot of disproportionate representation of minorities, and we see them underrepresented in the specialized programs. Where I worked was an urban area, so we had officers of every race. We also had a number of officers who spoke Spanish, which was critical because we had a large Hispanic population in our area. We absolutely needed officers who came with the interpretation skills and were able to converse with clients and work with them just as we did with our English-speaking clients.

To learn more about Jessica Johnston's experiences as a Probation Officer watch the Practitioner's Perspective video in SAGE Vantage.

serving prison terms or participating in ISPs, many offenders have chosen prison.[49] Consider the alternatives now facing offenders in one western state:

- *ISP:* The offender serves two years under this alternative. During that time, a probation officer visits the offender two or three times per week and phones on the other days. The offender is subject to unannounced searches of his or her home for drugs and has his or her urine tested regularly for alcohol and drugs. The offender must strictly abide by other conditions as set by the court: not carrying a weapon, not socializing with certain persons, performing community service, and being employed or participating in training or education. In addition, he or she will be strongly encouraged to attend counseling and/or other treatment, particularly if he or she is a drug offender.
- *Prison:* The alternative is a sentence of two to four years, of which the offender will serve only about three to six months. During this term, the offender is not required to work or to participate in any training or treatment but may do so voluntarily. Once released, the offender is placed on a two-year routine parole supervision and must visit his or her parole officer about once a month.[50] Although evidence of the effectiveness of this program is lacking, it has been deemed a public relations success.[51] Intensive supervision is usually accomplished by severely reducing caseload size per probation or parole officer, leading to increased contact between officers and clients or their significant others (such as spouse or parents).[52]

This point also can be illustrated with a case in Colorado, in which a felon convicted of sexual exploitation of a child asked the court to send him to prison, rather than serving out 20 years on ISP.[53] The court agreed to revoke his probation, and he was sentenced to 12 years in prison. At the hearing, the offender argued that he did not want to be in the "probation/treatment" trap any longer.[54]

House arrest/home confinement: detention of offenders in their own homes; compliance is often monitored electronically.

House Arrest

Since the late 1980s, **house arrest** (also known as **home confinement**) has become increasingly common. With house arrest, offenders receive a "sentence" of detention in their own homes, and their compliance is often monitored electronically. The primary motivation for using this intermediate sanction is a financial one: the conservation of scarce resources. Note that house arrest is often used in tandem with electronic monitoring, discussed in the next section.

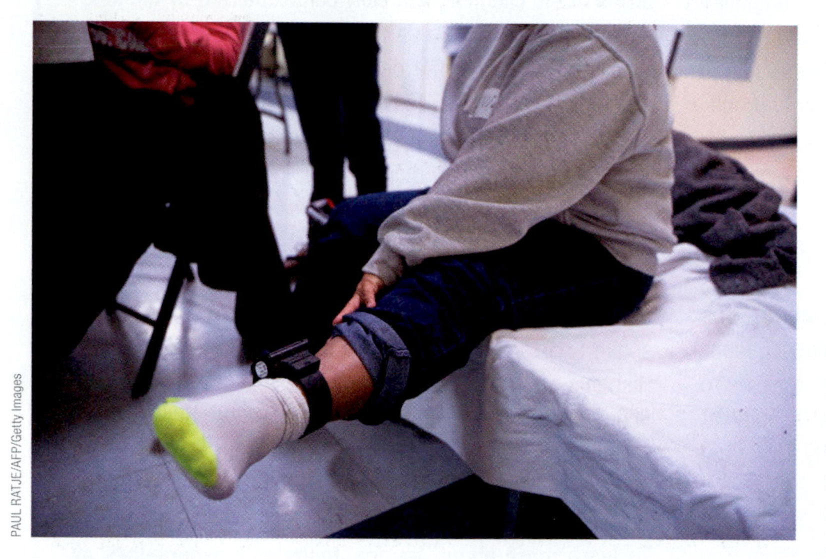

PAUL RATJE/AFP/Getty Images

Electronic monitoring systems are particularly useful for high-risk offenders, especially sex offenders for whom use of a global positioning system (GPS) is desirable.

Many people apparently feel that house arrest is not effective or punitive enough for offenders. Indeed, one study reported that only about 20 percent of the public think prisoners could be safely released into the community, even if under supervision;[55] and nearly half (44 percent) of the public feel that house arrest is not very effective or not effective at all.[56] Does house arrest work? Looking at a sample of 528 adult felony offenders who had been released from house arrest, Jeffery Ulmer found that the sentence combination associated with the least likelihood of rearrest was house arrest/probation.[57] The combinations of house arrest/work release and house arrest/incarceration were also significantly associated with decreased chances of rearrest compared to traditional probation. Furthermore, whenever any other sentence option was paired with house arrest, that sentence combination significantly reduced the chances of rearrest and the severity of rearrest charges.[58] House arrest puts the offender in touch with opportunities and resources for rehabilitative services (substance abuse or sex offender counseling, anger management classes, and so on), which supports the contention that for intermediate sanctions of any type to reduce recidivism, they must include a rehabilitative emphasis.[59]

AP Photo/Mike Groll

Correctional boot camps, also called shock incarceration, have been used in jails and prisons to place offenders in a quasi-military program to instill discipline and thus reduce recidivism, prison and jail populations, and operating costs.

Electronic Monitoring

Electronic monitoring (EM) or supervision can be used for a variety of offenders, but it is particularly useful for high-risk offenders, especially sex offenders for whom use of a global positioning system (GPS) is desirable.[60] The Urban Institute, in conjunction with the District of Columbia Crime Policy Institute, reports that EM reduces arrests by 24 percent, on average, for program participants.[61] Estimates suggest that EM costs about $750 per day and saves society $5,300 per program participant.[62]

Some early forms of EM technology are still in use today. There are two basic types of EM devices: active and passive. Active forms are continuous signaling devices attached to the offender that monitor his or her presence at a particular location. A central computer accepts reports from the receiver-dialer over telephone lines, compares them with the offender's curfew schedule, and alerts correctional officials to unauthorized absences.[63] Simpler systems consist of only two basic components: a transmitter and a portable receiver. The transmitter, which is strapped to the offender's ankle or wrist or worn around the neck, emits a radio signal that travels about one city block. By driving past the offender's residence, his or her place of employment, or wherever he or she is supposed to be, the officer can verify the offender's presence with the handheld portable receiver.[64] The passive type of EM involves the use of programmed devices that contact the offender periodically to verify his or her presence. One system uses voice verification technology. Another system uses satellite technology; the subject wears an ankle bracelet and carries or wears a portable tracking device about the size of a small lunchbox and weighing 3.5 pounds. A GPS satellite constellation is able to establish an offender's whereabouts within 150 feet of his or her location 24 hours a day.[65] However, technology is evolving and offender tracking systems continue to improve. The National Institute of Justice recently released a report specifying national minimum performance requirements and testing methods for EM devices to enhance public safety.[66]

Electronic monitoring: use of electronic devices (bracelets or anklets) to emit signals when a convicted offender (usually on house arrest) leaves the environment in which he or she is to remain.

Shock Probation/Parole

Shock probation/parole is another less costly intermediate alternative to incarceration that is supported by many correctional administrators. Typically used for first-time or early-career lower-level offenders who are eligible (drug crimes, larceny/theft), this form of corrections combines a brief exposure to incarceration with subsequent release. It allows sentencing judges to reconsider the original sentence to prison and, upon motion, to recall the inmate after a few months in prison and place him or her on probation, under conditions deemed appropriate. The idea is that the "shock" of a short stay in prison will give the offender a taste of institutional life and will make such an indelible impression that he or she will be deterred from future crime and will avoid the negative effects of lengthy confinement.[67] Like other alternatives to incarceration, shock probation can result in a cost savings (compared to longer term incarceration) and can provide the offender with community resources once on supervised probation.

Shock probation/parole: a situation in which individuals are sentenced to jail or prison for a brief period, to give them a taste or "shock" of incarceration and, it is hoped, turn them into more law-abiding citizens.

Boot Camps/Shock Incarceration

Boot camp: a short-term jail or prison program that puts offenders through a rigorous physical and mental regimen designed to instill discipline and respect for authority.

Correctional **boot camps**, also called shock incarceration, were first implemented as an intermediate sanction in 1983.[68] Early versions of boot camps placed offenders in a quasi-military program of 36 months' duration similar to a military basic training program. The goal was to reduce recidivism, prison and jail populations, and operating costs. Offenders generally served a short institutional sentence and then were put through a rigorous regimen of drills, strenuous workouts, marching, and hard physical labor. To be eligible, inmates generally had to be young, nonviolent offenders.

Unfortunately, early evaluations of boot camps generally found that participants did no better than other offenders without this experience.[69] The National Institute of Justice, in an overview of correctional boot camp research, reports that boot camps are generally not effective in reducing recidivism.[70] Only boot camps that were carefully designed, targeted the right offenders, and provided rehabilitative services and aftercare were deemed likely to save the state money and reduce recidivism.[71] As a result of these findings, the number of boot camps declined; by the year 2000, only 51 prison boot camps remained.[72] So-called second-generation boot camps have evolved over time, however, and added components such as alcohol and drug treatment and social skills training (some even include postrelease electronic monitoring, house arrest, and random urine tests); some boot camps have also substituted an emphasis on educational and vocational skills for the military components.[73] As an example of the latter, Pennsylvania's Motivational Boot Camp—one that combined militaristic exercise with multilayered treatment—had disappointing results with recidivism when boot camp participants were compared to those released from prison; however, more encouraging results were found regarding recidivism for participants who were employed or were repeat offenders. On average, a boot camp sentence reduced a prison stay by one year.[74]

Day reporting center: a structured corrections program requiring offenders to check in at a community site on a regular basis for supervision, sanctions, and services.

Day Reporting Centers

Another intermediate sanction that has recently gained popularity among correctional administrators and policy makers is the **day reporting center** (DRC). DRCs are places where offenders report with some frequency (usually once or twice a day); treatment services (job

Juvenile Justice Journal

Aftercare and Reentry

As with adult offenders, youths who are about to reenter the community after being in a custodial facility sorely need aftercare services. Indeed, up to two-thirds of all youths will be rearrested and one-third will be reincarcerated within a few years after release, so it is imperative that support services be in place to facilitate this reentry. Services should include wraparound support services that are collaborations between family members and a transition specialist to ensure that the transition is a successful one. Many of these youths have depression, schizophrenia, other mental disorders, or learning disabilities, so it is important that aftercare includes helping them to develop skills for participating in the workforce, school, and/or independent living; interacting appropriately with others; and developing a positive self-image and the ability to set personal goals.

Sources: Daniel P. Mears and Jeremy Travis, "Youth Development and Reentry," *Youth Violence and Juvenile Justice* 2, no. 1 (January 2004), pp. 3–20; also see David Altschuler and Shay Bilchik, "Critical Elements of Juvenile Reentry in Research and Practice," Justice Center, April 21, 2014, http://csgjusticecenter.org/youth/posts/critical-elements-of-juvenile-reentry-in-research-and-practice/.

training and placement, counseling, and education) are usually provided at the DRCs, either by the agency running the program or by other human services agencies.[75]

The purposes of DRCs are to heighten control and surveillance of offenders placed on community supervision, increase offender access to treatment programs, give offenders more proportional and certain sanctions, and reduce prison or jail crowding. One study of the effect of DRCs on recidivism, however, found no significant reduction in the rate of rearrest.[76]

Halfway Houses

Another approach to rehabilitation and incarceration is the **halfway house** (often called a community correctional center, a residential rehabilitation center, or, at the federal level, a residential reentry center). Here, convicted individuals who are serving alternative sentences or who have been released early from jail or prison live in a state-run or for-profit facility in hopes of transitioning to life outside. Like most reentry programs that attempt to transition offenders back into communities, the research on halfway house effectiveness is mixed. Some studies show declines in recidivism, while others show no effect on reoffending.[77] The idea of housing certain convicts in a less punitive setting was emphasized in 1961 by then–U.S. attorney general Robert Kennedy as a means of grooming young offenders for a law-abiding life.[78] Generally, such facilities are to provide a safe, structured, supervised environment, as well as employment counseling, job placement, financial management assistance, and other programs and services.[79]

Halfway house: a community center or home staffed by professionals or volunteers designed to provide counseling to ex-prisoners as they transition from prison to the community.

Furloughs

Another means by which a prison or jail inmate can leave a correctional facility is via furlough. For jail inmates, the purpose of furlough is commonly to allow—on a limited-time basis—them to obtain medical or mental health treatment, attend the birth of a child or a relative's funeral, and so on. For prison inmates, the release is also typically for a limited-time basis and for one of the following reasons: to transfer from one correctional facility to another or to home confinement, to attend to a family crisis or other urgent situation, or to engage in structured group programs (usually for vocational, medical, religious, educational, or recreational reasons) for a short period of time.[80]

CONFRONTING RECIDIVISTS: THE RISK-NEED-RESPONSIVITY (RNR) MODEL

How can the cycle of reoffending and reincarceration be stopped? Should an offender be confined in prison, placed on community supervision, or both? Several decades of research and experience have provided specific programs—practices and principles—that, when implemented, can answer this age-old question.

First, it is imperative that an offender's future risk for reoffending be assessed and that it be possible to do so. As an example, in 2009 the Washington state legislature required its department of corrections to develop and use an instrument that has the "highest degree of predictive accuracy" for assessing an offender's risk of reoffense. After considering a number of such instruments, five assessment tools were selected and used for this purpose.[81] Next, what kinds of programs should be used with these offenders who are in jails and prisons, and on probation and parole? Such approaches are known collectively as the risk-need-responsivity (RNR) model. In short, they operate on the assumption that using trained personnel to identify individual offenders' risks and needs—and then responding to those needs with the best combination of services and supervision—can lead to a significant reduction in recidivism.[82] Offenders' risk levels are determined by certain factors or characteristics that can be changed through intervention: one's antisocial behavior, close association with known offenders, poor family and/or marital relationships, bad school or work relationships, and substance abuse.

Figure 11.3 /// Center for Advancing Correctional Excellence Criminal Cognitions Scale, RNR Simulation Tool

Please indicate how well each statement describes your current thinking.

Externalization of Blame

	Please indicate how well each statement describes your current thinking.			
	Strongly Disagree	Disagree	Agree	Strongly Agree
Bad childhood experiences are partly to blame for my current situation.	❍	❍	❍	❍
I feel like what happens in my life is mostly determined by powerful people.	❍	❍	❍	❍
Because of my history I get blamed for a lot of things I did not do.	❍	❍	❍	❍
Sometimes I cannot control myself.	❍	❍	❍	❍
I am just a "born criminal."	❍	❍	❍	❍

Notions of Entitlement

	Please indicate how well each statement describes your current thinking.			
	Strongly Disagree	Disagree	Agree	Strongly Agree
When I want something, I expect people to deliver.	❍	❍	❍	❍
I will never be satisfied until I get all that I deserve.	❍	❍	❍	❍
I expect people to treat me better than other people.	❍	❍	❍	❍
I insist on getting the respect that is due me.	❍	❍	❍	❍
I deserve more than other people.	❍	❍	❍	❍

Devaluing Authority

	Please indicate how well each statement describes your current thinking.			
	Strongly Disagree	Disagree	Agree	Strongly Agree
Most of the laws are good.	❍	❍	❍	❍
Most police officers/guards abuse their power.	❍	❍	❍	❍
People in authority are usually looking out for my best interest.	❍	❍	❍	❍
If a police officer/guard tells me to do something, there's usually a good reason for it.	❍	❍	❍	❍
People in positions of authority generally take advantage of others.	❍	❍	❍	❍

Immediate Gratification

	Please indicate how well each statement describes your current thinking.			
	Strongly Disagree	Disagree	Agree	Strongly Agree
The future is unpredictable and there is no point planning for it.	❍	❍	❍	❍
Even though I got caught, it was still worth the risk.	❍	❍	❍	❍
Why plan to save for something if you can have it now?	❍	❍	❍	❍
I think it is better to enjoy today than worry about tomorrow.	❍	❍	❍	❍
I do not like to be tied down to a regular work schedule.	❍	❍	❍	❍

Insensitivity to Impact of Crime

	Please indicate how well each statement describes your current thinking.			
	Strongly Disagree	Disagree	Agree	Strongly Agree
My crime(s) did not really harm anyone.	❍	❍	❍	❍
A theft is all right as long as the victim is not physically injured.	❍	❍	❍	❍
Victims of crime usually get over it with time.	❍	❍	❍	❍
When you commit a crime the only one affected is the victim.	❍	❍	❍	❍
Society makes too big of a deal about my crime(s).	❍	❍	❍	❍

Source: Adapted from Jeffrey Stuewig, Emi Furukawa, Sarah Kopelo, June Price Tangney, Sarah Kopelovich, Patrick J. Meyer, and Brandon Cosby, "Reliability, Validity, and Predictive Utility of the 25-Item Criminogenic Cognitions Scale (CCS)," *Criminal Justice and Behavior* 39 (October 2012): pp. 1340–1360.

Once these risks are identified, specific treatment interventions can be put in place to reduce offenders' likelihood of continuing criminality.[83] Screening and assessment instruments are used for these purposes within the corrections field. Dozens of studies have shown, unequivocally, that using RNR principles of offender rehabilitation significantly reduces recidivism.[84] Although a one-size-fits-all model has been used in the past, recent advancements in this field have suggested the need for culturally-specific and gender-specific tools to enhance RNR effectiveness.[85] Figure 11.3 shows the kinds of questions used in an RNR assessment tool.

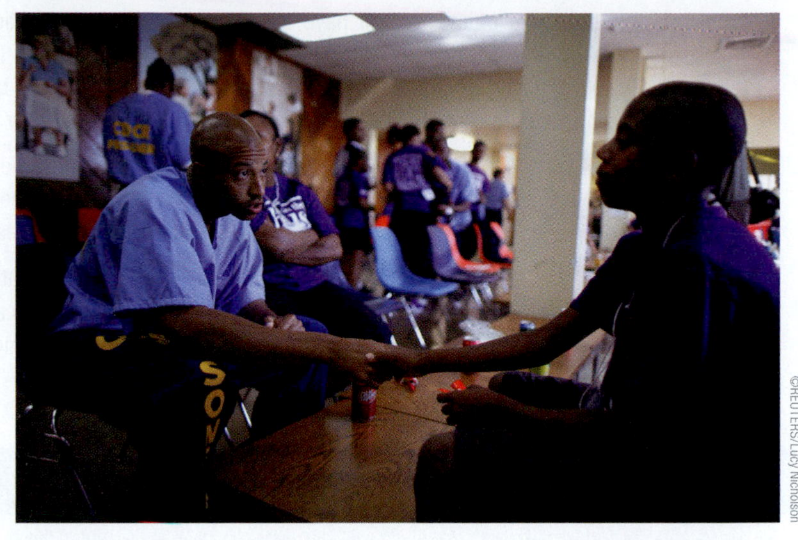

Restorative justice maintains that crime affects not only the victim but the entire community as well, so it involves meetings including the offender, the victim, and members of the community in order to devise a rehabilitative plan to make them whole again.

RESTORATIVE JUSTICE

In the view of many American citizens, a major shortcoming in the criminal justice system is that it considers crime as an act against the *state*, rather than against *people*, and thus removes human victims and their voices from the entire process. They believe that, even though costs of prisons and jails have skyrocketed (now estimated at a minimum of $71 billion),[86] we are not safer. Meanwhile, most offenders while away their time in prisons and jails without any encouragement or incentive to take responsibility for their crimes and are not changed upon release, leaving their victims dissatisfied with the entire experience.

Restorative justice works to remove the barriers between victims and offenders, who many times have a strong desire to meet each other and to discuss the criminal event that brought them together. It is victim driven, emphasizing ways to transform offenders by learning of the harms they perpetrated against the victim and community and making them more accountable (and thus less likely to recidivate). It can involve victims and offenders having a dialogue by meeting face-to-face or merely exchanging letters, and any type of crime. Although difficult for offenders to do, it can be for them (and their victims) a very satisfying and rewarding experience that is critical to restoration and healing.

Restorative justice: the view that crime affects the entire community, which must be healed and made whole again through the offender's remorse, community service, restitution to the victim, and other such activities.

A national organization, Restorative Justice International, provides a wealth of information about restorative justice efforts both in the United States and abroad, as well as a short, informative video (see http://www.restorativejusticeinternational.com/about/).

/// IN A NUTSHELL

- Probation involves a court allowing a convicted person to remain at liberty in the community, subject to certain conditions and restrictions; similarly, parole is the conditional release of a prisoner but before the prisoner's full sentence has been served. Both allow the offender to remain in the community to take advantage of treatment or counseling, maintain family and employment ties, and so on.
- If one is allowed to be placed on probation, the probationer must still abide by certain conditions. A technical violation occurs when, for example, the probationer fails to pay court costs or fines, misses a probation meeting, uses alcohol, or violates curfew. A substantive violation occurs when the probationer commits a new crime.
- Because they may lose their freedom, probationers are allowed to have counsel present at the probation revocation hearing; other due process rights are afforded as well.
- Parolees must also abide by certain conditions that the court and parole officers put in place. If they do not comply, parole may be revoked and the parolee returned to prison. Parolees also have certain due process rights prior to revocation.
- Probation officers perform duties such as reporting to the court any violations of probation, performing presentence investigations, arresting those who violate the terms of their probation, performing searches, seizing evidence, and arranging for drug testing.

- Parole officers supervise offenders who have been released from prison through personal contact with offenders and their families. They help parolees adjust back into society, developing a plan for employment, housing, health care, education, drug screening, and other activities.
- Whether probation and parole officers should be armed has been a debated topic in corrections; there is no national or standard policy for these personnel regarding weapons, and officers themselves are not in agreement about being armed.
- A number of intermediate sanctions are used in attempting to further rehabilitative efforts and to reduce jail and prison populations. These include intensive supervision probation and parole, house arrest, electronic monitoring, shock probation/parole, boot camps/shock incarceration, day reporting centers, halfway houses, and furloughs.
- To avoid reincarceration, offenders' future risk for reoffending must be assessed; approaches used are known as the risk-need-responsivity (RNR) model.
- Restorative justice holds that the first priority of justice processes is to assist victims, whereas the second priority is to restore the community.

/// KEY TERMS & CONCEPTS

Review key terms with eFlashcards at **edge.sagepub.com/peakbrief**.

Alternatives to incarceration 247
Boot camp 264
Caseload 259
Community corrections 246
Day reporting center 264
Electronic monitoring 263
Halfway house 265
Home arrest/home confinement 264
Intensive supervision probation and parole 262
Intermediate sanctions 262
Parole officer 260
Probation 248
Probation officer 260
Restorative justice 269
Revocation 253
Shock probation/parole 265
Substantive violation 253
Technical violation 253

/// REVIEW QUESTIONS

Test your understanding of chapter content. Take the practice quiz **at edge.sagepub.com/peakbrief**.

1. What is meant by the term *community corrections?*
2. What historical events led to modern-day probation? Parole?
3. What are the rights granted by the U.S. Supreme Court to people who are on probation and parole in general? When the state wishes to revoke their probation/parole status and send them to prison?
4. What are some examples of technical and substantive conditions that correctional agencies commonly apply to probationers and parolees?
5. Why does the criminal justice system use alternatives to incarceration?
6. How would you describe the primary functions of probation and parole officers, and the impact of high caseloads in terms of how these officers are able to achieve their goals?
7. How would you describe the arguments for and against arming probation and parole officers?
8. What are the purposes and functions of intermediate sanctions, including intensive supervision, house arrest, electronic monitoring, shock probation/parole, shock incarceration/boot camps, day reporting centers, halfway houses, and furloughs?
9. How does the risk-need-responsivity model work with offenders to stop the cycle of reoffending and reincarceration?
10. What is the meaning and purpose of the restorative justice concept, and how does it function?

/// LEARN BY DOING

1. You graduated recently with a criminal justice degree and are now employed as a state probation and parole officer. You are asked by a criminal justice professor at a nearby college to guest lecture in an introduction to criminal justice course concerning the primary challenges of working in probation and parole. Develop a presentation for the class.

2. Assume it is your first day working as a state probation and parole officer, and your new training officer says the following:

> Hi, nice to meet you. I'm Chuck, the training advisor. Here's what I tell everyone on their first day. First, the real world is different from what you've been told in college classes. Politicians promise the public that they will get tough on crime, so first off they spend money for more police officers. No one gets elected by promising to build more courts, or add more probation and parole officers. Eventually having more police means the courts get backlogged, which in turn crowds the prisons and the jails, puts more people on probation, and forces the parole board to grant more early releases. Meanwhile, our average caseload increases by 50 percent. We also have more drug and sex offenders than we know what to do with. We spend too much time bailing water out of the boat, with no one steering, so we are just drifting in circles. To vent the frustrations, about once a week the gang and I hold "choir practice" at a bar down the street. After about five or six beers and some carousing and loud singing, then this job looks a lot better.

   a. Should Chuck be retained as a training officer? Why or why not?

   b. How would changes in politics affect the corrections system directly and indirectly?

   c. Do you believe Chuck's comments are accurate concerning the police getting so much new funding—and the subsequent impacts on courts and corrections?

   d. Why do crowded jails and prisons make the job of parole officers more difficult?

   e. If you were Chuck's supervisor and heard others discussing their frequent "choir practices" at the bar, would you attempt to discontinue such gatherings or leave the situation alone?

3. Your criminal justice instructor has assigned a class debate concerning the use of incarceration versus alternatives to incarceration (e.g., probation and parole, electronic monitoring, house arrest, halfway houses). What will be your argument?

/// STUDY SITE

SAGE edge™

Get the tools you need to sharpen your study skills. SAGE edge offers a robust online environment featuring an impressive array of free tools and resources.

Access practice quizzes, eFlashcards, video, and multimedia at **edge.sagepub.com/peakbrief**

EXTRAORDINARY PROBLEMS AND CHALLENGES

PART 5

Chapter 12: On the Crime Policy and Prevention Agenda: Drug Abuse, Sex Trafficking, Terrorism, and Immigration

This part consists of a single chapter that provides an in-depth view of four particularly challenging problems and policy issues confronting society and the criminal justice system today: drug abuse, sex trafficking, terrorism, and immigration.

Genaro Molina/Los Angeles Times/Getty Images

CHAPTER 12

ON THE CRIME POLICY AND PREVENTION AGENDA

Drug Abuse, Sex Trafficking, Terrorism, and Immigration

The human story does not always unfold like a mathematical calculation on the principle that two and two make four. Sometimes in life they make five or minus three; and sometimes the blackboard topples down in the middle of the sum and leaves the class in disorder.

—Winston Churchill

LEARNING OBJECTIVES

As a result of reading this chapter, you will be able to

1. Describe the current drug epidemic in the United States, particularly the status of opioid use and marijuana legislation, and what is being done by first responders, courts, and legislators toward dealing with illicit drug abuse
2. Define and explain major issues facing the United States and its criminal justice system relating to handling of sex trafficking, to include current legislation, programs, and proposals to address those challenges
3. Delineate the types of terrorism—including cyberterrorism, bioterrorism, school shootings, and gun violence in general—and some approaches for combating these issues
4. Explain the controversial nature and recent patterns of illegal immigration, what is known about the link between illegal immigration and crime, dealing with asylum seekers, and the debate surrounding construction of a border wall

ASSESS YOUR AWARENESS

Test your knowledge of selected criminal justice policy issues by responding to the following seven true-false items; check your answers after reading this chapter.

1. Currently, the prescription drug problem (particularly involving opioids) in the United States has all but driven heroin and cocaine use out of existence.
2. Putting drug addicts in jail does little to stop prescription drug abuse, so many police agencies are cooperating with initiatives that put substance abusers into treatment programs.
3. At the federal level, the government has softened in its views of enforcement of marijuana laws, even while marijuana remains classified as a Schedule I substance under the Controlled Substances Act.
4. Victims of sex trafficking buy into the traffickers' promise of a better life in the United States, only to find they have been lied to, forced into servitude, or otherwise victimized.
5. Terrorist attacks have become increasingly focused on civilian targets, with terrorists using more traditional methods of violence such as employing guns and hostage-taking, rather than large-scale bombings.
6. Because crimes by and arrests of juveniles have declined so much in the United States in recent years, there are no related policy and prevention aspects or concerns.
7. The rates of crimes committed by illegal immigrants in the United States are extraordinarily high, driving the current fervor for closing the borders.

<< Answers can be found on page 293.

Now more than ever, Americans need to be able to rely on crime policies and practices that will guard against all types of crimes and vulnerabilities. Perhaps nowhere is that vulnerability more apparent than in Las Vegas, Nevada. Each year more than 36 million people are drawn to this valley, the self-proclaimed "Entertainment Capital of the World," where they can enjoy 18 of the world's 25 largest hotels and countless entertainment acts and events. For those reasons, Las Vegas has long been viewed as a prime terrorist target and hotbed for all manner of crimes. Given that, under the veneer of Vegas's glitter and fun is an underlying layer of security that may be unprecedented, and it starts with what is termed a *fusion center*.

Although there are more than 75 state and local fusion centers identified by the U.S. Department of Homeland Security,[1] the one in Las Vegas is unique. Inside its walls are 14 different federal, state, and local law enforcement agencies collaborating 24/7 to keep residents and tourists safe, focusing on all crimes and hazards, not just terrorism. Personnel in the center work in either an intelligence collections section or a crime analysis section to connect the dots between crimes that may look unrelated but

iStock/f11photo

Las Vegas's status as a world destination, with more than 40 million visitors per year and about 40 casinos on its famed Strip, stands as a prime terrorist target.

could be precursors to a bigger event, collecting and sharing information about suspicious activities. The intelligence collections section follows up on suspicious activity reports, collects information in the field, and conducts surveillance. The crime analysis group looks at all types of crime occurring in the valley, from robberies to rapes and murders, to analyze trends. But the two sections often work together. For example, upon seeing a 700 percent increase in thefts of 20-pound propane tanks, the crime analysis section wasn't sure if the thefts were terrorism related (the tanks could be used to make bombs). The intelligence collections section began interviewing people who were caught stealing propane tanks and determined they were not terrorism related at all. Rather, people were selling them to recycling yards and using them to heat homes. Looking at police reports to identify trends or activities is obviously key to the center's success.[2]

Although fusion centers are often seen as dealing only with the specter of terrorism, they work with all types of real and potential problems that can and do plague the United States. This chapter explores some of those pressing issues.

INTRODUCTION

What can be done with the current opioid epidemic in the United States, and how might law enforcement and legislative enactments assist in addressing it? What does the future hold for recreational use of marijuana? How can we protect individuals from becoming victims of sex traffickers? What should be done about terrorists who now attack using little help and with inexpensive, widely available weapons and explosives made from everyday ingredients—and, increasingly, computers?[3] How should the United States resolve the tension between national security and protecting civil liberties for all? What is the answer to the long-standing debate surrounding immigration and border protection? There is no shortage of crime and policy challenges in America today—issues that affect and must be confronted by its criminal justice system.

This chapter briefly examines these four "hot-button" issues in terms of the problems and policy questions they pose. Each has deservedly received widespread attention and publicity, and been the subject of great concern, but they give no indication of abating in the foreseeable future.

AMERICA'S DRUG EPIDEMIC

Heroin, Cocaine, and Opioid "Pill Mills"

Opioids: a class of drugs that includes heroin, synthetic opioids such as fentanyl, and pain relievers available legally by prescription.

So-called pill mills run by physicians who are prescribing **opioids** (a class of drugs naturally found in the opium poppy plant) illegally and/or excessively, and residents who engage in "doctor shopping" to feed an addiction, have been garnering most of the nation's attention. For example, in Reno, Nevada, federal drug agents in 2016 raided a car dealership and arrested nine persons for operating an opioid distribution ring, including the prescribing physician, who was sentenced to 10 years in prison. More than $8 million in drugs were put on the streets in about two years' time.[4] At the same time, the levels of heroin and cocaine use in the United States remain significant: According to the National Survey on Drug Use and Health, about 948,000 Americans reported using heroin in the previous year, a number that has been on the rise since 2007. This trend appears to be driven largely by young adults, aged 18–25, among whom there have been the greatest increases.[5] Similarly, cocaine use in the United States gives no indication of slowing. Cocaine is the second most common cause of fatal drug overdose, nearly a million people try it for the first time each year, it can be mixed with potent opioids, and its use can easily undergo a resurgence in the United States at any time.[6]

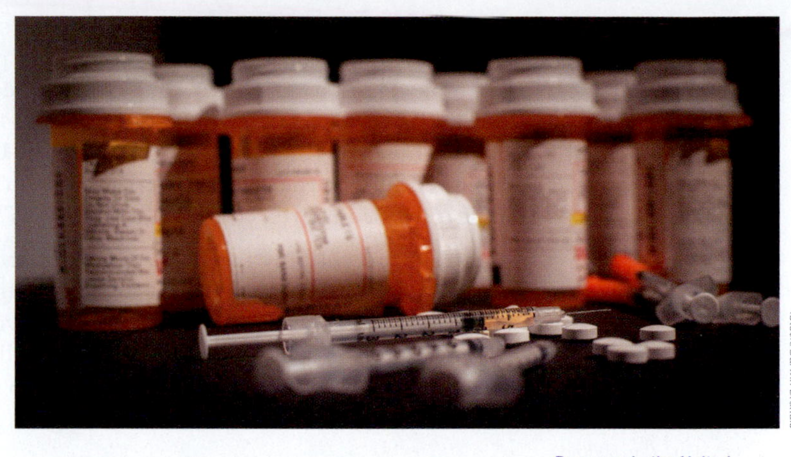

iStock/Darwin Brandis

Drug use in the United States, fueled by heroin, cocaine, and opioid "pill mills," has reached epidemic proportions.

The most alarming form of synthetic opioid is fentanyl, which is similar to morphine but 50–100 times more potent. Given its strength, fentanyl is now the drug most often involved in fatal overdoses. Its illegal form is made in labs and sold as a powder, on blotter paper, or put in eye droppers or nasal sprays or pills; mixing it with other drugs (such as heroin, cocaine, or meth) makes it a cheaper option for obtaining a high.[7]

This epidemic has reached staggering levels, with about 47,600 people dying from opioid overdoses in 2018[8] (on top of the 49,000 deaths in 2017 and the 64,000 in 2016). In a recent year, 11.4 million people misused opioid prescriptions, 2.1 million people had an opioid use disorder, and 28,500 deaths were attributed to synthetic opioids other than methadone (see Figure 12.1 for more statistics); indeed, in late 2017 the federal government declared this epidemic a public health emergency.[9] Moreover, for the first time, in early 2019 it was announced that Americans had greater odds of dying from an accidental opioid overdose than from a motor vehicle accident.[10]

Chasing "El Chapo"

Joaquin "El Chapo" (translating to "The Shorty") Guzman was extradited to the United States and put on federal trial in early 2019 in Brooklyn, New York. After more than two

Figure 12.1 /// The Opioid Epidemic by the Numbers

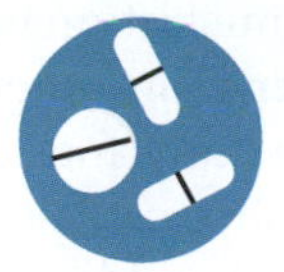

130+
People died every day from opioid-related drug overdoses[3] (estimated)

11.4 m
People misused prescription opioids[1]

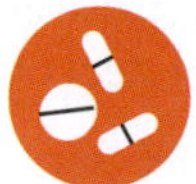

47,600
People died from overdosing on opioids[2]

2.1 million
People had an opioid use disorder[1]

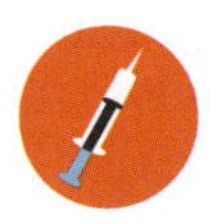

886,000
People used heroin[1]

81,000
People used heroin for the first time[1]

2 million
People misused prescription opioids for the first time[1]

15,482
Deaths attributed to overdosing on heroin[2]

28,466
Deaths attributed to overdosing on synthetic opioids other than methadone[2]

Source:

1. 2017 National Survey on Drug Use and Health. Mortality in the United States, 2016
2. NCHS Data Brief No. 293, December 2017
3. NCHS, National Vital Statistics System. Estimates for 2017 and 2018 are based on provisional data.

decades of eluding police and accomplishing several escapes from prison, the Mexican drug lord—who rose from working in the opium poppy fields at age 15 to being kingpin of the infamous Sinaloa cartel—finally saw his career brought to an end. Guzman led an organization that, over three decades, trafficked $14 billion in drugs to the United States, caused 119,000 murders (including thousands in the United States), and contributed to $1 trillion in U.S. spending on the war on drugs.[11] So much cash is reaped in this trade that, as one author put it, "they don't even count it, they *weigh* it."[12]

After 56 prosecution witnesses testified for more than 200 hours, Guzman was convicted in February 2019 of multiple counts of distributing large amounts of narcotics internationally. His conviction should be viewed as a hollow victory, though. Guzman will likely die in a U.S. supermax prison, but the Sinaloa cartel is still known to be operating as the most powerful cartel in Mexico.[13] Indeed, two weeks prior to Guzman's conviction, enough fentanyl for 100 million lethal doses was seized at the Nogales port of entry.

What Can Be Done?

First, go after profits; in August 2019 an Oklahoma jury fined a major drug company $572 million for helping to fuel its epidemic (and more than 1,500 similar lawsuits have been filed).

Members of Congress are attempting to stem the tide of addiction in several ways: by encouraging research on new nonaddictive pain medications, electronically linking all nationwide efforts to fight the epidemic, stemming the influx of illicit drugs into the country, establishing grants to help fight addiction, directing the U.S. Food and Drug Administration to set up programs for the safe return of unused opioids, and ensuring that medical professionals have access to a consenting patient's complete health history when deciding how to treat a patient (to avoid overprescribing opioids).[14]

In addition, realizing that putting drug addicts in jail does little or nothing to stop drug use, many police agencies are cooperating with initiatives that put substance abusers into treatment rather than jail, while also supporting an "early warning system" of rapid testing of confiscated substances and directing aid to people who have overdosed on opioids. Regarding the latter, when encountering a fatal overdose, expediting the testing of the substance that was used can help public health agencies to more quickly identify any overdose clusters or new batches of particularly dangerous substances. Furthermore, often being first on the scene, officers and other first responders are being trained in the administration of medications such as **Narcan/naloxone** (though fentanyl, being much stronger than other opioids, might require multiple doses); a Police Executive Research Forum study found that this approach has already saved 3,500 lives in 276 agencies surveyed.[15] Police can also provide drug treatment in jails; incarceration can provide an opportunity to help addicted persons undergo treatment. **Suboxone** can be administered in the jails to relieve withdrawal and cravings for opioids without causing the "high" of heroin or dangerous side effects.[16]

Narcan/naloxone/suboxone: a medication that can be used to block the effects of opioids, particularly for overdoses.

Finally, specialty problem-solving drug courts (discussed in Chapter 7) can assist persons who are addicted to opioids by providing medication-assisted treatment (MAT). However, many drug courts do not recommend (or even allow) the use of such treatment for opioid dependence. One survey of drug courts found that although 98 percent of their drug court participants were opioid dependent, only about half of the courts offered any form of MAT to participants.[17]

The Status of Medical and Recreational Marijuana Use

Although the number is subject to rapid change, 33 states and the District of Columbia currently have legalized marijuana in some form. The District of Columbia and 10 states—Alaska, California, Colorado, Maine, Massachusetts, Michigan, Nevada, Oregon, Vermont, and Washington—have adopted the most expansive laws legalizing marijuana for recreational use.[18] In addition, at least nine states all across the map are looking at legislation that would legalize marijuana in the near future.[19]

Note that **marijuana decriminalization** and **marijuana legalization** are two different concepts and approaches. Decriminalization involves a state repealing or amending its marijuana laws to make certain acts criminal (such as sales or trafficking), but some aspects are no longer subject to prosecution (e.g., individuals with small amounts of marijuana for personal consumption). Marijuana legalization typically means that one will not be arrested for marijuana use if he or she follows the state laws as to age, place, and amount for consumption (however, one might still be arrested for selling or trafficking if such are prohibited under the law).

Marijuana decriminalization: where a state has repealed or amended its marijuana laws to make certain acts criminal but no longer subject to prosecution (e.g., individuals caught with small amounts of marijuana for personal consumption may not be prosecuted and, thus, avoid a criminal record).

Marijuana legalization: typically, where one will not be arrested for marijuana use if he or she follows the state laws as to age, place, and amount for consumption (however, one might still be arrested for selling or trafficking if such are prohibited under the law).

At the federal level, the government has softened in its views of enforcement of marijuana laws, even while marijuana remains classified as a Schedule I substance under the Controlled Substances Act. In early 2019, U.S. Attorney General William Barr indicated that he favors a more lenient, albeit federalist, approach to marijuana laws, and for cannabis to be legalized nationwide rather than having states continue to go against the federal prohibition.[20]

Today it is estimated that 61 percent of Americans favor legalizing adult-use marijuana, while 88 percent favor legalizing medical marijuana nationwide. The police certainly stand to benefit from this trend, with an estimated $10 billion to $20 billion being spent each year on marijuana enforcement. Given that more than 600,000 marijuana arrests have been made during each of the past several years (about 9 in 10 for possession only), at an estimated cost of $750 per arrest, the police would surely have many resources to divert elsewhere.[21] Similarly, courts and corrections agencies would realize tremendous savings if not having to adjudicate and house such offenders.

SEX TRAFFICKING

Definition and Victimization

The U.S. Department of Justice defines **sex trafficking** as a commercial sex act that is "induced by force, fraud, or coercion, or in which the person induced to perform such act has not attained 18 years of age." It can also include "recruitment, harboring, transportation, provision, or obtaining of a person for labor or services, through the use of force, fraud, or coercion."[22] Today sex trafficking is a $99 billion annual industry that encompasses about 4.8 million victims around the world, mostly women and children.[23]

Sex trafficking: the use of humans for the purpose of sexual slavery or commercial sexual exploitation.

The United States is known as a destination country for transnational trafficking networks that bring foreign nationals into the country for purposes of both sexual and labor exploitation. Trafficking victims in the United States are primarily from Asia, Latin America, Eastern Europe, and Africa.[24]

What Can Be Done?

No single agency can prevent sex trafficking or protect its victims while also prosecuting traffickers. First and foremost, relationships must be developed between law enforcement and social services agencies before sex trafficking can be addressed and victims assisted. In particular, the two agencies need to be able to delegate tasks and clearly define their respective roles.

Sex trafficking typically occurs behind the closed doors of private homes or under the radar of legitimate businesses; crime rings have also become quite astute in the methods of smuggling persons across borders and hiding victims of trafficking as they move them between cities. Victims believe the promise of traffickers of a better life in the United States, only to find they have been lied to, forced into servitude, or otherwise victimized. These victims need assistance and benefits, such as those provided by the Trafficking Victims Protection Act of 2000 (TVPA), which specified severe punishment for traffickers and afforded qualifying victims shelter, rehabilitation programs, freedom from deportation, and other benefits.[25]

The T Visa

T visa: a type of visa that allow victims of human trafficking and their immediate family members to remain and work temporarily in the United States (provided certain conditions are met).

In conjunction with the aforementioned TVPA and other laws created to protect victims of trafficking, the **T visa** was also created in 2000. It allows some victims of sex trafficking and their immediate family members to remain and work temporarily in the United States if they agree to assist law enforcement in testifying against the perpetrators. Such legislation was necessary because victims of sex trafficking in the United States are usually undocumented and subject to deportation; therefore, while a criminal case is developing against the victim's trafficker(s), the victim may now be allowed to legally remain here to assist with the prosecution. Furthermore, three years after obtaining a T visa, victims may apply for permanent resident status and even file for immediate family members to join them legally in the United States.[26] In addition, such victims may qualify for federal and state benefits and programs, such as food stamps, Medicaid, and cash assistance.[27]

Additional Efforts: The Diagnostic Center

Notwithstanding all of the legislative and enforcement efforts now in place, effectively addressing sex trafficking can still be extremely challenging for communities that do not possess established practices for identifying and protecting sex trafficking victims and investigating and prosecuting cases. Now, however, the U.S. Department of Justice, Office of Justice Programs, provides assistance in what is termed a **diagnostic center**. Established in 2012, diagnostic centers employ data-driven strategies for combating sex trafficking.

Diagnostic center: Launched by the U.S. Department of Justice, the center provides technical assistance resources for law enforcement executives and policy makers for combatting all types of crimes, including sex trafficking.

With such efforts as those described here, and a variety of training programs and other resources in the police toolkit,[28] there have been some impressive results:

- In Los Angeles County, 510 people were arrested and 56 persons rescued in a crackdown on trafficking by regional task forces.[29]
- A state trooper, a pastor, and a convicted sex offender were among 1,000 people arrested in a national operation run by the National Johns Suppression Initiative and law enforcement agencies across the country.[30]
- A national operation resulted in 120 arrests and 84 children rescued as part of the Innocence Lost National Initiative; during this operation a three-month old girl and her five-year old sister were recovered after a family friend made a deal with an undercover officer to sell both children for sex for $600.[31]

POLICING TERRORISM

Definitions and Types

The FBI succinctly defines terrorism as an act "perpetrated by individuals and/or groups inspired by or associated with primarily U.S.-based movements that espouse extremist ideologies of a political, religious, social, racial, or environmental nature."[32] More broadly, terrorism as defined in the U.S. Code can be both domestic and international in nature and is any act that appears to be intended to intimidate or coerce a civilian population; to influence the policy of a government by intimidation or coercion; or to affect the conduct of a government by mass destruction, assassination, or kidnapping (see Figure 12.2). School shootings (often involving mass killings) fall within those parameters and thus are included in this discussion of terrorism.[33]

International terrorism: acts of terrorism perpetrated in a foreign country by terrorists who are not native to that country.

Terrorism, as defined previously, occurs around the world, as suicide bombings and shootings, kidnappings, assassinations, and beheadings each year will attest. International terrorist groups have become more adept at recruiting "foot soldiers" through the Internet and other means. These attacks have become increasingly focused on civilian targets, with terrorists using more traditional methods of violence such as deploying guns and hostage-taking, rather than large-scale bombings. **International terrorism** pointedly demonstrates the need

Figure 12.2 /// FBI Most Wanted Poster

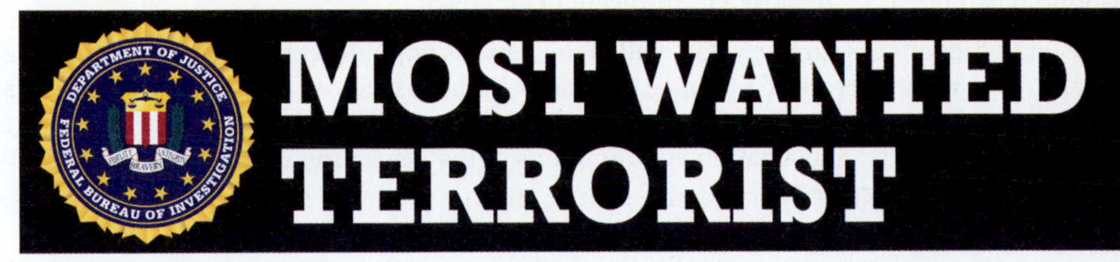

IBRAHIM SALIH MOHAMMED AL-YACOUB

Conspiracy to Kill U.S. Nationals; Conspiracy to Murder U.S. Employees; Conspiracy to Use Weapons of Mass Destruction Against U.S. Nationals; Conspiracy to Destroy Property of the U.S.; Conspiracy to Attack National Defense Utilities; Bombing Resulting in Death; Use of Weapons of Mass Destruction Against U.S. Nationals; Murder While Using Destructive Device During a Crime of Violence; Murder of Federal Employees; Attempted Murder of Federal Employees

DESCRIPTION

Date(s) of Birth Used: October 16, 1966

Hair: Black

Height: 5'4"

Build: Unknown

Sex: Male

Languages: Arabic

Place of Birth: Tarut, Saudi Arabia

Eyes: Brown

Weight: 150 pounds

Complexion: Olive

Citizenship: Saudi Arabian

Scars and Marks: None known

REWARD

The Rewards For Justice Program, United States Department of State, is offering a reward of up to $5 million for information leading directly to the apprehension or conviction of Ibrahim Salih Mohammed Al-Yacoub.

REMARKS

Al-Yacoub is an alleged member of the terrorist organization, Saudi Hizballah.

CAUTION

Ibrahim Salih Mohammed Al-Yacoub has been indicted in the Eastern District of Virginia for the June 25, 1996, bombing of the Khobar Towers military housing complex in Dhahran, Kingdom of Saudi Arabia.

SHOULD BE CONSIDERED ARMED AND DANGEROUS

If you have any information concerning this person, please contact your local FBI office or the nearest American Embassy or Consulate.

Field Office: Washington D.C.

Source: https://www.fbi.gov/wanted/wanted_terrorists/ibrahim-salih-mohammed-al-yacoub/@@download.pdf

for the international law enforcement community to become much more knowledgeable about terrorists' methods, how to predict and possibly prevent future attacks, and how to respond when terrorists do strike in the event of an attack.

These and other attacks similarly have called the nation's law enforcement agencies to action, to learn more about terrorists' methods here and abroad, as well as how to predict and possibly prevent future attacks. In addition, they have had to adopt a broader view of protecting the homeland (the organization and functions of the Department of Homeland Security were discussed in Chapter 4).

Right-Wing Extremists and Lone-Wolf Terrorists

In December 2015, a husband and wife wearing military-style clothing and black masks entered a Christmas party for employees of the San Bernardino, California, county health department and opened fire with two assault-style weapons, killing 14 people in what was the deadliest terrorist attack in the United States since 9/11. The couple were parents and college graduates, solidly middle class, with no criminal records, but they were also typical homegrown **extremists** wanting to defend or spread their beliefs.

Extremists: persons who hold extreme or fanatical political or religious views, which can lead to the commission or advocacy of extreme action.

Then, in October 2018, Robert Bowers, harboring anti-Semitic views, carried out a mass shooting at the Tree of Life Synagogue near Pittsburgh, Pennsylvania. Bowers was charged with 63 counts, including obstructing religious beliefs resulting in death as well as violations of the Matthew Shepard and James Byrd Jr. Hate Crimes Prevention Act.[34]

Also in October 2018, 56-year-old Cesar Sayoc, who drove a white van with stickers depicting some of President Trump's critics and targets over their images, was arrested for sending 14 pipe bombs through the mail to prominent Democrats around the nation (none of which detonated).[35] Officials admitted the difficulty of trying to detect dangerous items such as these, given that the U.S. Postal Service handles millions of packages each day.[36]

Although none of these individuals was known to be formally affiliated with a foreign terrorist group, they could easily turn to an extremist community online, for training, to espouse their views, and to try to incite others. Such extremists—**homegrown violent extremists** (HVEs)—warrant particular concern today. They are self-radicalized and "encourage, endorse, condone, justify, or support the commission of a violent criminal act" in order "to achieve political, ideological, religious, social, or economic goal." HVEs "can include U.S.-born citizens, naturalized citizens, green card holders or other long-term residents, foreign students, or illegal immigrants" wishing to commit terrorist acts inside Western countries or against Western interests abroad.[37] Some might have been inspired by calls by the Islamic State of Iraq and the Levant (ISIL) for individual jihadists in the West to retaliate for U.S.-led airstrikes on ISIL.[38]

Homegrown violent extremists: persons who are inspired by foreign terrorist organizations and radicalized in the countries in which they are born, raised, or reside.

A related concern is the **lone-wolf terrorist** who undergoes **radicalization** (see the next section)—a single individual driven to hateful attacks based on a particular set of beliefs without a larger group's knowledge or support.

Lone-wolf terrorist: one who plans and commits acts of terrorism without the support of a larger terrorist group.

Radicalization: a process whereby one adopts radical positions on political or social issues.

The Radicalization of Zachary Chesser

Zachary Chesser was an average high school student in northern Virginia. He participated in his high school's gifted and talented program, joined his high school break-dancing team, was an avid soccer player with aspirations of getting a scholarship to play in college, and worked part time at a video rental store.

In the summer of 2008, the 18-year-old Chesser converted to Islam and quickly became radicalized, solely on the Internet. He began posting views that supported Islamist terrorist groups, watching sermons by Anwar al-Awlaki, and he exchanged emails with the cleric about joining Al-Shabab. Within weeks, he had quit his job because he "objected to working at a place that rented videos featuring naked women" and became increasingly hostile to his parents.

Soon Chesser had committed himself solely to using his computer and graphics skills to contribute to and promote violent extremist messages. He also attempted to travel to Somalia with his wife to join Al-Shabab but was unsuccessful when his mother-in-law hid his wife's passport. Next, he uploaded a video to YouTube in which he threatened the creators of the television show *South Park* after an episode depicted the Prophet Muhammad dressed in a bear costume. He then attempted to join Al-Shabab once again but was held for questioning at the airport. A few days after being questioned, Chesser was arrested for attempting to provide material support to a terrorist organization. He pleaded guilty to three federal felony charges—communicating threats, soliciting violent jihadists to desensitize law enforcement, and attempting to provide material support to a designated foreign terrorist organization—and was sentenced to 25 years in federal prison.[39]

Zachary Chesser, a U.S. citizen, used his computer and graphics skills to promote violent extremist messages and was ultimately arrested for attempting to provide material support to a terrorist organization.

Cyberterrorism: The Asian Threat

Our foreign adversaries are very much aware that U.S. businesses, public agencies, citizens, and governments are so dependent on computer systems to run everything that hackers who can infiltrate those systems can be as effective and damaging as a military effort using troops and weaponry. Which foreign power is now the biggest threat against the United States with respect to cyberattacks or **cyberterrorism**? Although Russia was recently at or near the top of the list for its suspected efforts to disrupt the 2016 U.S. presidential elections, the most challenging and potentially disastrous type of **cybercrime**—actually, cyberespionage—now being perpetrated against the United States is by China's state-supported cyber warfare. Its three categories of cyber warfare—involving specialized military forces, military-authorized cyber warfare specialists, and civilians—have over the past dozen years stolen data from the U.S. Navy (concerning submarine warfare), NASA, IBM, and other agencies and companies and compromised many of the nation's most sensitive advanced weapons systems, including missile defense technology and combat aircraft.[40] China's interests have shifted recently to stealing financial and tech-related secrets from corporations, universities, government departments, and biotech firms.[41] Not to be overlooked, however, is the rising threat from North Korea, where the economy has long struggled and a group of hackers is believed to be actively targeting U.S. businesses and "critical infrastructure." North Korean hackers have tried to infiltrate finance, telecommunications, energy, and defense sectors around the world.[42]

Cyberterrorism: a premeditated, politically motivated attack against computer systems, computer programs, or data.

Cybercrime: the use of computers to commit crimes, such as embezzlement, diversion of bank monies to other accounts, and hacking personal information.

According to INTERPOL, cybercrime is one of the fastest growing areas of crime and includes attacks against computer hardware and software; financial crimes and corruption; and abuse, in the form of grooming or "sexploitation," especially crimes against children.[43] Unquestionably, many people around the world are now working full time trying to hack into online data, and the threats Americans face have never been more real or severe. According to the Privacy Rights Clearinghouse, for the 9,100 data breaches that were made public since 2005, about 11.6 billion records were breached.[44] Furthermore, each breach is estimated to cost American businesses an average of nearly $8 million, and that figure is increasing.[45]

There are also concerns about threats posed to U.S. nuclear reactors, banks, subways, and pipeline companies. The specter of electricity going out for days and perhaps weeks, the gates of a major dam opening suddenly and flooding complete cities, or pipes in a chemical plant rupturing and releasing deadly gas are nightmare scenarios that keep homeland security professionals awake at night.

Bioterrorism

Another means of attack by terrorists involves the use of chemical/biological agents, or **bioterrorism**. Poisons have been used for several millennia; recent attacks using chemical/biological agents including toxins, viruses, or bacteria such as anthrax, ricin, and sarin have underscored their potential dangers and uses by terrorists today.

Bioterrorism: the use of biological material, such as anthrax or botulin, to commit an act of terrorism.

Practitioner's Perspective

Director or IT and Electronic Security

Name: Dennis Bachman
Position: Director of IT and Electronic Security, Kansas Racing and Gaming Commission
Location: Kansas

What is your career story? I've been in law enforcement for 42 years now. I started with the Topeka Police Department as a patrol officer, progressed up through the ranks—detective, field sergeant—until I was injured and had to leave. Fortunately, I had obtained my degree while I was working. From there, I went to another role in the criminal justice system, as a probation officer, and started out doing presentence investigations. From that position, I went on to be chief of one of the regions in Kansas, to the Third Judicial District. I went back to the Third Judicial District and became a supervising probation officer. At this point, I had developed an expertise in handling high-risk offenders working with gang members. Because of that, the state parole board requested that I come over and handle high-risk offenders for them, so I went into the Department of Corrections. In that role, and because of my background of working as a law enforcement officer, I also worked on fugitive apprehension and was cross-deputized as a special deputy U.S. marshal in their Fugitive Apprehension Unit. After I'd been out in the field doing that, I realized I wanted to find work that was a little bit more subdued. I had the background in information technology [IT] and started using that specialty in the area of regulatory work and gaming. I started out working with tribal gaming and then worked in casino gaming with the state.

What are some challenges and misconceptions you face in this position? One of the misconceptions in my field is that people don't think my role is necessary because they think you can't cheat the slot machines. The fact is you can, and people try, and they do. We catch them. That's our specialty, figuring out how we catch them. There are a lot of ways of cheating slot machines. There are a lot of reasons that we work with the machines before they get put on the floor.

Do you see any common trends in this position? One of the ways that our field is going to change is how it's going to handle the Millennials. The way the game works now, they're actually set up more for people my age and older. The Millennials are more gaming oriented. They are going to change gaming to accommodate their expertise, but at the same time, to get a continuous flow of play that requires placing coins in the machine. Mobile gaming is also a new trend. People don't want to be tied to the machine. They like being able to move around and go from spot to spot while playing. We are going to see a rise in wireless gaming.

What advice would you give to someone either wishing to study, or now studying, criminal justice and wanting to become a practitioner in this position? One of the things I would say—and I would do this in any part of the criminal justice system—is to keep information technology in your training. When I was in probation and parole both, I used databases. I built them, they helped me a lot all the way through. When I was with the Topeka Police Department, I was in criminal intelligence—built databases, worked with different systems, so having that background helped me all the way through my career. And right now, I'm seeing a lot of people coming in, and they're having problems with some very basic IT work. So I would stress that you're going to need these skills, and having a good foundation for IT will help you.

To learn more about Dennis Bachman's experiences as a Director of IT and Electronic Security watch the Practitioner's Perspective video in SAGE Vantage.

Chemical weapons—including several types of gases—suffocate the victim immediately or cause massive burning. Biological weapons are slower acting, spreading a disease such as anthrax or smallpox through a population before the first signs are noticed. Many experts believe it is only a matter of time before chemical/biological weapons are used like explosives have been to date.[46] All that is required is for a toxin to be cultured and put into a spray form that can be weaponized and disseminated into the population. Fortunately, such bioterrorism agents are extremely difficult for all but specially trained individuals to make in large quantities and in the correct dosage; they are also difficult to transport because live organisms are delicate.

What Can Be Done?

Federal Efforts Against Terrorism and Cyber Warfare

Police have several possible means of attempting to prevent or address acts of terrorism. On a practical level, there is the federal Department of Homeland Security (DHS, discussed in Chapter 4), which has more than 240,000 employees in jobs that range from aviation and border security to emergency response, from cybersecurity analyst to chemical facility inspector.[47]

Furthermore, the Federal Emergency Management Administration (FEMA) administers a National Incident Management System (NIMS), which provides a framework for agencies of all levels to work effectively together to prepare for, prevent, respond to, and recover from domestic incidents.[48] With respect to cyber warfare, American governments, businesses, and citizens have long had a fortress mentality, putting faith in firewalls, patching software, guarding passwords—approaches that have failed, as we know from major hacks. Governments and businesses are now working harder to guard their systems from the inside rather than protecting against hacks from the outside, such as by creating information security departments, hiring "white-hat" consultants (hacking-savvy persons who can detect weak points), and hardening their systems.[49]

At the federal level is the U.S. Cyber Command, which protects U.S. Department of Defense networks and conducts "full spectrum military cyberspace operations."[50] At the state and local levels, three approaches have been undertaken: First, a growing number of state governments are using multiagency groups to tackle cybersecurity and to analyze real and emerging threats. Some such alliances include the FBI, the DHS, and academic institutions, in hopes of determining ways to speed up coordinated responses to cyberattack. Second, employees are being trained to make them more conscious of security overall and to reduce the kinds of mistakes that can bring about an intrusion, cause an attack, or allow certain types of fraud. Third, interestingly enough, is that both states and local governments can now purchase cyber insurance to help cover the cost of investigating an attack, notifying individuals impacted by the breach, and recovering the theft of government funds.[51]

From a local policing standpoint, the minimal approach is to have trained personnel gathering digital evidence; this has even been termed the "future of policing." Not only must these personnel know how to secure a computer's memory cache, activity logs, and Internet history, but they also must maximize speed in getting to and securing the crime scene. Every second a computer is left unattended, it loses data stored in its memory cache and crucial evidence is permanently lost. Police themselves must also know how to hack, from conducting information-gathering reconnaissance to covering their tracks by restoring the criminal's computer to its preattack state.[52]

Posse Comitatus Act of 1878: legislation that prohibits the federal government from using the armed forces for the purpose of law enforcement, except in cases authorized by the Constitution or an act of Congress.

Domestic terrorism: the commission of terrorist acts in the perpetrator's own country against his or her fellow citizens.

USA PATRIOT Act: broadly written legislation that affords law enforcement with new tools to detect and prevent terrorism.

Legislative Measures

Several types of federal legislation have been enacted to help criminal justice agencies in the nation's fight against terrorism. First, although the **Posse Comitatus Act of 1878** prohibits using the military to execute the laws domestically, the military may be called on to provide personnel and equipment for special support activities, such as responding to **domestic terrorism** events involving weapons of mass destruction.[53]

Immediately following the 9/11 attacks, Congress passed the **USA PATRIOT Act** (Uniting and Strengthening America by Providing Appropriate Tools Required to Intercept and Obstruct

Terrorism Act of 2001). The act dramatically expanded the federal government's ability to investigate Americans without establishing probable cause for "intelligence purposes" and to conduct searches if there are "reasonable grounds to believe" there may be national security threats. Federal agencies such as the FBI and others are given access to financial, mental health, medical, library, and other records. The act was reauthorized in March 2006, providing additional tools: the "roving wiretap" portion and the "sneak and peek" section. The first allows the government to get a wiretap on every phone a suspect uses, while the second allows federal investigators to access library, business, and medical records without a court order. In June 2015, Congress extended the act through 2019 but amended it to stop the National Security Agency from continuing its mass phone data collection program.[54]

Military Commissions Act: legislation authorizing trial by military commission for violations of the law of war, and for other purposes.

Furthermore, in October 2006 President George W. Bush signed the **Military Commissions Act** (MCA).[55] Under the MCA, the president is authorized to establish military commissions to try unlawful enemy combatants. The commissions are authorized to sentence defendants to death, and defendants are prevented from invoking the Geneva Conventions as a source of rights during commission proceedings. The law also strips detainees of the right to file *habeas corpus* petitions in federal court and also allows hearsay evidence to be admitted during proceedings, so long as the presiding officer determines it to be reliable.[56]

Related Issues: School Shootings and Gun Violence

Extent and Nature

As Daniel Victor wrote in the *New York Times* in May 2018, "It's never long until the next school shooting in America."[57] Sad but, unfortunately, true; according to the Center for Homeland Defense and Security, there have been more than 1,350 school shootings in the United States since 2000 (see Figure 12.3). Of those, a little more than half (51 percent) occurred outside, on school property, and about 44 percent were inside the school building. Incidents occurred about equally by day of week, being slightly higher in number on Fridays, and were more common in January/February and September/October. Finally, most shootings tend to be perpetrated by a small (.22 caliber) handgun.[58]

Figure 12.3 /// K–12 School Shooting Database, Incidents by Year

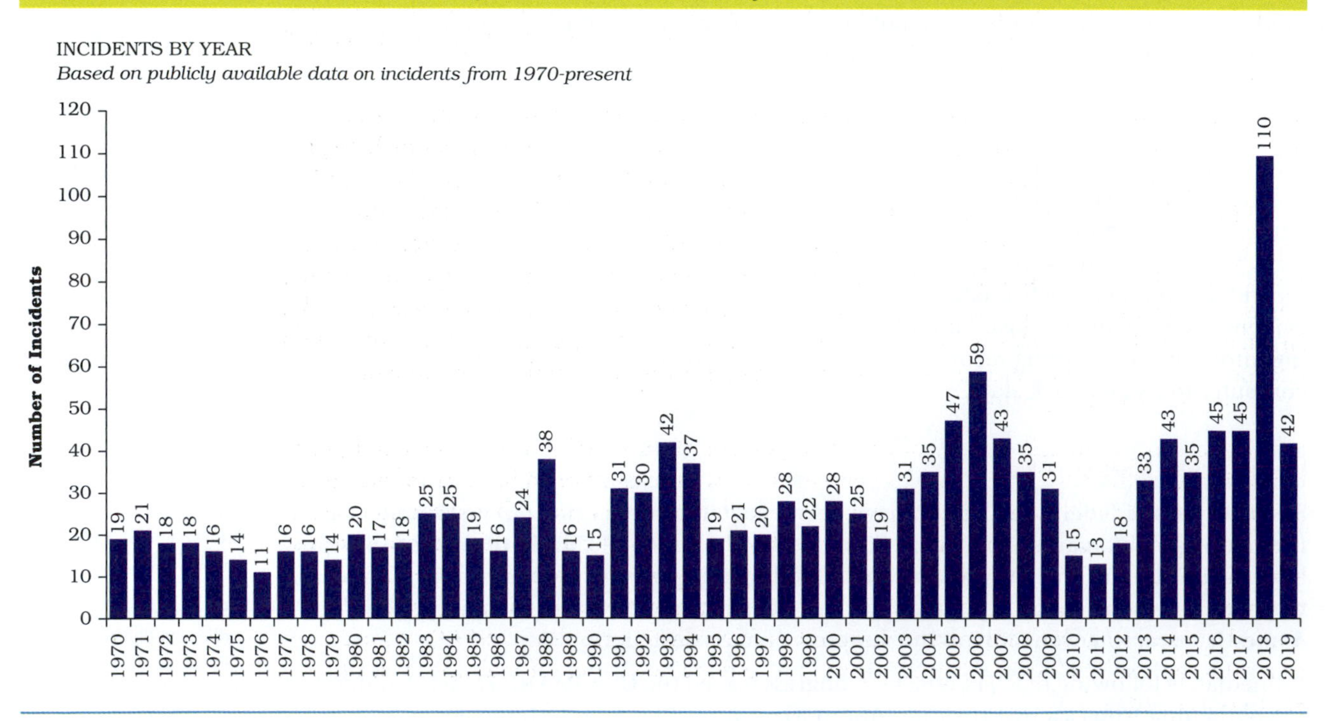

With respect to police response to an active-shooter situation, much has changed in the decade since the April 1999 mass murders by two shooters at Columbine High School in Littleton, Colorado. Thirteen people were killed while on-scene officers waited 45 minutes for an elite SWAT team to arrive. Police have greatly modified their protocols for dealing with such critical incidents. The most immediate change in protocol calls for police across the country to react swiftly to an active-shooter situation; responding officers are being trained to rush toward gunfire and—if necessary—even step over victims in order to stop the **active shooter** before more lives are lost. Such training is grounded on the assumption that, in a mass shooting, a gunman kills a person every 15 seconds. (Prior to Columbine, the protocol was for police to take a contain-and-wait strategy, intended to prevent officers and bystanders from getting killed; first responders would establish a perimeter to contain the situation and then wait for the special-weapons team to go in and neutralize the shooters(s).) Finally, another change in tactics wrought by Columbine is that special-weapons teams now typically have armed medics and rescue teams trained to remove wounded persons under fire.

Active shooter: an individual seeking to kill or attempting to kill people in a confined and populated area.

With respect to prevention, it is known that after school shootings occur, so-called red flags are always discovered concerning the shooter, so the key to prevention is to find those red flags before shootings occur. The key is information sharing: People who become concerned about a fellow student, a coworker, or a neighbor need to share information. K–12 schools and universities have student privacy laws, but every law has a safety provision that lets someone exchange information if it is a matter of public safety. Police, school personnel, parents, and other stakeholders must find all the flags, to include talking to the person's neighbors, coworkers, or whoever else might be able to reveal something.[59]

Closely related to the subject of school shootings is that of gun violence. For our Founding Fathers, the right to "keep and bear arms" with a "well-regulated militia" was "so central to the notion of liberty that it came second in the Bill of Rights only to the freedom to think and speak."[60] But they could not have foreseen a time when each year more than 42,000 people would die or be injured through firearms violence (including about 700 children)[61] and more than 300 mass shootings would occur.[62]

In the United States today, approximately 265 million guns are privately owned (although only 4 percent of Americans hunt); each day those guns are used to kill 97 persons, and guns are used in 97 percent of all homicides.[63] Yet—although two-thirds of Americans say they want stricter laws regarding firearms sales, nearly one-half support a ban on assault-style weapons, and more than 80 percent support requiring background checks for private sales and at gun shows[64]—there seems to be little appetite among lawmakers to attempt to enact stricter gun control laws.

Widely publicized, senseless mass shootings have prompted what may be a turning point in gun control laws and related violence. Signs of a growing gun control movement include protests by high school students, legislation by politicians, and boycotts of corporate America. In addition, for the first time, gun control advocates outspent the National Rifle Association during the 2018 midterm elections, and new gun control measures were enacted in several states (including increasing the minimum age for purchasing a firearm, requiring waiting periods, and allowing temporary confiscation of weapons from people deemed a safety risk). In early 2019, the U.S. House of Representatives held its first hearing on gun control in a decade, and some financial firms refused to do business with gun makers or companies that sell firearms and ammunition.[65]

Studies of prison inmates show that about 80 percent of these offenders used a handgun to commit their offenses. Nearly all (96 percent) who were legally prohibited from gun ownership acquired their guns from a supplier not required to conduct a background check.[66] The message is clear: (1) Stricter gun ownership laws would have made firearm possession illegal for many state prison inmates who used a gun to commit a crime; (2) guns in the hands of high-risk individuals present a serious threat to public safety (given that the United States has an extraordinarily high rate of firearm homicides); and (3) requiring all gun sales to be subject to a background check would make it more difficult for these offenders to obtain guns.[67]

Approaches to gun violence include background checks, preventing mentally ill people from purchasing guns, tracking gun sales, banning assault-style weapons, and barring gun purchases by people on watch or no-fly lists.

What Can Be Done?

Several approaches can be taken toward mitigating the harms caused by firearms. First are legislative enactments: background checks, preventing people with mental illness from purchasing guns, creating a federal database to track gun sales, banning assault-style weapons (with bump stocks), and barring gun purchases by people on watch or no-fly lists.[68] In addition to watching for red flags and sharing information, as mentioned earlier, the following recommendations by the Police Executive Research Forum could be given due consideration:

- *Gun owners must secure their guns in their homes:* Doing so can dramatically decrease the risk of domestic violence homicide, suicide, and mass shootings. Child access protection statutes should be enacted for the safe storage of firearms in homes and vehicles, and laws enacted that allow family members or friends to petition the courts to order the temporary removal of firearms from individuals who may be at risk of harming themselves or others.
- *Target the small number of offenders who are responsible for most gun violence:* Most gun crimes take place in a relatively small number of compact geographic areas; enforcement strategies should include identifying and deterring these small groups of gun criminals, and police should vigorously investigate nonfatal shootings and gun possession cases.
- *Ballistics technology should be fully utilized to connect guns to shooters:* The same weapon is frequently used in multiple gun crimes, and today's technology makes it easier and faster to connect weapons that are used in multiple crimes and to trace them to their original purchaser.[69]

Whether or not public school teachers (and university professors) should be armed has become a flashpoint of controversy in the discussion of gun control.

Should School Teachers Be Armed?

An area of debate—especially in the aftermath of a school shooting—concerns whether or not public school teachers should be armed. On one side are those who support such a policy as a "last defense" against armed attackers; on the other are those who believe it would bring about more problems than it would solve. The Gun-Free School Zones Act, enacted in 1990 (later overturned, but reenacted by Congress in 1996), makes knowingly possessing a firearm in a school zone a federal offense (with certain exceptions); however, the law also allows states to authorize certain individuals to carry firearms on school grounds. At present, at least nine states have policies that specifically allow school employees—other than security personnel—to carry guns on school property, and the number is greater if including those states that grant school districts power to make their own decisions and do not limit carrying to school employees. Furthermore, half of the states have policies allowing schools or districts to give permission to "individuals" to carry guns.[70]

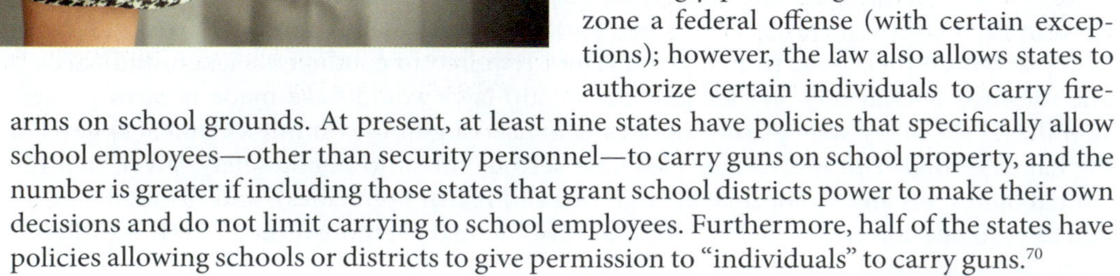

Juvenile Justice Journal

Causes and Treatment of Juvenile Offending

In earlier chapters we looked at juvenile justice from a variety of perspectives. Here we briefly consider this subject in terms of extent of crime, treatment, and implications for policy and prevention.

There is good news concerning juvenile arrest trends in the United States: Reaching its highest level in 1996, juvenile arrests have since declined 72 percent by 2017. Why this long-running reduction in juvenile arrests? Two reasons are commonly offered: First, the drop in juvenile violent crime arrests virtually coincides with the drop in juvenile commitments to residential facilities (studies show that juveniles who are kept in the community recidivate less often than previously detained youths). Second, states are increasingly using alternative forms of punishment over incarceration (discussed in Chapter 11). Other possible reasons commonly offered are the increasing immigrant population—a group that tends to have lower crime rates—and the fact that marijuana laws are becoming more lax. However, given that about 620,000 juveniles are still arrested each year in the United States, there remains cause for concern.

Several theories have been offered by experts to explain juvenile crime; however, no single theory has been universally accepted. Nevertheless, experts agree that a correlation exists between juvenile crime and the following factors:

- *Family dysfunction:* Family background is one of the most potent influences on juvenile development. Juveniles who live in unstable homes and social environments are deemed to be at-risk because of their vulnerability to detrimental influences and the influence of such environments on antisocial behavior.
- *Drug use and deviance:* Alcohol and tobacco are the drugs of choice for many juveniles.
- *Socioeconomic class:* Children from poor and working-class backgrounds are much more likely than other children to engage in delinquent behavior. Large numbers of the urban poor are caught in a chronic generational cycle of poverty, low educational achievement, teenage parenthood, unemployment, and welfare dependence.
- *Educational experiences:* Academic achievement is considered to be one of the principal stepping stones toward success in American society. Ideally, opportunities for education, mentoring, and encouragement to excel would be equally available for all children. Unfortunately, that is not the situation.

With respect to policy and prevention aspects of juvenile crime, one recommendation is to stop prosecuting persons under age 18 as adults, which often happens no matter how minor the crime. Raising the age at which youthful offenders can be treated as adults can have a huge impact on their lives—and recidivism—because teens who are channeled into adult prisons do not have access to rehabilitative services that the juvenile justice system provides; furthermore, adult prisons can be extremely dangerous for teens.

In addition—and taking into consideration the Juvenile Justice and Delinquency Prevention Act, discussed in Chapter 6, which hopes to address racial and ethnic disparities and calls for sound separation and jail removal for youth awaiting trial as adults—policy makers also need to address the tough-on-crime statutes that result in automatic transfer of juveniles to adult courts and to continue to increase the use of community-based alternatives to incarceration.

Sources: "Juvenile Incarceration Rate Has Dropped in Half. Is Trend Sustainable?" Justice Policy Institute, November 10, 2015, http://www.justicepolicy.org/news/9854; also see Manny Araujo, "Study: Number of Juvenile Arrests Declining," *Eureka Times-Standard*, November 7, 2016, http://www.times-standard.com/article/NJ/20161007/NEWS/161009838; list of factors adapted from Gus Martin, *Juvenile Justice: Process and Systems* (Thousand Oaks, Calif.: Sage, 2005), pp. 64–67; and Teresa Wiltz, "How 'Raise the Age' Laws Might Reduce Recidivism," Pew Charitable Trusts, May 31, 2017, https://www.pewtrusts.org/en/research-and-analysis/blogs/stateline/2017/05/31/how-raise-the-age-laws-might-reduce-recidivism.

IMMIGRATION: WHO SHOULD BE COMING TO AMERICA?

Many Questions, Few Answers

Today, of the 326 million people in the United States, an estimated 11 million are undocumented immigrants, 5.9 million of whom are under age 18. Most (56 percent) have come from Mexico, while 11 percent arrived here from Guatemala and El Salvador. In four states—Texas, California, Nevada, and New Jersey—such persons make up more than five percent of the total population.[71] However, **immigration** is a national issue—with political, economic, and social elements—and not one that only impacts a few states.

Immigration: the act of relocating to and living permanently in a foreign country.

Undocumented immigrant: one who is foreign born and does not have a legal right to be or remain in the United States.

What should be done with those **undocumented immigrants** who are already here? And in the future, should all such persons who wish to migrate to the United States be allowed to cross its borders? Or, rather, should all such persons be banned? Or, should only *some* be admitted onto our soil? Should the United States erect walls to separate it from its neighbors? Deny protection to victims of gang and domestic violence who are from other countries and seek asylum here?

These are complex social, political, and humanitarian questions, each of them leading to dozens more issues and questions. At times it seems unlikely that one could find any two U.S. politicians who could agree on the answers to these policy questions. Indeed, immigration has become one of the most vexing and polarizing challenges of our time—and our law enforcement agencies are often caught in the middle while the political battles are being waged.

Making the issue even more complicated is that U.S. citizens themselves cannot seem to agree on whether immigration is good or bad. On the one hand, it is argued that the nation is richer in culture, diversity, and civilization as it is reshaped by people from around the world. Conversely, immigration also has illegal and dangerous aspects that threaten the nation's security (as illustrated in the earlier discussion of terrorism). Nor do we agree on whether immigration helps (through reduced labor costs) or hurts (via displaced U.S.-born workers and lower wages) the U.S. economy. The U.S. labor industry certainly has no problem with hiring undocumented immigrants, with an estimated 8 million undocumented immigrants now working in all sorts of occupations.[72]

©iStockphoto.com/rikirisnandar

Whether or not the United States should open its borders—and make it richer in culture, diversity, and civilization, while also possibly threatening the nation's security—has also, like gun control, become a flashpoint of controversy.

Not even police executives seem to support stricter immigration enforcement. For example, a 40-member task force of the National Immigration Forum has argued that immigration enforcement is primarily a federal responsibility. They believe that immigration enforcement at the local level diverts limited resources from already financially strapped agencies, hinders community policing efforts, and damages relationships with immigrant communities.[73]

Under the law, immigration enforcement is a duty of all federal, state, and local law enforcement officers and agents; in practice, however, immigration enforcement decisions are made from the top of our federal government, with the executive branch (i.e., the president) having much of the control. As examples, in past presidential administrations, law enforcement agencies had "prosecutorial discretion" in determining when and whether to enforce immigration laws and deport certain criminal groups. Under the recent Trump administration policy, however, nearly all such discretion was removed when a 2017 executive order directed the enforcement of immigration laws against "all removable aliens."[74]

The Wall Conundrum

The question of whether or not there should be a southwestern **border wall** (alternatively termed a fence or barrier) has been at the heart of a major political divide in the United States concerning immigration, and it has been so since the 2016 presidential campaign. After President Trump's election, U.S. Customs and Border Protection (CBP) proposed a plan for a wall to extend 316 miles (resulting in a total length of 970 miles) that would cost $18 billion (for construction as well as ongoing personnel roads, towers, technology, and maintenance).[75] Later, the cost was downgraded to $5.7 billion for a 100-mile wall along the southern border.[76] Then, in February 2019—following a 35-day federal government shutdown and a declaration of national emergency (resulting in several legal challenges[77])—another political compromise was reached: President Trump signed a bipartisan spending bill allocating $1.38 billion for 55 miles of border barriers—far less than the amounts CBP and the president had initially requested. In fact, over time eight different prototypes of walls were designed and proposed for this purpose.[78]

Border wall: some form of separation barrier that runs along an international border.

Although President George W. Bush performed what he called "the first step toward immigration reform" by signing the $35 billion Secure Fence Act into law in 2006, authorizing the construction of 700 miles of double-layer fencing along the southwestern border (and deploying National Guard troops there, as did President Barack Obama),[79] whether or not there will ever be a wall along the entire 2,000-mile southwestern border remains to be seen.

The Controversy Over Criminality

Another controversial and often-used argument for a strong border policy and hardening of immigration enforcement is that undocumented immigrants commit more crimes than U.S.-born citizens. Widely publicized murders committed by undocumented immigrants—such as the murders of four people in northern Nevada in early 2019,[80] an Iowa college student in mid-2018,[81] and a 32-year-old woman in San Francisco in 2015[82]—have fomented angst and ire among many Americans who believe, as one national news commentator put it, that "illegal immigration kills Americans."[83] Further, because the San Francisco woman was shot by a known criminal who had slipped into the United States illegally on multiple occasions, the killing prompted a national debate about sanctuary cities.

At the same time, a number of immigration experts and academic researchers have long held that immigrants commit fewer crimes. But what is the truth? The research is frustrated by the fact that local police do not list the immigration status of those persons arrested, meaning it is impossible to determine with certainty how many crimes are committed by legal immigrants, undocumented immigrants, and native-born citizens.

Attempts to learn the "truth" have compared incarceration rates for both native-born citizens and undocumented immigrants, violent crime rates both at times when national immigration trends were high and low, and numbers of persons removed from the United States after committing crimes or being identified as gang members. All of these attempts to research the question have flaws, according to the Center for Immigration Studies. The center argues that the crime rates of such persons don't matter. What *does* matter is what we do with those undocumented immigrants who are committing crimes and causing problems.[84]

Those Who Would Seek Asylum

We are a world of refugees; more than 40 million people across the globe have recently been forced to flee their homes as a result of war, famine, and political instability.[85] Another recent issue concerns **asylum seekers**—now ranging from about 2,000 to 10,000 applications per month (many are from Mexico and Central America, but the most are from China, as political and religious dissidents[86])—wishing to migrate to the United States to escape violence, poverty, political volatility, and persecution of all kinds in their home countries.[87] In late 2018 and early 2019, 7,000 active-duty troops were sent to southwestern border communities

Asylum seekers: persons fleeing their home countries (often due to persecution on the basis of race, religion, nationality, political affiliation, and so on) in search of sanctuary in another country, where they hope to remain permanently.

to guard fences, install razor wire, and halt asylum seekers' northward march toward the United States from Central America.[88] Meanwhile, the U.S. Department of Homeland Security offered support to its Mexican counterparts who were attempting to block the caravans of people heading north.

Under laws enacted in 1948 for "displaced persons" who needed refuge following World War II, asylum could be granted for people seeking to enter the United States when certain criteria were met. Under current practice, people who meet with immigration authorities and express a fear of returning to their home country will be interviewed (to judge whether or not their fears are well grounded); however, it is when a person is *not* granted asylum that this issue becomes charged with controversy. With a current backlog of nearly 800,000 cases, there simply are not enough judicial and enforcement personnel to handle these numbers, so a "catch-and-release" system evolved where people can stay in the country while awaiting their court date.[89]

As with the other issues discussed in this chapter, refugee and asylum policy is driven largely by the nation's chief executive—who gets to decide whether to commit resources to working with asylum seekers or instead devote those resources to immigration enforcement.[90]

What Can Be Done?

In the foreseeable future, there will likely remain much turmoil concerning immigration issues as they relate to law, policy, and enforcement at the federal level. So what can state and local law enforcement agencies do?

First, state and local agencies should continue to cooperate with federal immigration officials in order to protect our national security. This network is of critical importance for dealing with crimes and threats involving lone-wolf, homegrown terrorists as well as gangs, sex trafficking, smuggling, illicit drug offenses, and other serious crimes that are often tied to illegal immigration. Furthermore, federal agencies rely on and benefit greatly from intelligence developed at the state and local levels; therefore, these agencies should never turn a blind eye to immigration violations. Agencies operating intelligence units (often referred to as fusion centers) must continue to do their work.

Finally, as they will come into frequent contact with immigrants, police officers should receive adequate training and possess a working knowledge of immigration law and policy (see the accompanying "You Be the . . . Police Officer" box).[91]

You Be the . . .

Police Officer

U.S. Immigration and Customs Enforcement (ICE) is responsible for enforcing federal immigration laws as part of its homeland security mission. The so-called 287(g) program, contained in the Illegal Immigration Reform and Immigrant Responsibility Act of 1996,[92] is one of ICE's top partnership initiatives; it allows a state or local law enforcement entity to enter into a partnership with ICE immigration enforcement and allows certain selected and specially trained police officers to locate and apprehend undocumented immigrants in their communities.[93]

These deputized officers are allowed to interview individuals to ascertain their immigration status, check DHS databases for information or to enter information on individuals, issue detainers to hold individuals for ICE, file documents to begin the removal process, and make recommendations for detention and immigration bond.

Currently, ICE has 287(g) agreements with 78 law enforcement agencies in 20 states. ICE has trained and certified more than 1,500 state and local officers work in the program.[94]

/// IN A NUTSHELL

- International terrorist groups have become more adept at recruiting "foot soldiers" through the Internet and other means. These attacks have become increasingly focused on civilian targets, and great concern now exists for homegrown violent extremists (HVEs) as well as the lone offender who becomes radicalized.
- Cybercrime is one of the fastest growing areas of crime; the most challenging and potentially disastrous type of cybercrime now being perpetrated against the United States is China's state-supported cyber warfare.
- Many experts believe it is only a matter of time before chemical/biological weapons are used by terrorists; all that is required is for a toxin to be cultured and put into a spray form that can be weaponized and disseminated into the population.
- Police have several possible means of attempting to prevent or address acts of terrorism, including legislative measures such as the USA PATRIOT Act.
- There have been more than 1,350 school shootings in the United States since 2000; new policies and practices have been developed for responding to and preventing such acts in the future.
- Closely related to the subject of school shootings is that of gun violence; although the topic of gun control has long been controversial, it has received new attention due to mass shootings.
- Another long-standing issue is what should be done with undocumented individuals, including those wishing to come here and those who are already here. A tandem issue concerns what to do with those seeking asylum. The criminal justice system—and the police in particular—are caught in the middle of this political, economic, and social debate. Related to the question of immigration are questions concerning whether or not undocumented immigrants commit more crimes.

/// KEY TERMS & CONCEPTS

Review key terms with eFlashcards at **edge.sagepub.com/peakbrief**.

Active shooter 285
Asylum seekers 289
Bioterrorism 281
Border wall 289
Cybercrime 281
Cyberterrorism 281
Diagnostic center 278
Domestic terrorism 283
Extremists 280
Homegrown violent extremists 280
Immigration 288
International terrorism 278
Lone-wolf terrorist 280
Marijuana decriminalization 277
Marijuana legalization 277
Military Commissions Act 284
Narcan/naloxone/suboxone 276
Opioids 274
Posse Comitatus Act of 1878 283
Radicalization 280
Sex trafficking 277
T visa 278
Undocumented immigrant 288
USA PATRIOT Act 283

/// REVIEW QUESTIONS

Test your understanding of chapter content. Take the practice quiz at **edge.sagepub.com/peakbrief**.

1. What are today's most pressing issues concerning drug abuse, and what are some approaches that legislatures, police, and other agencies can take toward addressing them?
2. What is the legal status of medical and recreational marijuana use?
3. What are the major issues facing the United States and its criminal justice system with respect to handling of sex trafficking, to include current legislation, programs, and proposals to address those challenges?
4. What is a commonly accepted definition of terrorism?
5. Why are homegrown and lone-wolf extremists of concern today?
6. What kinds of measures might police employ against all types of terroristic acts?
7. What is cyberterrorism, and what are some of the kinds of activities cybercriminals engage in?
8. What is bioterrorism, and why have those who would employ it been largely unsuccessful?

9. What role is played by legislation (e.g., the USA PATRIOT Act, the Military Commissions Act, and the Posse Comitatus Act of 1878) in combatting terrorism?
10. What is the status of gun ownership and violence in our society? What are some methods police can use to control such violence, and what kinds of policy issues might be in play to effect stronger gun control measures?
11. What is the extent and status of immigration to the United States today? The status of federal law? The building of a wall? The granting of asylum?
12. What is known about the link between illegal immigration and crime?

/// LEARN BY DOING

1. How would you determine whether or not an opioid epidemic exists in your community? Consider what resources—data, agencies, publications, interviews, Internet searches, and so on—you would employ in this endeavor.
2. Assume that you are assigned a class presentation in which you will investigate what measures police and social services agencies in your area are taking to combat sex trafficking (to include such approaches as the T visa). How will you approach these questions?
3. Assume that shortly after 8 p.m., a lone male throws a pipe bomb amongst moviegoers in a theater complex. The initial 911 call reported numerous injuries, and that the suspect had left the theater running toward another retail mall. What kinds of resources should the police and other first responders deploy to the scene and beyond?
4. Terrorists often seek out public gathering areas and critical infrastructures (such as dams, nuclear reactors, chemical plants, tourist attractions, water systems, public safety and emergency services). Therefore, first responders should develop advance plans for identifying and protecting these assets *as well as* an action plan in the event a terrorist attack or other significant event occurs. What such events and critical infrastructures in your community would warrant such careful pre- and post-planning?
5. To better grasp the methods used and problems confronted by federal, state, and local law enforcement agencies regarding the immigration issues discussed in this chapter, you could do no better than to interview individuals who work in this arena on a daily basis—in this case, a homeland security specialist or fusion center employee, a member of the Border Patrol or ICE, and so on.

/// STUDY SITE

SAGE edge™

Get the tools you need to sharpen your study skills. SAGE edge offers a robust online environment featuring an impressive array of free tools and resources.

Access practice quizzes, eFlashcards, video, and multimedia at **edge.sagepub.com/peakbrief**

/// ASSESS YOUR AWARENESS ANSWERS

Chapter 1 1. True; 2. False; 3. False; 4. False; 5. True; 6. False

Chapter 2 1. True; 2. False; 3. False; 4. True; 5. True; 6. False; 7. True; 8. True; 9. False

Chapter 3 1. True; 2. False; 3. True; 4. False; 5. False; 6. False; 7. False

Chapter 4 1. True; 2. False; 3. True; 4. False; 5. True; 6. False; 7. True; 8. True; 9. False

Chapter 5 1. False; 2. False; 3. True; 4. False; 5. True; 6. True; 7. True; 8. True

Chapter 6 1. True; 2. True; 3. True; 4. True; 5. False; 6. True; 7. True; 8. True

Chapter 7 1. True; 2. True; 3. True; 4. False; 5. False; 6. False

Chapter 8 1. False; 2. True; 3. True; 4. False; 5. True; 6. True; 7. False

Chapter 9 1. False; 2. True; 3. False; 4. True; 5. True; 6. False; 7. True

Chapter 10 1. True; 2. False; 3. True; 4. False; 5. False; 6. False; 7. False; 8. True

Chapter 11 1. True; 2. False; 3. False; 4. True; 5. False; 6. True; 7. True

Chapter 12 1. False; 2. True; 3. True; 4. True; 5. True; 6. False; 7. False

/// GLOSSARY

Absolute ethics: the type of ethics where there are only two sides—good or bad, black or white; some examples would be unethical behaviors such as bribery, extortion, excessive force, and perjury, which nearly everyone would agree are unacceptable for criminal justice personnel.

Academy training: where police and corrections personnel are trained in the basic functions, laws, and skills required for their positions.

Accepted lying: police activities intended to apprehend or entrap suspects. This type of lying is generally considered to be trickery.

Acquittal: a court or jury's judgment or verdict of not guilty of the offenses charged.

Active shooter: an individual seeking to kill or attempting to kill people in a confined and populated area.

Actus reus: Latin for "guilty deed"—an act that accompanies one's intent to commit a crime, such as pulling out a knife and then stabbing someone.

Adjudication: the legal resolution of a dispute—for example, when one is declared guilty or not guilty—by a judge or jury.

Adversarial system: a legal system wherein there is a contest between two opposing sides, with a judge (and possibly a jury) sitting as an impartial arbiter, seeking truth.

Affidavit: any written document in which the signer swears under oath that the statements in the document are true.

Affirmative defense: the response to a criminal charge in which the defendant admits to committing the act charged but argues that for some mitigating reason he or she should not be held criminally responsible under the law.

Aggravating circumstances: elements of a crime that enhance its seriousness, such as the infliction of torture, killing of a police or corrections officer, and so on.

Alternatives to incarceration: a sentence imposed by a judge other than incarceration, such as probation, parole, shock probation, or house arrest.

Arraignment: a criminal court proceeding during which a formally charged defendant is informed of the charges and asked to enter a plea of guilty or not guilty.

Arrest: the taking into custody or detaining of one who is suspected of committing a crime.

Asylum seekers: persons fleeing their home countries (often due to persecution on the basis of race, religion, nationality, political affiliation, and so on) in search of sanctuary in another country, where they hope to remain permanently.

Bail: surety (e.g., cash or paper bond) provided by a defendant to guarantee his or her return to court to answer to criminal charges.

Bioterrorism: the use of biological material, such as anthrax or botulin, to commit an act of terrorism.

Booking: the clerical procedure that occurs after an arrestee is taken to jail, during which a record is made of his or her name, address, charge(s), arresting officers, and time and place of arrest.

Boot camp: a short-term jail or prison program that puts offenders through a rigorous physical and mental regimen designed to instill discipline and respect for authority.

Border wall: some form of separation barrier that runs along an international border.

Burden of proof: the requirement that the state must meet to introduce evidence or establish facts.

Capital punishment: a sentence of death, or carrying out same via execution of the offender.

Caseload: the number of cases awaiting disposition by a court, or the number of active cases or clients maintained by a probation or parole officer.

Causation: a link between one's act and the injurious act or crime, such as one tossing a match in a forest and igniting a deadly fire.

Chain of command: vertical and horizontal power relations within an organization, showing how one position relates to others.

Circuit courts: originally courts wherein judges traveled a circuit to hear appeals, now courts with several counties or districts in their jurisdiction; the federal court system contains 11 circuit courts of appeals (plus the District of Columbia and territories), which hear appeals from district courts.

Civil law: a generic term for all noncriminal law, usually related to settling disputes between private citizens, governmental, and/or business entities.

Classification (of inmates): an inmate security and treatment plan based on one's security, social, vocational, psychological, and educational needs while incarcerated.

Community corrections: probation, parole, and a variety of other measures that offer convicted offenders an alternative(s) to incarceration.

Community era: beginning in about 1980, a time when the police retrained to work with the community to solve problems by looking at their underlying causes and developing tailored responses to them.

Community policing and problem solving: a proactive management philosophy that involves police-community collaboration and a four-step process (scanning, analyzing, response, and assessment) to focus police activities and thus enable officers to respond more effectively to crime and disorder with arrests or other appropriate actions.

Conflict theory of justice: explains how powerful groups create laws to protect their values and interests in diverse societies.

Consensus theory of justice: explains how a society creates laws as a result of common interests and values, which develop largely because people experience similar socialization.

Constable: in England, favored noblemen who were forerunners of modern-day U.S. criminal justice functionaries; largely disappeared in the United States by the 1970s.

Contract services: a for-profit firm or individuals hired by an individual or company to provide security services.

Conviction: the legal finding, by a jury or judge, or through a guilty plea, that a criminal defendant is guilty.

Coroner: an early English court officer; today one (usually a physician) in the United States whose duty it is to determine cause of death.

County sheriff's office: a unit of county government with a sheriff (normally elected) and deputies whose duties vary but typically include policing unincorporated areas, maintaining county jails, providing security to courts in the county, and serving warrants and court papers.

Court of last resort: the last court that may hear a case at the state or federal level.

Courtroom work group: the criminal justice professionals who work together to move cases through the court system.

Crime control model: a model by Packer that emphasizes law and order and argues that every effort must be made to suppress crime and to try, convict, and incarcerate offenders.

Crime rate: the number of reported crimes divided by the population of the jurisdiction, and multiplied by 100,000 persons; developed and used by the FBI *Uniform Crime Reports*.

Crime scene: any location where a crime occurred and that may contain forensic evidence relating to and supporting a criminal investigation.

Crimes against persons: violent crimes, to include homicide, sexual assault, robbery, and aggravated assault.

Crimes against property: crimes during which no violence is perpetrated against a person, such as burglary, theft, and arson.

Criminal justice flow and process: the movement of defendants and cases through the criminal justice process, beginning with the commission of a crime, and including stages that involve actions of criminal justice actors working within police, courts, and correctional agencies.

Criminal law: the body of law that defines criminal offenses and prescribes punishments for their infractions.

Criminalistics: the interdisciplinary study of physical evidence related to crime; drawing on mathematics, physics, chemistry, biology, anthropology, and many other scientific fields.

Cybercrime: the use of computers to commit crimes, such as embezzlement, diversion of bank monies to other accounts, and hacking personal information.

Cyberterrorism: a premeditated, politically motivated attack against computer systems, computer programs, or data.

Day reporting center: a structured corrections program requiring offenders to check in at a community site on a regular basis for supervision, sanctions, and services.

Defendant: a person against whom a criminal charge is pending; one charged with a crime.

Defense: the response by a defendant to a criminal charge, to include denial of the criminal allegations in an attempt to negate or overcome the charges.

Defense attorney: one whose responsibility it is to see that the rights of the accused are upheld prior to, during, and after trial; the Sixth Amendment provides for "effective" counsel, among other constitutionally enumerated rights that defense attorneys must see are upheld.

Delay (trial): an attempt (usually by defense counsel) to have a criminal trial continued until a later date.

Deontological ethics: one's duty to act.

Detective/criminal investigator: a police officer who is assigned to investigate reported crimes, to include gathering evidence, completing case reports, testifying in court, and so on.

Determinate sentence: a specific, fixed-period sentence ordered by a court.

Deterrence: the effect of punishments and other actions to deter people from committing crimes.

Deviant lying: occasions when officers commit perjury to convict suspects or are deceptive about some activity that is illegal or unacceptable to the department or public in general.

Diagnostic center: Launched by the U.S. Department of Justice, the center provides technical assistance resources for law enforcement executives and policy makers for combatting all types of crimes, including sex trafficking.

Discovery: a procedure wherein both the prosecution and the defense exchange and share information as to witnesses to be used, results of tests, recorded statements by defendants, or psychiatric reports, so that there are no major surprises at trial.

Discretion: authority to make decisions in enforcing the law based on one's observations and judgment ("spirit of the law") rather than the letter of the law.

District courts: trial courts at the county, state, or federal level with general and original jurisdiction.

Diversion program: a sentencing alternative that removes a case from the criminal justice system, typically to move a defendant into another treatment program or modality.

DNA: deoxyribonucleic acid, which is found in all cells; used in forensics to match evidence (hair, semen) left at a crime scene with a particular perpetrator.

Domestic terrorism: the commission of terrorist acts in the perpetrator's own country against his or her fellow citizens.

Double jeopardy: the prosecution of an accused person twice for the same offense; prohibited by the Fifth Amendment except under certain circumstances.

Dual court system: the state and federal court systems of the United States.

Due process model: a model by Packer that advocates defendants' presumption of innocence, protection of suspects' rights, and limitations placed on police powers to avoid convicting innocent persons.

Eighth Amendment: in the Bill of Rights, it contains the protection against excessive bail and fines, as well as cruel and unusual punishment.

Electronic monitoring: use of electronic devices (bracelets or anklets) to emit signals when a convicted offender (usually on house arrest) leaves the environment in which he or she is to remain.

Entrapment: police tactics that unduly encourage or induce an individual to commit a crime he or she typically would not commit.

Ethics: a set of rules or values that spell out appropriate human conduct.

Exclusionary rule: the rule (see *Mapp v. Ohio,* 1961) providing that evidence obtained improperly cannot be used against the accused at trial.

Exigent circumstance: an instance in which quick, emergency action is required to save lives, protect against serious property damage, or prevent suspect escape or evidence destruction; in such cases, officers can enter a structure without a search warrant.

Exoneration: to absolve someone of criminal blame, or find someone not guilty.

Extremists: persons who hold extreme or fanatical political or religious views, which can lead to the commission or advocacy of extreme action.

Federal court system: the four-tiered federal system that includes the Supreme Court, circuit courts of appeal, district courts, and magistrate courts.

Federal law enforcement agencies: federal organizations that, for example, are charged with protecting the homeland (DHS); investigating crimes (FBI) and enforcing particular laws, such as those pertaining to drugs (DEA) or alcohol/tobacco/firearms/explosives (ATF); and guarding the courts and transporting prisoners (USMS).

Federalism: a type of government that divides powers between a national (federal) government and governments of smaller geographic territories, including states, counties, and cities.

Felony: a serious offense with a possible sentence of more than a year in prison.

Felony-murder rule: the legal doctrine that says that, if a death occurs during the commission of a felony, the perpetrator of the crime may be charged with murder in the first degree, regardless of the absence of intent, premeditation and deliberation, or malice aforethought.

Field training officer (FTO): one who is to oversee and evaluate the new police officer's performance as he or she transitions from the training academy to patrolling the streets.

Fifth Amendment: in the Bill of Rights, among other protections, it guards against self-incrimination and double jeopardy.

Forensic science: the study of causes of crimes, deaths, and crime scenes.

Fourth Amendment: in the Bill of Rights, it contains the protection against unreasonable searches and seizures and protects people's homes, property, and effects.

Grand jury: a body that hears evidence and determines probable cause regarding crimes and can return formal charges against suspects; use, size, and functions vary among the states.

Gratuities: the receipt of some benefit (a meal, gift, or some other favor) either for free or for a reduced price.

Halfway house: a community center or home staffed by professionals or volunteers designed to provide counseling to ex-prisoners as they transition from prison to the community.

Hierarchy rule: in the FBI *Uniform Crime Reports* reporting scheme, the practice whereby only the most serious offense of several that are committed during a criminal act is reported by the police.

Homegrown violent extremists: persons who are inspired by foreign terrorist organizations and radicalized in the countries in which they are born, raised, or reside.

House arrest/home confinement: detention of offenders in their own homes; compliance is often monitored electronically.

Idealistic contrast: the differences between juvenile and adult criminal justice processes, to include treatment and terminology.

Immigration: the act of relocating to and living permanently in a foreign country.

Incapacitation: rendering someone as unable to act or move about, either through incarceration or by court order.

Indeterminate sentence: a scheme whereby one is sentenced for a flexible time period (e.g., 5–10 years) so as to be released when rehabilitated or when the opportunity for rehabilitation is presented.

Initial appearance: a formal proceeding during which the accused is read his or her rights and informed of the charges and the amount of bail required to secure pretrial release.

Intensive supervision probation and parole: post-release supervision that usually includes much closer and stricter supervision, more contact with offenders, more frequent drug tests, and other such measures.

Intent, specific: a purposeful act or state of mind to commit a crime.

Intermediate court of appeals: a state court that stands between a trial court and a court of last resort; it typically has appellate jurisdiction only.

Intermediate sanctions: forms of punishment that are between freedom and prison, such as home confinement and day reporting.

International terrorism: acts of terrorism perpetrated in a foreign country by terrorists who are not native to that country.

INTERPOL: the only international crime-fighting organization; it collects intelligence information, issues alerts, and assists in capturing world criminals, and it has nearly 200 member countries.

Jail: a facility that holds persons who have been arrested for crimes and are awaiting trial, persons who have been convicted for misdemeanors and are serving a sentence (up to a year in jail), federal offenders, and others.

Judicial misconduct: inappropriate behavior by a judge.

Judicial selection (methods of): means by which judges are selected for the bench, to include election, a nominating commission, or a hybrid of these methods.

Jurisdiction, court: the authority of a court to hear a particular type of case, based on geography (city, state, or federal) and subject matter (e.g., criminal, civil, probate).

Justice of the peace (JP): a minor justice official who oversees lesser criminal trials; one of the early English judicial functionaries.

Kansas City Preventive Patrol Experiment: in the early 1970s, a study of the effects of different types of patrolling on crime—patrolling as usual in one area, saturated patrol in another, and very limited patrol in a third area; the results showed no significant differences.

Legal jurisdiction: the authority to make legal decisions and judgments, often based on geographic area (territory) or the type of case in question.

Lex talionis: Latin for "an eye for an eye, a tooth for a tooth"; retaliation or revenge that dates back to the Bible and the Middle Ages.

Lineup: a procedure in which police ask suspects to submit to a viewing by witnesses to determine the guilty party, based on personal and physical characteristics; information obtained may be used later in court.

Lone-wolf terrorist: one who plans and commits acts of terrorism without the support of a larger terrorist group.

Marijuana decriminalization: where a state has repealed or amended its marijuana laws to make certain acts criminal but no longer subject to prosecution (e.g., individuals caught with small amounts of marijuana for personal consumption may not be prosecuted and, thus, avoid a criminal record).

Marijuana legalization: typically, where one will not be arrested for marijuana use if he or she follows the state laws as to age, place, and amount for consumption (however, one might still be arrested for selling or trafficking if such are prohibited under the law).

Mass incarceration: a term generally referring to what is perceived as the United States's disproportionately high rates of imprisonment of young African American men; some believe it deters crime and incapacitates offenders, while others say that it weakens poor families and keeps them socially marginalized.

Mens rea: Latin for "guilty mind"—the purposeful intention to commit a criminal act.

Merit selection: a means of selecting judges whereby names of interested candidates are considered by a committee and recommendations are then made to the governor, who then makes the appointment; known also as the Missouri Plan.

Military Commissions Act: legislation authorizing trial by military commission for violations of the law of war, and for other purposes.

Misdemeanor: a lesser offense, typically punishable by a fine or up to one year in a local jail.

Mitigating circumstances: circumstances that would tend to lessen the severity of the sentence, such as one's youthfulness, mental instability, not having a prior criminal record, and so on.

Model Code of Judicial Conduct: adopted by the House of Delegates of the American Bar Association in 1990, it provides a set of ethical principles and guidelines for judges.

Motive: the reason for committing a crime.

Municipal police department: a police force that enforces laws and maintains peace within a specified city or municipality.

Narcan/naloxone/suboxone: a medication that can be used to block the effects of opioids, particularly for overdoses.

National Crime Victimization Survey: a random survey of U.S. households that measures crimes committed against victims; includes crimes not reported to police.

National Incident-Based Reporting System: a crime reporting system in which police describe each offense that occurs during a crime event as well as characteristics of the offender.

New generation/direct supervision jail: a jail that, by its architecture and design, eliminates many of the traditional features of a jail, allowing staff members greater interaction and control.

Noble cause corruption: a situation in which one commits an unethical act but for the greater good; for example, a police officer violates the Constitution in order to capture a serious offender.

Opioids: a class of drugs that includes heroin, synthetic opioids such as fentanyl, and pain relievers available legally by prescription.

Organization: an entity of two or more people who cooperate to achieve one or more objectives.

Organizational structure or chart: a diagram of the vertical and horizontal parts of an organization, showing its chain of command, lines of communication, division of labor, and so on.

Organized crime: crimes committed by members of illegal organizations.

Parole: early release from prison, with conditions attached and under supervision of a parole agency.

Parole officer: one who supervises those who are on parole.

Plaintiff: the party bringing a lawsuit or initiating a legal action against someone else.

Plea negotiation (or bargaining): a preconviction process between the prosecutor and the accused in which a plea of guilty is given by the defendant, with certain specified considerations in return—for example, having several charges or counts tossed out, and a plea by the prosecutor to the court for leniency or shorter sentence.

Police corruption: misconduct by police officers that can involve but is not limited to illegal activities for economic gain, gratuities, favors, and so on.

Policing role: the function of the police in contemporary society.

Policing styles: James Q. Wilson argued that there are three styles of policing: watchman, legalistic, and service.

Policy making: the act of creating laws or setting standards to govern the activities of government; the U.S. Supreme Court, for example, has engaged in policy making in several areas, such as affirmative action, voting, and freedom of communication and expression.

Political era: from the 1840s to the 1930s, the period of time when police were tied closely to politics and politicians, dependent on them for being hired and promoted, and for assignments—all of which raised the potential for corruption.

Posse Comitatus Act of 1878: legislation that prohibits the federal government from using the armed forces for the purpose of law enforcement, except in cases authorized by the Constitution or an act of Congress.

Preliminary hearing: a stage in the criminal process conducted by a magistrate to determine whether a person charged with a crime should be held for trial based on probable cause; does not determine guilt or innocence.

Pretrial motions/processes: any number of motions filed by prosecutors and defense attorneys prior to trial, for instance, to quash evidence, change venue, conduct discovery, challenge a search or seizure, raise doubts about expert witnesses, or exclude a defendant's confession.

Prison: a state or federal facility housing long-term offenders, typically felons, for a period greater than one year.

Prison industries: use of prison and jail inmates to produce goods or provide services for a public agency or private corporation.

Private police/security officers: all nonpublic officers, including guards, watchmen, private detectives, and investigators; they have limited powers and only the same arrest powers as regular citizens.

Privatization: the operation of existing prison facilities, or the building and operation of new prisons, by for-profit companies.

Probable cause: a reasonable basis to believe that a crime has been, or is about to be, committed by a particular person.

Probation officer: one who supervises the activities of persons on probation.

Probation: an alternative to incarceration in which the convict remains out of jail or prison and in the community and thus on the job, with family, and so on, while subject to conditions and supervision of the probation authority.

Procedural law: rules that set forth how substantive laws are to be enforced, such as those covering arrest, search, and seizure.

Proprietary services: in-house security services whose personnel are hired, trained, and supervised by the company or organization.

Prosecuting attorney: one who brings prosecutions, representing the people of the jurisdiction.

Prosecution: the bringing of charges against an individual, based on probable cause, so as to bring the matter before a court.

Public defender: an attorney whose full-time job is to represent indigent defendants.

Public order crimes: offenses that violate a society's shared norms.

Punishment (and its purposes): penalties imposed for committing criminal acts, to accomplish retribution, deterrence, incapacitation, and/or rehabilitation.

Radicalization: a process whereby one adopts radical positions on political or social issues.

Reasonable doubt: the standard used by jurors to arrive at a verdict—whether or not the government (prosecutor) has established guilt beyond a reasonable doubt.

Reasonable suspicion: suspicion that is less than probable cause but more than a mere hunch that a person may be involved in criminal activity.

Reform era: also the professional era, from the 1930s to 1980s, when police sought to extricate themselves from the shackles of politicians, and leading to the crime-fighter era—with greater emphases being placed on *numbers*—arrests, citations, response times, and so on.

Rehabilitation: attempts to reform an offender through vocational and educational programming, counseling, and so forth, so that he or she is not a recidivist and does not return to crime and prison.

Relative ethics: the gray area of ethics that is not so clear-cut, such as releasing a serious offender in order to use him later as an informant.

Restorative justice: the view that crime affects the entire community, which must be healed and made whole again through the offender's remorse, community service, restitution to the victim, and other such activities.

Retribution: punishment that fits the crime, that is "equitable" for the offense.

Revocation: the court's revoking probation or parole status for the purpose of returning an offender to prison (usually for not following the conditions of probation or parole, or for committing a new offense).

Right-wrong test: the test of legal insanity, asking whether the defendant understood the nature and quality of his or her act and, if so, if he or she understood it was wrong.

Sanction: a penalty or punishment.

School-to-prison pipeline: the notion that certain policies and practices push schoolchildren, particularly those who are most at risk, out of classrooms and into the juvenile and adult criminal justice and prison systems.

Search and seizure: in the Fourth Amendment, the term refers to an officer's searching for and taking away evidence of a crime.

Sentencing guidelines: an instrument developed by the federal government that uses a grid system to chart seriousness of the offense, criminal history, and so forth and thus allows the court to arrive at a more consistent sentence for everyone.

Sex trafficking: the use of humans for the purpose of sexual slavery or commercial sexual exploitation.

Sheriff: the chief law enforcement officer of a county, typically elected and frequently operating the jail as well as law enforcement functions.

Shock incarceration: used in jails and prisons, a quasi-military program for offenders to instill discipline.

Shock probation/parole: a situation in which individuals are sentenced to jail or prison for a brief period, to give them a taste or "shock" of incarceration and, it is hoped, turn them into more law-abiding citizens.

Sixth Amendment: in the Bill of Rights, it guarantees the right to a speedy and public trial by an impartial jury, the right to effective counsel at trial, and other protections.

Sixth sense: in policing, the notion that an officer can "sense" or feel when something is not right, as in the way a person acts, talks, and so on.

Slippery slope: the idea that a small first step can lead to more serious behaviors, such as the receipt of minor gratuities by police officers believed to eventually cause them to desire or demand receipt of items of greater value.

Special-purpose state agencies: specially trained units for particular investigative needs, such as those for violations of alcoholic beverage laws, fish and game laws, organized crime, and so on.

Speedy Trial Act of 1974: later amended, a law originally enacted to ensure compliance with the Sixth Amendment's provision for a speedy trial by requiring that a federal case be brought to trial no more than 100 days following the arrest.

Standing: a legal doctrine requiring that one must not be a party to a lawsuit unless he or she has a personal stake in its outcome.

Stare decisis: Latin for "to stand by a decision"—a doctrine referring to court precedent, whereby lower courts must follow (and render the same) decisions of higher courts when the same legal issues and questions come before them, thereby not disturbing settled points of law.

State bureau of investigation: a state agency that is responsible for investigating crimes involving state statutes; they may also be called in to assist police agencies in serious criminal matters, and often publish state crime reports.

State court system: civil or criminal courts in which cases are decided through an adversarial process; typically including a court of last resort, an appellate court, trial courts, and lower courts.

State police: a state agency responsible for highway patrol and other duties as delineated in the state's statutes; some states require their police to investigate crimes against persons and property.

State prison: a correctional facility that houses convicted felons.

Status offense: a crime committed by a juvenile that would not be a crime if committed by an adult; examples would be purchasing alcohol and tobacco products, truancy, and violating curfew.

Substantive law: the body of law that spells out the elements of criminal acts.

Substantive violation: an allegation that one was arrested for a new criminal offense while serving probation.

Supermax prison: a penal institution that, for security purposes, affords inmates very few, if any, amenities and a great amount of isolation.

T visa: a type of visa that allow victims of human trafficking and their immediate family members to remain and work temporarily in the United States (provided certain conditions are met).

Tasks of policing (four basic): enforce the law, perform welfare tasks, prevent crime, and protect the innocent.

Technical violation: in probation and parole, when one violates certain conditions that must be obeyed to remain out of prison, such as violating curfew, using drugs or alcohol, or not maintaining a job.

Terry Stop: also known as a "stop and frisk"; when a police officer briefly detains a person for questioning and then frisks ("pats down") the person if the officer reasonably believes he or she is carrying a weapon.

Three-strikes law: a crime control strategy whereby an offender who commits three or more violent offenses will be sentenced to a lengthy term in prison, usually 25 years to life.

Trial process: all of the steps in the adjudicatory process, from indictment or charge to conviction or acquittal.

Undocumented immigrant: one who is foreign born and does not have a legal right to be or remain in the United States.

Uniform Crime Reports: published annually by the FBI, each report describes the nature of crime as reported by law enforcement agencies; includes analyses of Part I crimes.

U.S. Supreme Court: the court of last resort in the United States, also the highest appellate court; it consists of nine justices who are appointed for life.

USA PATRIOT Act: broadly written legislation that affords law enforcement with new tools to detect and prevent terrorism.

Utilitarianism: in ethics, as articulated by John Stuart Mill, a belief that the proper course of action is that which maximizes utility—usually defined as that which maximizes happiness and minimizes suffering.

Victim impact statements: information provided prior to sentencing by the victims of a crime (or, in cases of murder, the surviving family members) about the impact the crime had on their lives; allowed by the U.S. Supreme Court.

Warden: the chief administrator of a federal penitentiary or state prison.

Warrant, arrest: a document issued by a judge directing police to immediately arrest a person accused of a crime.

Warrant, search: a document issued by a judge, based on probable cause, directing police to immediately search a person, a premises, an automobile, or a building for the purpose of finding illegal contraband felt to be located therein and as stated in the warrant.

Wedding cake model of criminal justice: a model of the criminal justice process whereby a four-tiered hierarchy exists, with a few celebrated cases at the top, and lower tiers increasing in size as the seriousness of cases declines and informal processes (use of discretion) become more likely to occur.

Whistleblower Protection Act: a federal law prohibiting reprisal against employees who reveal information concerning a violation of law, rules, or regulations; gross mismanagement or waste of funds; an abuse of authority; and so on.

White-collar crime: crimes committed by wealthy or powerful individuals in the course of their professions or occupations.

/// NOTES

/// CHAPTER 1

1. Jennifer Gonnerman, "Before the Law," *The New Yorker,* September 29, 2014, https://www.newyorker.com/magazine/2014/10/06/before-the-law.
2. Benjamin Weiser, "Kalief Browder's Suicide Brought Changes to Rikers. Now It Has Led to a $3 Million Settlement.," *New York Times,* January 24, 2019, https://www.nytimes.com/2019/01/24/nyregion/kalief-browder-settlement-lawsuit.html.
3. Federal Bureau of Investigation, "Table 2. Crime in the United States by Community Type, 2017," *Crime in the United States—2017* (Washington, D.C.: Uniform Crime Reporting Program), https://ucr.fbi.gov/crime-in-the-u.s/2017/crime-in-the-u.s.-2017/tables/table-2.
4. Jennifer Bronson, "Justice Expenditures and Employment Extracts, 2015—Preliminary," Statistical Tables (Washington, D.C.: U.S. Department of Justice, Office of Justice Systems, Bureau of Justice Statistics, June 2018), https://www.bjs.gov/index.cfm?ty=pbdetail&iid=6310.
5. YourDictionary, "Justice Quotes," *Webster's New World Dictionary of Quotations* (Hoboken, N.J.: Wiley, 2010), http://www.yourdictionary.com/quotes/justice.
6. James Austin, "'Three Strikes and You're Out': The Likely Consequences on the Courts, Prisons, and Crime in California and Washington State," *St. Louis University Public Law Review* 14, no. 1 (1994): 239–257.
7. See, for example, Joe Domanick, *Cruel Justice: Three Strikes and the Politics of Crime in America's Golden State* (Berkeley: University of California Press, 2004).
8. Austin, "'Three Strikes and You're Out,'" pp. 239–257.
9. Scott Ehlers, Vincent Schiraldi, and Jason Ziedenberg, *Still Striking Out: Ten Years of California's Three Strikes* (Washington, D.C.: Justice Policy Institute, 2004), http://www.justicepolicy.org/research/2028.
10. Brent Staples, "California Horror Stories and the 3-Strikes Law," *New York Times,* November 24, 2012, http://www.nytimes.com/2012/11/25/opinion/sunday/california-horror-stories-and-the-3-strikes-law.html.
11. Ehlers et al., *Still Striking Out.*
12. Ibid.
13. John J. Sloan III and Lynne M. Vieraitis, "'Striking Out' as Crime Reduction Policy: The Impact of 'Three Strikes' Laws on Crime Rates in U.S. Cities," *Justice Quarterly* 21, no. 2 (2004): 207–239.
14. Staples, "California Horror Stories and the 3-Strikes Law."
15. Ibid.
16. Alexander B. Smith and Harriet Pollack, *Criminal Justice: An Overview* (New York: Holt, Rinehart and Winston, 1980), p. 9.
17. Ibid., p. 10.
18. Ibid., p. 366.
19. Thomas J. Bernard, *The Consensus-Conflict Debate: Form and Content in Social Theories* (New York: Columbia University Press, 1983), p. 78.
20. Thomas Hobbes, *Leviathan* (New York: E. P. Dutton, 1950), pp. 290–291.
21. Laws that prohibit conducting certain types of business on Sundays, sometimes called "Blue Laws," are still in effect in some locations throughout the United States. For a historical overview of these laws, see David N. Laband and Deborah Hendry Heinbuch, *Blue Laws: The History, Economics, and Politics of Sunday Closing Laws* (New York: Free Press, 1987).
22. Justin McCarthy, "Two in Three Americans Now Support Legalizing Marijuana," Gallup News, October 22, 2018, https://news.gallup.com/poll/243908/two-three-americans-support-legalizing-marijuana.aspx.
23. Jean-Jacques Rousseau, "A Discourse on the Origin of Inequality," in G. D. H. Cole (ed.), *The Social Contract and Discourses* (New York: E. P. Dutton, 1946), p. 240.
24. Bernard, *Consensus-Conflict Debate*, pp. 83, 85.
25. Roe v. Wade, 410 U.S. 113 (1973).
26. See Gallup, "Abortion," http://www.gallup.com/poll/1576/abortion.aspx.
27. U.S. Sentencing Commission, *Special Report to the Congress: Cocaine and Federal Sentencing Policy* (Washington, D.C.: Richard P. Conaboy et al., February 1995), http://www.ussc.gov/sites/default/files/pdf/news/congressional-testimony-and-reports/drug-topics/199502-rtc-cocaine-sentencing-policy/EXECSUM.pdf.
28. Herbert L. Packer, *The Limits of the Criminal Sanction* (Stanford, Calif.: Stanford University Press, 1968).
29. Herbert L. Packer, *Two Models of the Criminal Process*, 113 U. PA. L. Rev. 1, 2 (1964).
30. The President's Commission on Law Enforcement and Administration of Justice, *The Challenge of Crime in a Free Society* (Washington, D.C.: U.S. Government Printing Office, 1967), p. 5.
31. See Office of Juvenile Justice and Delinquency Prevention, "Upper Age of Original Juvenile Court Jurisdiction," *Statistical Briefing Book: Juvenile Justice System Structure and Process* (Washington, D.C.: U.S. Department of Justice, 2012), http://www.ojjdp.gov/ojstatbb/structure_process/qa04101.asp.
32. Samuel Walker, *Sense and Nonsense About Crime and Drugs*, 4th ed. (Belmont, Calif.: Wadsworth, 1997), p. 15.
33. Adapted from Mike Broemmel, *The Wedding Cake Model Theory of Criminal Justice* (Bellevue, Wash.: eHow, n.d.), http://www.ehow.com/about_5143074_wedding-model-theory-criminal-justice.html.
34. Robert F. Kennedy, *The Enemy Within: The McClellan Committee's Crusade Against Jimmy Hoffa and Corrupt Labor Unions* (Jackson, Tenn.: Perseus Books, 1994), p. 324.
35. Benjamin S. Bloom (ed.), Max D. Englehart, Edward J. Furst, Walker H. Hill, and David R. Krathwohl, *Taxonomy of Educational Objectives: The Classification of Educational Goals.* Handbook I: Cognitive Domain (New York: Longman, 1956).

/// CHAPTER 2

1. Brian P. Block and John Hostettler, *Famous Cases: Nine Trials That Changed the Law* (Hook, Hampshire, U.K.: Waterside Press, 2002), pp. 9–12.
2. Henry Campbell Black, *Black's Law Dictionary,* 4th ed. (St. Paul, Minn.: West, 1951).
3. Law Library of Congress, "Case Law (or Common Law)," *American Memory,* http://memory.loc.gov/ammem/awhhtml/awlaw3/common_law.html.
4. Texas Politics, "State Constitutions," *The Texas Constitution Today,* http://texaspolitics.laits.utexas.edu/7_3_1.html.
5. Carissa Byrne Hessick, "Motive's Role in Criminal Punishment," *Southern California Law Review 80*, no. 89 (2006): 89.
6. See *Arizona Criminal Code*, generally, at http://www.azleg.gov/arizonarevisedstatutes.asp?title=13.
7. Ibid.
8. See *Nevada Revised Statutes*, generally, at https://www.leg.state.nv.us/NRS/.
9. *People v. Anderson*, 70 Ca1.2d 15 (1968).
10. *U.S. v. Brown*, 518 F.2d 821 (1975).
11. See, for example, James R. Elkins, "Depraved Heart Murder," *West Virginia Homicide Jury Instructions Project* (Morgantown: West

Virginia University College of Law, Spring 2006), http://myweb.wvnet.edu/~jelkins/adcrimlaw/depraved_heart_murder.html.

12. Federal Bureau of Investigation, "Rape," in *Crime in the United States—2017* (Washington, D.C.: Uniform Crime Reporting Program), https://ucr.fbi.gov/crime-in-the-u.s/2017/crime-in-the-u.s.-2017/topic-pages/rape.
13. Federal Bureau of Investigation, "Robbery," in *Crime in the United States—2017* (Washington, D.C.: Uniform Crime Reporting Program), https://ucr.fbi.gov/crime-in- https://ucr.fbi.gov/crime-in-the-u.s/2017/crime-in-the-u.s.-2017/topic-pages/robbery.
14. Federal Bureau of Investigation, "Aggravated Assault," in *Crime in the United States—2017* (Washington, D.C.: Uniform Crime Reporting Program), https://ucr.fbi.gov/crime-in-the-u.s/2017/crime-in-the-u.s.-2017/topic-pages/aggravated-assault.
15. Federal Bureau of Investigation, "Burglary," in *Crime in the United States—2017* (Washington, D.C.: Uniform Crime Reporting Program), https://ucr.fbi.gov/crime-in-the-u.s/2017/crime-in-the-u.s.-2017/topic-pages/burglary.
16. For Nevada's larceny-theft statute, see *Nevada Revised Statutes*, Chapter 205.220; for Iowa's larceny statutes, see *Iowa Code*, Chapter 714.2, "Degrees of Theft," http://law.justia.com/codes/iowa/2011/titlexvi/subtitle1/chapter714/714–2.
17. Federal Bureau of Investigation, "Larceny-Theft," in *Crime in the United States—2017* (Washington, D.C.: Uniform Crime Reporting Program), https://ucr.fbi.gov/crime-in-the-u.s/2017/crime-in-the-u.s.-2017/topic-pages/larceny-theft.
18. Federal Bureau of Investigation, "Vehicle Theft," in *Crime in the United States—2017* (Washington, D.C.: Uniform Crime Reporting Program), https://ucr.fbi.gov/crime-in-the-u.s/2017/crime-in-the-u.s.-2017/topic-pages/motor-vehicle-theft.
19. Federal Bureau of Investigation, "Arson," in *Crime in the United States—2017* (Washington, D.C.: Uniform Crime Reporting Program), https://ucr.fbi.gov/crime-in-the-u.s/2017/crime-in-the-u.s.-2017/topic-pages/arson.
20. United Nations Office on Drugs and Crime, *Global Report on Trafficking in Persons,* 2014, https://www.unodc.org/documents/data-and-analysis/glotip/GLOTIP_2014_full_report.pdf.
21. Edwin H. Sutherland, "White Collar Criminality," *American Sociological Review* 5 (February 1940): 1–12.
22. Frank E. Hagan, *Introduction to Criminology: Theories, Methods, and Criminal Behavior,* 8th ed. (Los Angeles, Calif.: Sage Publications, 2013), p. 281.
23. Ibid., p. 288.
24. Ibid., p. 320.
25. Robert Lenzner, "Bernie Madoff's $50 Billion Ponzi Scheme," *Forbes*, December 12, 2008, http://www.forbes.com/2008/12/12/madoff-ponzi-hedge-pf-ii-in_rl_1212croesus_inl.html.
26. "Martha Stewart's Conviction Upheld," CBS News, February 11, 2009, http://www.cbsnews.com/2100–207_162–1183526.html.
27. Charles B. Fleddermann, "The Ford Pinto Exploding Gas Tank," in *Engineering Ethics*, 2nd ed. (Upper Saddle River, N.J.: Prentice Hall, 2004), pp. 72–73.
28. Bureau of Justice Statistics, *Victims of Identity Theft, 2014* (September 2015), p. 1, https://www.bjs.gov/content/pub/pdf/vit14.pdf.
29. Herbert Edelhertz, *The Nature, Impact and Prosecution of White-Collar Crime,* Report No. ICR 70–1 (Washington, D.C.: U.S. Department of Justice, 1970).
30. Federal Bureau of Investigation, *What We Investigate*, "Organized Crime," https://www.fbi.gov/investigate/organized-crime.
31. See "Statistics," http://www.twainquotes.com/Statistics.html.
32. Federal Bureau of Investigation, "Incidents and Offenses," *Hate Crime Statistics—2017* (Washington, D.C.: U.S. Department of Justice), https://ucr.fbi.gov/hate-crime/2017/topic-pages/incidents-and-offenses.
33. Madeline Masucci, *Hate Crime Victimization, 2004–2015* (Washington, D.C.: U.S. Department of Justice, Office of Justice Programs, Bureau of Justice Statistics, February 2014), https://www.bjs.gov/content/pub/pdf/hcv0415.pdf.
34. For access to the publications, see Federal Bureau of Investigation, *Uniform Crime Reports,* https://www.fbi.gov/services/cjis/ucr/publications.
35. Federal Bureau of Investigation, "Table 1," *Crime in the United States—by Volume and Rate per 100,000 Inhabitants, 1998–2017* (Washington, D.C.: U.S. Department of Justice), https://ucr.fbi.gov/crime-in-the-u.s/2017/crime-in-the-u.s.-2017/topic-pages/tables/table-1.
36. Ibid.
37. Federal Bureau of Investigation, "Murder," *Crime in the United States—2017* (Washington, D.C.: Uniform Crime Reporting Program), https://ucr.fbi.gov/crime-in-the-u.s/2017/crime-in-the-u.s.-2017/topic-pages/murder.
38. Federal Bureau of Investigation, "Offense Definitions," *Crime in the United States—2017,* https://ucr.fbi.gov/crime-in-the-u.s/2017/crime-in-the-u.s.-2017/topic-pages/offense-definitions.
39. Federal Bureau of Investigation, "FBI Releases 2015 Crime Statistics," *Press Releases* (Washington, D.C.: U.S. Department of Justice), https://www.fbi.gov/news/pressrel/press-releases/fbi-releases-2015-crime-statistics.
40. See, for example, Nathan James and Logan Rishard Council, "How Crime in the United States Is Measured," *Congressional Research Service Report for Congress,* January 3, 2008, pp. 17–20, http://www.policyarchive.org/handle/10207/bitstreams/18912.pdf.
41. See Federal Bureau of Investigation, *Uniform Crime Reporting Handbook* (Washington, D.C.: U.S. Department of Justice), p. 10, http://www.fbi.gov/about-us/cjis/ucr/additional-ucr-publications/ucr_handbook.pdf/view.
42. Federal Bureau of Investigation, "FBI Releases 2015 Crime Statistics from the National Incident-Based Reporting System, Encourages Transition," *Press Releases* (Washington, D.C.: U.S. Department of Justice), https://www.fbi.gov/news/pressrel/press-releases/fbi-releases-2015-crime-statistics-from-the-national-incident-based-reporting-system-encourages-transition.
43. Federal Bureau of Investigation, "FBI Releases 2017 NIBRS Crime Data as Transition to NIBRS 2021 Continues," *Press Releases* (Washington, D.C.: U.S. Department of Justice), https://www.fbi.gov/news/pressrel/press-releases/fbi-releases-2017-nibrs-crime-data-as-transition-to-nibrs-2021-continues.
44. Bureau of Justice Statistics, "Data Collection: National Crime Victimization Survey (NCVS)," http://bjs.ojp.usdoj.gov/index.cfm?ty=dcdetail&iid=245.
45. National Archive of Criminal Justice Data, "National Crime Victimization Survey Resource Guide," http://www.icpsr.umich.edu/icpsrweb/NACJD/NCVS/.
46. National Archive of Criminal Justice Data, "Accuracy of NCVS Estimates," https://www.icpsr.umich.edu/icpsrweb/NACJD/NCVS/accuracy.jsp.
47. See *State v. Thompson*, 865 p.2d 1125 (Mont. 1993) and Mont. Code. Ann. §§45–501–1-3. The principal was acquitted of sexual assault because the statute did not contemplate purely psychological force. The Montana legislature later amended the statute to clarify this issue in anticipation of future cases.
48. Defendant Peterson was charged with manslaughter. At trial, the judge instructed the jury that self-defense is *not* available as a defense when someone acts as the aggressor, stands one's ground when other options are available, and actually provokes a conflict. Peterson was convicted; he appealed on the grounds that his shooting the driver should be excused because he acted in self-defense. The appellate court agreed with the trial judge: One cannot claim to have acted in "self-defense" by a *self-generated necessity to kill* another person. The evidence demonstrated that Peterson instigated the confrontation, and his failure to retreat was also a factor that the jury could take into account. *U.S. v. Peterson*, 483 F. 2d 1222 (D.C. Cir. 1973).

/// CHAPTER 3

1. Stewart was an associate justice of the U.S. Supreme Court from 1958 to 1981.
2. Adapted from Pamela Manson, "Utah Nurse Reaches $500,000 Settlement in Dispute Over Her Arrest for Blocking Cop From Drawing Blood From Patient," *Salt Lake Tribune,* November 1, 2017, https://www.sltrib.com/news/2017/10/31/utah-nurse-arrested-for-blocking-cop-from-drawing-blood-from-patient-receives-500000-settlement/.
3. Jodi S. Cohen, "Chicago Cop Under Investigation Again Over Social Media Posts," ProPublica Illinois, January 25, 2018, https://www.propublica.org/article/chicago-police-officer-john-catanzara-investigation.
4. This scenario is based loosely on David Gelman, Susan Miller, and Bob Cohn, "The Strange Case of Judge Wachtler," *Newsweek,* November 23, 1992, pp. 34–35. Wachtler was later arraigned on charges of attempting to extort money from the woman and threatening her 14-year-old daughter (it was later determined that the judge had been having an affair with the woman, who had recently ended the relationship). After being placed under house arrest with an electronic monitoring bracelet, the judge resigned from the court, which he had served with distinction for two decades.
5. ACLU/Florida, "Civil Rights Groups Call for Federal Investigation Into Solitary Confinement Abuse in Florida Prisons," March 11, 2016, https://www.aclufl.org/en/press-releases/civil-rights-groups-call-federal-investigation-solitary-confinement-abuse-florida a copy of the letter is available at: http://aclufl.org/resources/letter-to-doj-calling-for-investigation-into-abuse-of-solitary-in-florida-prisons/.
6. Immanuel Kant, "Foundations of the Metaphysics of Morals," *The German Library 13* (New York: Continuum, 2006).
7. Richard Kania, "Police Acceptance of Gratuities," *Criminal Justice Ethics* 7 (1988): 37–49.
8. John Kleinig, *The Ethics of Policing* (New York: Cambridge University Press, 1996).
9. T. J. O'Malley, "Managing for Ethics: A Mandate for Administrators," *FBI Law Enforcement Bulletin* (April 1997): 20–25.
10. Thomas J. Martinelli, "Unconstitutional Policing: The Ethical Challenges in Dealing With Noble Cause Corruption," *The Police Chief* (October 2006): 150.
11. John P. Crank and Michael A. Caldero, *Police Ethics: The Corruption of Noble Cause* (Cincinnati, Ohio: Anderson, 2000), p. 75.
12. U.S. Department of Justice, National Institute of Justice, Office of Community Oriented Policing Services, *Police Integrity: Public Service With Honor* (Washington, D.C.: U.S. Government Printing Office, 1997), p. 62.
13. Ibid.
14. Lawrence W. Sherman, ed., *Police Corruption: A Sociological Perspective* (Garden City, N.Y.: Anchor, 1974), p. 1.
15. Herman Goldstein, *Policing a Free Society* (Cambridge, Mass.: Ballinger, 1977), p. 188.
16. Ibid.
17. William A. Westley, *Violence and the Police* (Cambridge, Mass.: MIT Press, 1970), pp. 113–114.
18. International Association of Chiefs of Police, "Law Enforcement Oath of Honor," https://www.theiacp.org/sites/default/files/all/i-j/IACP_Oath_of_Honor_En_8.5x11_Web.pdf.
19. Ibid.
20. David Carter, "Theoretical Dimensions in the Abuse of Authority," in Thomas Barker and David Carter (Eds.), *Police Deviance* (Cincinnati, Ohio: Anderson, 1994), pp. 269–290; also see Thomas Barker and David Carter, "Fluffing Up the Evidence and 'Covering Your Ass': Some Conceptual Notes on Police Lying," *Deviant Behavior* 11 (1990): 61–73.
21. Gary T. Marx, "Who Really Gets Stung? Some Issues Raised by the New Police Undercover Work," *Crime & Delinquency* (1982): 165–193.
22. *Illinois v. Perkins,* 110 S. Ct. 2394 (1990).
23. Barker and Carter, *Police Deviance.*
24. Thomas Barker, "An Empirical Study of Police Deviance Other than Corruption," in Barker and Carter, *Police Deviance,* pp. 123–138.
25. For an excellent analysis of how the acceptance of gratuities can become endemic to an organization and pose ethical dilemmas for new officers within, see Jim Ruiz and Christine Bono, "At What Price a 'Freebie'? The Real Cost of Police Gratuities," *Criminal Justice Ethics* (Winter/Spring 2004): 44–54. The authors also demonstrate through detailed calculations how the amount of gratuities accepted can reach up to 40 percent of an annual officer's income—and is therefore no minor or inconsequential infraction of rules that can be left ignored or unenforced.
26. Quoted in David Burnham, "Police Aides Told to Rid Commands of All Dishonesty," *New York Times,* October 29, 1970, p. 1.
27. From the Washoe County Sheriff's Office, Reno, Nevada.
28. Edward Tully, "Misconduct, Corruption, Abuse of Power: What Can the Chief Do?" http://www.neiassociates.org/mis2.htm (Part I) and http://www.neiassociates.org/misconductll.htm (Part II).
29. See "Officer Caught on Camera Buying Shoes for Homeless Man," *USA Today,* December 3, 2015, https://www.usatoday.com/story/news/humankind/2015/12/03/officer-caught-camera-buying-homeless-man-shoes/76727994/; Anthony de Stefano, "Larry DePrimo, NYPD Cop, Buys Homeless Man Boots," *Huffington Post,* November 2017, https://www.huffingtonpost.com/2012/11/29/larry-deprimo-nypd-cop-gives-homeless-boots_n_2209178.html.
30. See "NY Police Apologize to Victim With Food, Favors," Policeone.com, May 14, 2014, https://www.policeone.com/police-heroes/articles/6999864-NY-police-apologize-to-victim-with-food-favors/.
31. "Arizona Cop Praised for Giving Bike, Riding Lessons to Teen," Policeone.com, May 13, 2013, https://www.youtube.com/watch?v=DcQ8M3hCeWc; "Teen Who Walked 2 Hours to Work Surprised by Cop Who Buys Him Bike," KTVU News, September 21, 2016, http://www.ktvu.com/news/mobile-app-ktvu/teen-who-walked-2-miles-to-work-surprised-by-cop-who-buys-him-bike.
32. See this and other such instances in Lorraine Burger, "5 Cases That Fight the Stigma of 'Unethical' Policing," Policeone.com, April 10, 2014, https://www.policeone.com/police-jobs-and-careers/articles/7063313-5-cases-that-fight-the-stigma-of-unethical-policing/.
33. Roscoe Pound, "The Causes of Popular Dissatisfaction With the Administration of Justice," address before the annual convention of the American Bar Association, August 29, 1906. *American Lawyer* 14 (1906): 445.
34. John P. MacKenzie, *The Appearance of Justice* (New York: Scribner, 1974).
35. Pierro Calamandrei, quoted in Frank Greenberg, "The Task of Judging the Judges," *Judicature* 59 (May 1976): 464.
36. For thorough discussions and examples of these areas of potential ethical shortcomings, see Jeffrey M. Shaman, Steven Lubet, and James J. Alfini, *Judicial Conduct and Ethics,* 3rd ed. (San Francisco: Matthew Bender, 2000).
37. Ibid.
38. Ibid., p. vi.
39. Tim Murphy, "Test Your Ethical Acumen," *Judges' Journal* 8 (1998): 34.
40. American Judicature Society, *Judicial Conduct Reporter* 16 (1994): 2–3.
41. Shaman, Lubet, and Alfini, *Judicial Conduct and Ethics,* p. viii.
42. *Berger v. U.S.,* 295 U.S. 78 (1935).
43. See *Dunlop v. U.S.,* 165 U.S. 486 (1897), involving a prosecutor's inflammatory statements to the jury.
44. 386 U.S. 1 (1967). In this case, the Supreme Court overturned the defendant's conviction after determining that the prosecutor "deliberately misrepresented the truth."
45. Elliot D. Cohen, "Pure Legal Advocates and Moral Agents: Two Concepts of a Lawyer in an Adversary System," in Michael C. Braswell, Belinda R. McCarthy, and Bernard J. McCarthy (Eds.), *Justice, Crime and Ethics,* 2nd ed. (Cincinnati, Ohio: Anderson, 1996), p. 168.

46. Ibid.
47. Ibid., pp. 131–167.
48. Cynthia Kelly Conlon and Lisa L. Milord, *The Ethics Fieldbook: Tools for Trainers* (Chicago: American Judicature Society, n.d.), pp. 23–25.
49. Administrative Office of the United States Courts, *Code of Conduct for United States Judges*, http://www.uscourts.gov/RulesAndPolicies/CodesOfConduct/CodeConductUnitedStatesJudges.aspx; further guidance appears in Federal Judicial Center, *Guide to Judiciary Policies and Procedures* (vol. 2): *Maintaining the Public Trust: Ethics for Federal Judicial Law Clerks* (Washington, D.C.: Author, 2002).
50. Criminal Justice Research, "Correctional Ethics," http://criminal-justice.iresearchnet.com/criminal-justice-ethics/correctional-ethics/
51. Ibid., p. 28.
52. Elizabeth L. Grossi and Bruce L. Berg, "Stress and Job Dissatisfaction Among Correctional Officers: An Unexpected Finding," *International Journal of Offender Therapy and Comparative Criminology* 35 (1991): 79.
53. Irving L. Janis, "Group Dynamics Under Conditions of External Danger," in Darwin Cartwright and Alvin Zander (Eds.), *Group Dynamics: Research and Theory* (New York: Harper & Row, 1968).
54. Ibid.
55. American Correctional Association, "Code of Ethics," http://www.aca.org/ACA_Prod_IMIS/ACA_Member/About_Us/Code_of_Ethics/ACA_Member/AboutUs/Code_of_Ethics.aspx?hkey=61577ed2-c0c3-4529-bc01-36a248f79eba
56. CBS News, March 30, 1977; see John R. Jones and Daniel P. Carlson, *Reputable Conduct: Ethical Issues in Policing and Corrections*, 2nd ed. (Upper Saddle River, N.J.: Prentice Hall, 2001), p. 76.
57. Adapted from Florida Regional Community Policing Institute, Ethical Issues and Decisions in Law Enforcement (March 2005): 37–39, http://cop.spcollege.edu/Training/Ethics/EN/ethicsInstructorMarch2005.pdf.
58. Kleinig, *The Ethics of Policing.*
59. Adapted from Gail Diane Cox, "Judges Behaving Badly (Again)," *National Law Journal* 21 (May 3, 1999): 1–5.

/// CHAPTER 4

1. There are a number of articles available that focus on "federalizing" the local police, meaning granting them more authority (to enforce federal immigration laws, and so on); here, however, we consider whether or not there should be a single, national police agency; but see: Glenn Harlan Reynolds, "Want a lawless police force? Federalize it." *USA Today,* May 3, 2015, https://www.usatoday.com/story/opinion/2015/05/03/baltimore-federal-drugs-evidence-fbi-column/26830873/; Lance Eldridge, The federalization of local law enforcement, PoliceOne, January 3, 2011, https://www.policeone.com/patrol-issues/articles/3139476-The-federalization-of-local-law-enforcement/.
2. Bruce Smith, *Rural Crime Control* (New York: Columbia University, 1933), p. 40.
3. Ibid.
4. Ibid.
5. Ibid., pp. 182–184.
6. Ibid., pp. 188–189.
7. Ibid., pp. 218–222.
8. Ibid., pp. 245–246.
9. David R. Johnson, *American Law Enforcement History* (St. Louis, Mo.: Forum Press, 1981), pp. 18–19.
10. Leon Radzinowicz, *A History of English Criminal Law and Its Administration From 1750* (vol. IV): *Grappling for Control* (London: Stevens and Son, 1968), pp. 20–21.
11. C. Germann, Frank D. Day, and Robert R. J. Gallati, *Introduction to Law Enforcement and Criminal Justice* (Springfield, Ill.: Charles C. Thomas, 1962), p. 63.
12. Johnson, *American Law Enforcement History*, p. 26.
13. James F. Richardson, *Urban Police in the United States* (London: Kennikat Press, 1974), pp. 47–48.
14. Johnson, *American Law Enforcement History*, p. 26.
15. Herman Goldstein, *Policing a Free Society* (Cambridge, Mass.: Ballinger, 1977).
16. August Vollmer, "Police Progress in the Past Twenty-Five Years," *Journal of Criminal Law and Criminology* 24 (1933): 161–175.
17. Alfred E. Parker, *Crime Fighter: August Vollmer* (New York: Macmillan, 1961).
18. Nathan Douthit, "August Vollmer," in Carl B. Klockars (Ed.), *Thinking About Police: Contemporary Readings* (New York: McGraw-Hill, 1983), p. 102.
19. Ibid.
20. Ibid.
21. Richardson, *Urban Police in the United States*, pp. 139–143.
22. Herman Goldstein, *Policing a Free Society* (Cambridge, Mass.: Ballinger, 1977).
23. Kenneth J. Peak and Ronald W. Glensor, *Community Policing and Problem Solving: Effectively Addressing Crime and Disorder*, 7th ed. (New York: Pearson, 2016), pp. 10–11.
24. Elaine Cumming, Ian Cumming, and Laura Edell, "Policeman as Philosopher, Guide, and Friend," *Social Problems* 12 (1965), 285; T. Bercal, "Calls for Police Assistance," *American Behavioral Scientist* 13 (1970): 682; Albert J. Reiss Jr., *The Police and the Public* (New Haven, Conn.: Yale University Press, 1971).
25. Bureau of Justice Statistics, "Federal Law Enforcement Statistics, 2015–2016: Summary," January 2019, pp. 1, 3. https://www.bjs.gov/content/pub/pdf/fjs1516_sum.pdf.
26. U.S. Department of Homeland Security, "Creation of the Department of Homeland Security," http://www.dhs.gov/creation-department-homeland-security.
27. National Priorities Project, "U.S. Security Spending Since 9/11," February 28, 2013, https://www.nationalpriorities.org/analysis/2013/homeland-security-spending-since-911/.
28. U.S. Department of Homeland Security, "The Lifesaving Missions of CBP," https://www.dhs.gov/news/2018/08/20/life-saving-missions-cbp.
29. Department of Homeland Security, "On a Typical Day in Fiscal Year 2018, CBP . . . ," https://www.cbp.gov/newsroom/stats/typical-day-fy2018.
30. U.S. Immigration and Customs Enforcement, "Who We Aren't," https://www.ice.gov/about.
31. Ibid., "ICE: Enforcement and Removal Operations," http://www.ice.gov/about/offices/enforcement-removal-operations/.
32. Ibid., "Homeland Security Investigations," http://www.ice.gov/hsi; also see ICE, "Become a Criminal Investigator," http://www.ice.gov/careers/occupation/investigator.
33. Transportation Security Administration, "Jobs at TSA," https://www.tsa.gov/about/jobs-at-tsa.
34. See "Are TSA Screeners Really Law Enforcement 'Officers'?" KDVR, November 4, 2013, http://kdvr.com/2013/11/04/are-tsa-screeners-really-law-enforcement-officers/.
35. See U.S. Coast Guard, "Missions," https://www.work.uscg.mil/Missions/.
36. U.S. Secret Service, "Careers Built on Integrity," https://www.secretservice.gov/join/careers/
37. U.S. Secret Service, "About," https://www.secretservice.gov/about/history/; also see https://www.secretservice.gov/about/faqs/.
38. Federal Protective Service, *Annual Report, Fiscal Year 2015,* p. 20.,https://www.dhs.gov/sites/default/files/publications/Federal%20Protective%20Service%20Annual%20Report%20508%20Compliant%20FY2015.pdf. This report is the latest available.
39. U.S. Department of Justice, Federal Bureau of Investigation, "A Brief History," https://www.fbi.gov/history/brief-history.
40. U.S. Department of Justice, Federal Bureau of Investigation, "Special Agent Selection Process All You Need to Know to Apply," https://www.fbijobs.gov/sites/default/files/how-to-apply.pdf.

41. U.S. Department of Justice, Federal Bureau of Investigation, "About," https://www.fbi.gov/about.
42. "FBI Seeks Sweeping New Powers," The Nation, August 22, 2008, www.thenation.com/article/fbi-seeks-sweeping-new-powers.
43. Bureau of Alcohol, Tobacco, Firearms and Explosives, "ATF's History," http://www.atf.gov/about/history/.
44. Bureau of Alcohol, Tobacco, Firearms and Explosives, "Fact Sheet—Facts and Figures for Fiscal Year 2017," https://www.atf.gov/resource-center/fact-sheet/fact-sheet-facts-and-figures-fiscal-year-2017.
45. U.S. Department of Justice, Drug Enforcement Administration, "DEA History," http://www.justice.gov/dea/about/history.shtml.
46. U.S. Department of Justice, Drug Enforcement Administration, "DEA Mission Statement," http://www.dea.gov/about/mission.
47. Bureau of Justice Statistics, "Federal Law Enforcement Statistics, 2015–2016: Summary," January 2019, https://www.bjs.gov/content/pub/pdf/fjs1516_sum.pdf.
48. See U.S. Marshals Service, "Facts and Figures," https://www.usmarshals.gov/duties/factsheets/facts.pdf.
49. U.S. Marshals Service, *The FY 1993 Report to the U.S. Marshals* (Washington, D.C.: U.S. Department of Justice, 1994), pp. 188–189; also see U.S. Marshals Service, "Fact Sheets: Facts and Figures," Office of Public Affairs, April 15, 2011 http://www.usmarshals.gov/duties/factsheets/facts-2011.html.
50. Central Intelligence Agency, "CIA Vision, Mission, & Values," https://www.cia.gov/about-cia/cia-vision-mission-values/index.html; also see CIA, "About CIA," https://www.cia.gov/about-cia/index.html.
51. Central Intelligence Agency, "Careers & Internships," https://www.cia.gov/careers/opportunities/cia-jobs/index.html.
52. Internal Revenue Service, "Financial Investigations: Criminal Investigation (CI)," http://www.irs.gov/uac/Financial-Investigations—Criminal-Investigation-(CI).
53. Ibid.; also see IRS, *Criminal Investigation: Annual Report, 2018,* p. 31, https://www.irs.gov/pub/irs-utl/2018_irs_criminal_investigation_annual_report.pdf.
54. INTERPOL, "About INTERPOL," https://www.interpol.int/en/Who-we-are/What-is-INTERPOL; also see Michael Fooner, *Interpol: Issues in World Crime and International Criminal Justice* (New York: Plenum Press, 1989), p. 179.
55. Brian A. Reaves, *Census of State and Local Law Enforcement Agencies, 2013,* Bureau of Justice Statistics, May 2015, http://www.bjs.gov/content/pub/pdf/lpd13ppp.pdf. This is the most recent report available.
56. Missouri State Highway Patrol, Career Recruitment Division, "Duties of a Trooper," http://www.mshp.dps.missouri.gov/MSHPWeb/PatrolDivisions/HRD/Trooper/troopCareer.html.
57. State of Hawaii Department of Public Safety, http://dps.hawaii.gov/about/.
58. National Association of Attorneys General, http://www.naag.org/naag/about_naag/faq/what_does_an_attorney_general_do.php.
59. Bureau of Justice Statistics, *National Sources of Law Enforcement Employment Data*, October 4, 2016, p. 1, https://www.bjs.gov/content/pub/pdf/nsleed.pdf.
60. Reaves, *Census of State and Local Law Enforcement Agencies, 2013*, p. 2, Table 1, http://www.bjs.gov/content/pub/pdf/lpd13ppp.pdf.
61. Ibid.
62. Ibid.; see also Brian Reaves, *Local Police Departments, 2013: Equipment and Technology.* U.S. Department of Justice, Bureau of Justice Statistics, July 2015, http://www.bjs.gov/content/pub/pdf/lpd13et.pdf.
63. Andrea Burch, *Sheriffs' Offices, 2007.* U.S. Department of Justice, Bureau of Justice Statistics, December 2012, http://www.bjs.gov/content/pub/pdf/s007st.pdf.
64. Clemens Bartollas, Stuart J. Miller, and Paul B. Wice, *Participants in American Criminal Justice: The Promise and the Performance* (Englewood Cliffs, N.J.: Prentice Hall, 1983), pp. 51–52.
65. Gregory M. White, Resident Agent in Charge, U.S. Department of Homeland Security, U.S. Immigration and Customs Enforcement, personal communication, October 29, 2009.
66. Ibid.
67. Ibid.
68. See, for example, Lisa Riordan Seville, Hannah Rappleye and Andrew W. Lehren, "22 Immigrants Died in ICE Detention Centers During the Past 2 Years," NBC News, January 6, 2019, https://www.nbcnews.com/politics/immigration/22-immigrants-died-ice-detention-centers-during-past-2-years-n954781; Cristobal Ramon and Raven Quesenberry, "Police, Jails, and Immigrants: How Do Immigrants and the Immigration Enforcement System Interact With Local Law Enforcement?" Bipartisan Policy Center, February 23, 2018, https://bipartisanpolicy.org/library/police-jails-and-immigrants-how-do-immigrants-and-the-immigration-enforcement-system-interact-with-local-law-enforcement/; Tal Kopan, "States Taking More of a Lead on Immigration, Citing Federal Inaction," CNN, August 17, 2017, https://www.cnn.com/2017/08/07/politics/states-immigration-policies/index.html.
69. Saul D. Astor, "A Nation of Thieves," *Security World* 15 (September 1978).
70. Bureau of Justice Statistics, *Criminal Victimization—2017*, December 2018, pp. 1, 5,https://www.bjs.gov/content/pub/pdf/cv17.pdf.
71. Bureau of Labor Statistics, "Occupational Employment Statistics, July 2018," https://www.bls.gov/ooh/protective-service/security-guards.htm.
72. Niall McCarthy, "Private Security Outnumbers the Police in Most Countries Worldwide," *Forbes,* August 31, 2017, https://www.forbes.com/sites/niallmccarthy/2017/08/31/private-security-outnumbers-the-police-in-most-countries-worldwide-infographic/#6358df2a210f.
73. "10 Things You May Not Know About the Pinkertons," History Channel, October 23, 2015, https://www.history.com/news/10-things-you-may-not-know-about-the-pinkertons.
74. George F. Cole and Christopher E. Smith, *The American System of Criminal Justice,* 11th ed. (Belmont, Calif.: Thomson Wadsworth, 2007), p. 253.
75. Law Enforcement—Private Security Consortium, *Operation Partnership Trends and Practices in Law Enforcement and Private Security Collaborations,* August 2009, pp. 34–40, https://www.nationalpublicsafetypartnership.org/clearinghouse/Content/ResourceDocuments/OperationPartnership-TrendsandPracticesinLawEnforcementandPrivateSecurityCollaborations.pdf.
76. Shoshana Walter and Ryan Gabrielson, "America's Gun-Toting Guards Armed With Poor Training, Little Oversight," Center for Investigative Reporting, December 9, 2014, https://www.revealnews.org/article/americas-gun-toting-security-guards-may-not-be-fit-for-duty/; also see Byard Duncan, "5 Ways the Armed-Guard Industry Is Out of Control," Reveal, May 4, 2015, https://www.revealnews.org/article/heres-whats-wrong-with-the-us-armed-security-industry/; First Security, "Should Your Security Guards Be Armed or Not?" https://www.firstsecurityservices.com/should-your-security-guards-be-armed-or-not/.

/// CHAPTER 5

1. Justin Jouvenal, "To Find Alleged Golden State Killer, Investigators First Found His Great-Great-Great-Grandparents," *Washington Post,* April 30, 2018, https://www.washingtonpost.com/local/public-safety/to-find-alleged-golden-state-killer-investigators-first-found-his-great-great-great-grandparents/2018/04/30/3c865fe7-dfcc-4a0e-b6b2-0bec548d501f_story.html; also see Joseph Serna and Benjamin Oreskes, "Why Did It Take So Long to Arrest the Golden State Killer Suspect? Interagency Rivalries, Old Technology, Errors and Bad Luck," *Los Angeles Times,* May 25, 2018, http://www.latimes.com/local/lanow/la-me-ln-golden-state-killer-case-20180525-story.html.

2. William A. Westley, *Violence and the Police* (Cambridge, Mass.: MIT Press, 1970).
3. Quoted in V. A. Leonard and Harry W. More, *Police Organization and Management*, 3rd ed. (Mineola, N.Y.: Foundation Press, 1971), p. 128.
4. Roger G. Dunham and Geoffrey P. Alpert, *Critical Issues in Policing: Contemporary Readings*, 5th ed. (Long Grove, Ill.: Waveland, 2005), p. 12.
5. Ibid., p. 111.
6. Steven M. Cox, *Police: Practices, Perspectives, Problems* (Boston: Allyn and Bacon, 1996), p. 61.
7. See Albert Reiss, *The Police and the Public* (New Haven, Conn.: Yale University Press, 1971), p. 96.
8. Jerome H. Skolnick and David H. Bayley, *The New Blue Line: Police Innovation in Six American Cities* (New York: Free Press, 1986), p. 4.
9. James Q. Wilson, *Varieties of Police Behavior* (Cambridge, Mass.: Harvard University Press, 1968), pp. 140–226.
10. Michael B. Sauter and Charles Stockdale, "The Most Dangerous Jobs in the US Include Electricians, Firefighters and Police Officers," *USA Today*, January 8, 2019, https://www.usatoday.com/story/money/2019/01/08/most-dangerous-jobs-us-where-fatal-injuries-happen-most-often/38832907/.
11. National Law Enforcement Officers Memorial Fund, "Deaths, Assaults and Injuries," 2019, https://nleomf.org/facts-figures/deaths-assaults-and-injuries.
12. Officer Down Memorial Page, "Honoring Officers Killed in 2019," https://www.odmp.org/search/year/2019.
13. Quoted in Kevin Krajick, "Does Patrol Prevent Crime?" *Police Magazine* 1 (September 1978): 4–16.
14. Herman Goldstein, "Police Discretion: The Ideal Versus the Real," *Public Administration Review* 23, no. 3 (1963): 140–148.
15. Carl B. Klockars and Stephen D. Mastrofski, "Police Discretion: The Case of Selective Enforcement," in Carl B. Klockars and Stephen D. Mastrofski (Eds.), *Thinking About Police: Contemporary Readings*, 2nd ed. (Boston: McGraw-Hill, 1991), p. 330.
16. David H. Bayley and Egon Bittner, "Learning the Skills of Policing," in Roger G. Dunham and Geoffrey P. Alpert (Eds.), *Critical Issues in Policing: Contemporary Readings* (Prospect Heights, Ill.: Waveland, 1989), pp. 87–110.
17. Kenneth Culp Davis, *Police Discretion* (St. Paul, Minn.: West, 1975), p. 73.
18. Kenneth Culp Davis, *Discretionary Justice* (Urbana: University of Illinois Press, 1969), p. 222.
19. Klockars and Mastrofski, "Police Discretion," p. 331.
20. Susan Rahr, quoted in President's Task Force on 21st Century Policing, *Interim Report of the President's Task Force on 21st Century Policing* (Washington, D.C.: Office of Community Oriented Policing Services, April 2015), http://www.cops.usdoj.gov/pdf/taskforce/Interim_TF_Report.pdf, p. 9.
21. Dean Esserman, quoted in ibid., p. 1.
22. Adapted from ibid., pp. 5–10.
23. Peter R. DeForest, R. E. Gaensslen, and Henry C. Lee, *Forensic Science: An Introduction to Criminalistics* (New York: McGraw-Hill, 1983), p. 29.
24. Quoted in Leonard Roy Frank, ed., *Random House Webster's Quotationary* (New York: Random House, 1999), p. 761.
25. Marc H. Caplan and Joe Holt Anderson, *Forensic: When Science Bears Witness* (Washington, D.C.: U.S. Government Printing Office, 1984), p. 2.
26. Charles R. Swanson, Neil C. Chamelin, Leonard Territo, and Robert W. Taylor, *Criminal Investigation*, 9th ed. (Boston: McGraw-Hill, 2006), p. 10.
27. DeForest, et al., *Forensic Science*, p. 29.
28. Solomon Moore, "Progress Is Minimal in Clearing DNA Cases," *New York Times*, October 24, 2008, http://www.nytimes.com/2008/10/25/us/25dna.html.
29. DeForest et al., *Forensic Science*, p. 11.
30. John S. Dempsey and Linda S. Forst, *An Introduction to Policing*, 6th ed. (Independence, Ky.: Cengage, 2012), p. 470.
31. University of Michigan Law School, The National Registry of Exonerations, "Exonerations by State," http://www.law.umich.edu/special/exoneration/Pages/Exonerations-in-the-United-States-Map.aspx.
32. For an excellent overview of differential treatment of juveniles by the police, see Chapter 8 of Steven M. Cox, Jennifer M. Allen, and Robert D. Hanser, *Juvenile Justice: A Guide to Theory, Policy, and Practice*, 9th ed.) (Thousand Oaks, Calif.: Sage), 2017.

/// CHAPTER 6

1. *McCulloch v. Maryland*, 17 U.S. 316 (1819).
2. John Adams, "Novanglus Papers," in *The Works of John Adams*, ed. Charles Francis Adams (Boston: Little Brown and Company, 1851), p. 106.
3. David Neubauer, *America's Courts and the Criminal Justice System*, 9th ed. (Belmont, Calif.: Wadsworth, 2008), pp. 294–300.
4. *Draper v. United States*, 358 U.S. 307 (1959).
5. *Illinois v. Gates*, 462 U.S. 213 (1983).
6. *U.S. v. Sokolow*, 109 S.Ct. 1581 (1989).
7. *People v. Defore*, 242 N.Y. 214, 150 N.E. 585 (1926).
8. *Weeks v. United States*, 232 U.S. 383 (1914).
9. *Mapp v. Ohio*, 367 U.S. 643 (1961).
10. Ibid.
11. *Utah v. Strieff*, 579 U.S. ___, 136 S.Ct. 2056 (2016).
12. *Payton v. New York*, 445 U.S. 573 (1980).
13. *United States v. Reed*, 935 F. 2d 641 (4th Cir.), cert. denied, 502 U.S. 960 (1991).
14. *Delaware v. Prouse*, 440 U.S. 648 (1979).
15. *Michigan Department of State Police v. Sitz*, 110 S.Ct. 2481, 110 L.Ed.2d 412 (1990).
16. *Pennsylvania v. Muniz*, 110 S.Ct. 2638, 110 L.Ed.2d 528 (1990).
17. *Maryland v. Pringle*, 124 S.Ct. 795 (2004).
18. *Illinois v. Lidster*, 124 S.Ct. 885 (2004).
19. *Kentucky v. King*, 131 S.Ct. 1849 (2011).
20. *California v. Greenwood*, 486 U.S. 35 (1988).
21. *Florida v. Bostick*, 59 LW 4708 (June 20, 1991).
22. *California v. Hodari D.*, 59 LW 4335 (April 23, 1991).
23. *Chimel v. California*, 395 U.S. 752 (1969).
24. *Maryland v. Buie*, 58 LW 4281 (1990).
25. *Arizona v. Gant*, 07–542 (2009).
26. *Maryland v. King*, 569 U.S. 12 (2013).
27. *Riley v. California*, 573 U.S. ___ (2014).
28. *Terry v. Ohio*, 319 U.S. 1 (1968).
29. *Minnesota v. Dickerson*, 113 S.Ct. 2130 (1993).
30. *Maryland v. Wilson*, 117 S.Ct. 882 (1997).
31. *Missouri v. McNeely*, 569 U.S. ____ (2013).
32. *Birchfield v. North Dakota*, 579 U.S. ___ (2016).
33. *Carroll v. United States*, 267 U.S. 132 (1925).
34. *Harris v. United States*, 390 U.S. 234 (1968).
35. *Florida v. Jimeno*, 59 LW 4471 (May 23, 1991).
36. *Wyoming v. Houghton*, 119 S.Ct. 1297 (1999).
37. *Illinois v. Caballes*, 543 U.S. 405 (2005); *Florida v. Harris*, 568 U.S. 237 (2013): In a unanimous decision, the Supreme Court gave police authority to use dogs to uncover illegal drugs at traffic stops, upholding a police Labrador retriever's search of a truck that uncovered methamphetamine ingredients inside.
38. *Rodriguez v. United States*, 135 S.Ct. 1609 (2015).
39. *U.S. v. Jones*, 565 U.S. ___, 132 S.Ct. 945 (2012).
40. *Byrd n. United States*, 584 U.S. ____ (2018).
41. *New York v. Class*, 54 LW 4178 (1986).
42. *Oliver v. United States*, 466 U.S. 170 (1984).
43. *California v. Ciraolo*, 476 U.S. 207 (1986).

44. *Collins v. Virginia,* 584 U.S. ___ (2018).
45. *Florida v. Jardines,* 569 U.S. ___, 133 S. Ct. 1409 (2013): Justice Antonin Scalia's opinion stated that "to find a visitor knocking on the door is routine (even if sometimes unwelcome); to spot that same visitor exploring the front path with a metal detector, or marching his bloodhound into the garden before saying hello and asking permission, would inspire most of us to—well, call the police." Scalia said using the dog was no different from using thermal imaging technology from afar to peer inside homes without a warrant.
46. *Bumper v. North Carolina,* 391 U.S. 543 (1968).
47. *Stoner v. California,* 376 U.S. 483 (1964).
48. *City of Los Angeles v. Patel,* 576 U.S. ___ (2015).
49. *Georgia v. Randolph,* 126 S.Ct. 1515 (2006).
50. *Fernandez v. California,* 571 U.S. ____ (2014).
51. *Katz v. United States,* 389 U.S. 347 (1967).
52. *Berger v. New York,* 388 U.S. 41 (1967).
53. *Lee v. United States,* 343 U.S. 747 (1952).
54. John Kaplan, Jerome H. Skolnick, and Malcolm M. Feeley, *Criminal Justice: Introductory Cases and Materials,* 5th ed. (Westbury, N.Y.: Foundation Press, 1991), pp. 220–221.
55. *Miranda v. Arizona,* 384 U.S. 436 (1966)
56. *Edwards v. Arizona,* 451 U.S. 477 (1981).
57. *Arizona v. Roberson,* 486 U.S. 675 (1988).
58. *Berkemer v. McCarty,* 468 U.S. 420 (1984).
59. *Quarles v. New York,* 467 U.S. 649 (1984).
60. Adam Goodman, "How the Media Have Misunderstood Dzhokhar Tsarnaev's *Miranda* Rights," *The Atlantic,* April 22, 2013, http://www.theatlantic.com/national/archive/2013/04/how-the-media-have-misunderstood-dzhokhar-tsarnaevs-i-miranda-i-rights/275189/.
61. *Michigan v. Mosley,* 423 U.S. 93 (1975).
62. *Colorado v. Spring,* 479 U.S. 564 (1987).
63. *Connecticut v. Barrett,* 479 U.S. 523 (1987).
64. *Duckworth v. Eagan,* 109 S.Ct. 2875 (1989).
65. *Berghuis v. Thompkins,* 560 U.S. 370 (2010).
66. *U.S. v. Wade,* 388 U.S. 218 (1967).
67. *Kirby v. Illinois,* 406 U.S. 682 (1972).
68. *Salinas v. Texas,* 133 S.Ct. 2174 (2013). In a narrow 5–4 decision, the Court concluded that the Fifth Amendment's privilege against self-incrimination does not extend to defendants who decide to remain silent during questioning. A witness (someone who is being questioned but is not yet in custody as a suspect) must explicitly claim that protection if he or she wants to guard against self-incrimination.
69. *Foster v. California,* 394 U.S. 440 (1969).
70. E. F. Loftus, J. M. Doyle, and J. Dysert, *Eyewitness Testimony: Civil & Criminal,* 4th ed. (Charlottesville, Va: Lexis Law Publishing, 2008); E. F. Loftus, "The Malleability of Human Memory," *American Scientist* 67 (1979): 312–320; Gary L. Wells et al., "Eyewitness Identification Procedures: Recommendations for Lineups and Photospreads," *Law and Human Behavior* 22, no. 6 (1998): 603–647; Gary L. Wells, "Eyewitness Identifications: Systemic Reforms," *Wisconsin Law Review* 615 (2006): 628–629.
71. *State v. Henderson,* 27 A.3d 872 (2011).
72. *Powell v. Alabama,* 287 U.S. 45 (1932).
73. *Gideon v. Wainwright,* 372 U.S. 335 (1963).
74. *Argersinger v. Hamlin,* 407 U.S. 25 (1973).
75. *Luis v. United States,* 136 S.Ct. 1083 (2016).
76. *Escobedo v. Illinois,* 378 U.S. 478 (1964).

/// CHAPTER 7

1. Erwin C. Surrency, "The Courts in the American Colonies," *American Journal of Legal History* 11 (1967): 258.
2. Kermit Hall, *The Magic Mirror: Law in American History* (New York: Oxford University Press, 1989).
3. Surrency, "The Courts in the American Colonies," p. 258.
4. Freda Adler, Gerhard O. W. Mueller, and William S. Laufer, *Criminal Justice: An Introduction* (Boston: McGraw-Hill, 2006), pp. 325–326.
5. Stephen Whicher and R. Spiller, eds., *The Early Lectures of Ralph Waldo Emerson* (Philadelphia: University of Pennsylvania Press, 1953), p. 112.
6. *Tehan v. U.S. ex rel. Shott,* 382 U.S. 406 (1966), p. 416.
7. Howard Abadinsky, *Law and Justice: An Introduction to the American Legal System,* 4th ed. (Chicago: Nelson-Hall, 1999), p. 174.
8. Ibid., p. 170.
9. For the last compiled database on prisoner petitions, see U.S. Department of Justice, "Table 5.65: Petitions Filed in U.S. District Courts by Federal and State Prisoners," *Sourcebook of Criminal Justice Statistics Online,* https://www.albany.edu/sourcebook/pdf/t5652012.pdf.
10. Malcolm M. Feeley and Edward L. Rubin, *Judicial Policy Making and the Modern State: How the Courts Reformed America's Prisons* (New York: Cambridge University Press, 1998).
11. *Brown v. Plata,* 131 S.Ct. 1910 (2011).
12. Stephen L. Wasby, *The Supreme Court in the Federal System,* 3rd ed. (Chicago: Nelson-Hall, 1989), p. 5.
13. Abadinsky, *Law and Justice,* p. 171.
14. Ibid., p. 166.
15. Cassie Spohn and Craig Hemmens, *Courts: A Text/Reader* (Thousand Oaks, Calif.: Sage, 2008), p. 10.
16. National Center for State Courts, Conference of State Court Administrators, *Courts Statistics Project, CSP DataViewer, 2016 Trial Court Data,* http://www.courtstatistics.org/.
17. David W. Neubauer, *America's Courts and the Criminal Justice System,* 8th ed. (Belmont, Calif.: Thomson Wadsworth, 2005), pp. 402–403.
18. Ibid., p. 82.
19. Ibid., pp. 81, 83.
20. Nevada Appellate Courts, http://nvcourts.gov/CourtOfAppeals.aspx.
21. R. Schauffler, R. LaFountain, S. Strickland, K. Holt, & K. Genthon, *Examining the Work of State Courts: An Overview of 2015 State Court Caseloads* (Williamsburg, Va.: National Center for State Courts, 2016).
22. Neubauer, *America's Courts and the Criminal Justice System,* 8th ed., p. 85.
23. Ibid.
24. The U.S. Supreme Court reversed the decision of the lower court (i.e., the circuit court), saying the lack of an instruction on causation by itself in this case was not a violation of due process rights. The Court also said, notably, that "a person who is aware of, and consciously disregards a risk, must foresee the ultimate harm involved." See *Kibbe v. Henderson,* 534 F.2d 493 (1976). Kibbe's adversary (the appellee) in this case, Henderson, was the superintendent at Auburn Correctional Facility in New York. The phrase "exhausting all possible state remedies" is a legal doctrine meaning that the state's appeals courts are to be given the opportunity to correct any defects that occurred at trial before a party may pursue a case in federal court. This is basically the rule of comity, or courtesy—the federal courts defer to state courts to correct any defects prior to those claims being raised in a federal court. *Habeas corpus* (sometimes termed "the great writ") is a Latin term for "you have the body," and it is the inmate's means of asking a court to grant a hearing to determine whether or not he or she is being held illegally.
25. Ibid., p. 87
26. U.S. Courts, "Federal Judicial Caseload Statistics 2016," https://www.uscourts.gov/report-names/federal-judicial-caseload-statistics.
27. Federal Judicial Center, "Federal Judicial History," http://www.fjc.gov/history/home.nsf.
28. U.S. Courts, "Federal Judicial Caseload Statistics 2016."
29. Federal Judicial Center, "Federal Judicial History."

30. Supreme Court of the United States, "A Brief Overview of the Supreme Court," http://www.supremecourt.gov/about/briefoverview.aspx.
31. David W. Neubauer, *America's Courts and the Criminal Justice System,* 9th ed. (Belmont, Calif.: Thomson Wadsworth, 2008), p. 63.
32. Supreme Court of the United States, "Instructions for Admission to the Bar," http://www.supremecourtus.gov/bar/barinstructions.pdf.
33. See U.S. Courts, "U.S. Supreme Court Procedures," http://www.uscourts.gov/EducationalResources/ConstitutionResources/SeparationOfPowers/USSupremeCourtProcedures.aspx.
34. Ibid.
35. Supreme Court of the United States, "The Justices' Caseload," http://www.supremecourt.gov/about/justicecaseload.aspx; also see, as examples of written opinions, Supreme Court of the United States, "Opinions of the Court—2017," https://www.supremecourt.gov/opinions/slipopinion/17.
36. Ibid.; see the *Rules of the Supreme Court of the United States,* generally, at http://www.supremecourt.gov/ctrules/2010RulesoftheCourt.pdf.
37. Ram Subramanian et al., *Incarceration's Front Door: The Misuse of Jail in America* (New York: Vera Institute of Justice, 2015); Aimee Picchi, "Are America's Jails Used to Punish Poor People?" CBS News, February 11, 2015, http://www.cbsnews.com/news/how-jails-are-warehousing-those-too-poor-to-make-bail/.
38. Bureau of Justice Statistics, "Jail Inmates in 2016," February 2018, https://www.bjs.gov/content/pub/pdf/ji16_sum.pdf.
39. American Bar Association, "FAQs About the Grand Jury System," http://www.abanow.org/2010/03/faqs-about-the-grand-jury-system/.
40. Larry Buchanan et al., "What Happened in Ferguson?" *New York Times,* November 25, 2014, http://www.nytimes.com/interactive/2014/08/13/us/ferguson-missouri-town-under-siege-after-police-shooting.html.
41. For more information, search "grand jury" at https://www.americanbar.org/.
42. Arlen Specter, book review, 76 *Yale Law Journal* 604 (1967): 606–607.
43. John Nisbet, Burma *Under British Rule—and Before,* Vol. I (London: Archibald Constable, 1901), p. 177.
44. Cora L. Daniels and C. M. Stevans (Eds.), *Encyclopedia of Superstitions, Folklore, and the Occult Sciences of the World,* Vol. III (Milwaukee: J. H. Yewdale, 1903), p. 1243.
45. "Trial by Ordeal," http://www.absolute astronomy.com/topics/Trial_by_ordeal.
46. *Williams v. Florida,* 399 U.S. 78, 86 (1970).
47. *Apodaca v. Oregon,* 406 U.S. 404 (1972).
48. Vick and his attorneys bargained with federal prosecutors to avoid a trial. During this time, Vick tested positive for marijuana, in violation of his bail conditions, so the federal judge ordered stricter home-confinement conditions. Vick ultimately pleaded guilty to a variety of charges and was sentenced to, among other penalties, 21 months in federal prison at Leavenworth, Kansas, and extensive community service in the interest of stopping animal cruelty. While Vick was serving time at Leavenworth, Virginia authorities moved forward with their case, using Vick's guilty pleas in the federal case to bolster the state's evidence. Again, Vick pleaded guilty and struck a bargain for a suspended sentence conditioned on the successful serving of his federal sentence. Vick was released from prison in 2009 and ultimately returned to play in the NFL while devoting significant time to speaking out against animal cruelty.
49. *See Brady v. Maryland,* 373 U.S. 83 (1963).
50. Neubauer, *America's Courts and the Criminal Justice System,* 9th ed., p. 263.
51. National Association of Drug Court Professionals, "Did You Know?" https://www.nadcp.org/treatment-courts-work/; see also Douglas B. Marlowe, Carolyn D. Hardin, and Carson L. Fox, *Painting the Picture: A National Report on Drug Courts and Other Problem-Solving Court Programs in the United States,* June 2016, National Drug Court Institute, http://www.nadcp.org/sites/default/files/2014/Painting%20the%20Current%20Picture%202016.pdf.
52. Council of State Governments Justice Center, "Mental Health Courts," May 2015, http://csgjusticecenter.org/mental-health-court-project/.
53. Marlowe, Hardin, and Fox, *Painting the Picture: A National Report on Drug Courts and Other Problem-Solving Court Programs in the United States.*
54. Ibid.; see also National Association of Drug Court Professionals, "Problem Solving Courts" http://www.nadcp.org/learn/what-are-drug-courts/models/problem-solving-courts.
55. See Thomas Alexander Fyfe, *Charles Dickens and the Law* (London: Chapman and Hall, 1910), pp. 28–29.
56. Charles Dickens, *Bleak House* (London: Penguin Book, 1971; first published in 1853).
57. William Shakespeare, *Hamlet,* Act 3, Scene 1.
58. *Strunk v. U.S.,* 412 U.S. 434 (1973).
59. Speedy Trial Act of 1974, 18 U.S.C.S. §§ 3161–3174 (as amended, 1979).
60. *Barker v. Wingo,* 407 U.S. 514 (1972).
61. Ibid.
62. Ibid., p. 522.
63. "Casey Anthony Trial: Timeline of Key Events in the Murder Trial of the Florida Mother," ABC News, July 6, 2011, https://abcnews.go.com/US/casey-anthony-trial-timeline-key-events/story?id=13990853.
64. American Prosecutors Research Institute, *Basic Trial Techniques for Prosecutors,* May 2005, http://www.ndaa.org/pdf/basic_trial_techniques_ 05.pdf.
65. Justia, "Criminal Appeals Overview," http://www.justia.com/criminal/criminal-appeals/.
66. *Douglas v. California,* 372 U.S. 353 (1963).
67. *Ross v. Moffitt,* 417 U.S. 600 (1974).
68. Federal Judicial Center, *Annual Report,* 2015, https://www.fjc.gov/sites/default/files/2016/Annual%20Report%202015.pdf.
69. See U.S. Courts, "Chambers Online Automation Training (COAT)," http://uscourts.webapponline.com/.
70. Dwane Brown, "San Diego County Prosecutors Use App to Present Evidence," KPBS Public Broadcasting, May 29, 2015, http://www.kpbs.org/news/2015/may/29/san-diego-county-prosecutors-using-app-present-evi/.
71. Judge Herbert B. Dixon, Jr., "The Basics of a Technology-Enhanced Classroom," American Bar Association, November 1, 2017, https://www.americanbar.org/groups/judicial/publications/judges_journal/2017/fall/basics-technologyenhanced-courtroom/.
72. National Judicial College, "The Model Courtroom," http://www.judges.org/about/the-njc-experience/model-courtroom/.

/// CHAPTER 8

1. Kevin Grasha, "After More Than 3 Years, Here's Where the Tracie Hunter Case Stands," *Cincinnati Enquirer,* April 8, 2018, https://www.cincinnati.com/story/news/2018/04/08/after-more-than-3-years-heres-where-tracie-hunter-case-stands/493333002/.
2. Kara Foxx, "Judge Nadel Sentences Tracie Hunter to 6 Months in Jail," Fox 19 News (Cincinnati, Ohio), December 5, 2014, https://www.fox19.com/story/27557626/judge-nadel-sentences-tracie-hunter-to-6-months-in-jail/.
3. Hollie Silverman, "A Former Ohio Judge Was Dragged From Court After Her Sentencing," CNN, July 23, 2019, https://www.cnn.com/2019/07/23/us/ohio-judge-dragged-from-court/index.html.
4. Jong Son, "Top 5 List of Real-Life Judicial Misconduct," Ballotpedia, September 26, 2014, https://ballotpedia.org/Top_5_List_of_Real-Life_Judicial_Misconduct.
5. Samuel J. Brakel and Alexander D. Brooks, *Law and Psychiatry in the Criminal Justice System* (Buffalo, N.Y.: William S. Hein, 2001), p. 130.

6. Quoted in David Landy and Elliott Aronson, "The Influence of the Character of the Criminal and His Victims on the Decisions of Simulated Jurors," *Journal of Experimental Social Psychology* 5 (1969), pp. 141–142.
7. Abraham Blumberg, *Criminal Justice* (Chicago: Quadrangle Books, 1967), p. 120.
8. See, for example, Pennsylvanians for Modern Courts, "Choosing Judges," http://www.pmconline.org/node/25.
9. U.S. Department of Justice, Bureau of Justice Statistics, *State Court Organization, 2004* (Washington, D.C.: Author, October 2006), p. 23.
10. American Judicature Society, *Judicial Selection in the States: How It Works, Why It Matters* (Des Moines, Iowa: Author, 2008), p. 4.
11. Ibid.
12. James Podgers, "O'Connor on Judicial Elections: 'They're Awful. I Hate Them,'" *ABA Journal*, May 9, 2009, http://www.abajournal.com/news/oconnor_chemerinsky_sound_warnings_at_aba_conference_about_the_dangers_of_s/.
13. Ibid.
14. "O'Connor Says Judges Shouldn't Be Elected," Law.com, November 8, 2007, http://www.law.com/jsp/law/LawArticleFriendly.jsp?id=1194429842107.
15. *Williams v. Pennsylvania,* 136 S.Ct. 28 (2015). In a 5–3 ruling, the U.S. Supreme Court ruled that Judge Castille should have recused himself from the case. Justice Anthony Kennedy wrote in the majority opinion that "where a judge has had an earlier significant, personal involvement as a prosecutor in a critical decision in the defendant's case, the risk of actual bias in the judicial proceeding rises to an unconstitutional level." The court ruled that Castille's failure to recuse himself violated the defendant's Fourteenth Amendment due process protections.
16. Martha Neil, "Top Court Hears Judicial Influence Case, Leans Toward Stricter Recusal Standard," *ABA Journal,* March 3, 2009, http://www.abajournal.com/news/top_court_hears_judicial_influence_case_leans_toward_stricter_recusal_stand.
17. *Caperton v. A.T. Massey Coal Co., Inc.*, 556 U.S. _____ (2009).
18. Jeff Cranson, "The Price of Justice: High Court Wrestles With a Case That Should Spark Discussion in Michigan," *The Grand Rapids Press,* March 3, 2009, http://blog.mlive.com/talkingpolitics/2009/03/the_price_of_justice_high_cour.html.
19. Alice Bannon, Cathleen Lisk, and Peter Hardin, "Who Pays for Judicial Races," Brennan Center for Justice, 2017, https://www.brennancenter.org/publication/politics-judicial-elections.
20. Ibid.
21. Blumberg, *Criminal Justice,* p. 120.
22. Russell R. Wheeler and Howard R. Whitcomb, *Judicial Administration: Text and Readings* (Englewood Cliffs, N.J.: Prentice-Hall, 1977), p. 370.
23. Ibid., p. 372.
24. Ibid.
25. Ibid.
26. Ibid., p. 373.
27. William A. Batlitch, "Reflections on the Art and Craft of Judging," *The Judges' Journal* 43, no. 4 (Fall 2003), pp. 7–8.
28. Charles E. Patterson, "The Good Judge: A Trial Lawyer's Perspective," *The Judges' Journal* 43, no. 4 (Fall 2003), pp. 14–15.
29. See, for example, Allen K. Harris, "The Professionalism Crisis—The 'Z' Words and Other Rambo Tactics: The Conference of Chief Justices' Solution," 53 S.C. L. Rev. 549, 589 (2002).
30. *In re First City Bancorp of Tex., Inc.*, 282 F.3d 864 (5th Cir. 2002).
31. Marla N. Greenstein, "The Craft of Ethics," *The Judges Journal* 43, no. 4 (Fall 2003), p. 42.
32. Ty Tasker, "Sticks and Stones: Judicial Handling of Invective in Advocacy," *The Judges' Journal* 43, no. 4 (Fall 2003), pp. 17–18.
33. Collins T. Fitzpatrick, "Building a Better Bench: Informally Addressing Instances of Judicial Misconduct," *The Judges' Journal* 45, no. 2 (Winter 2005), pp. 16–20.
34. Ibid.
35. U.S. Courts, "Guide to Judiciary Policy," http://www.uscourts.gov/sites/default/files/v0102b-ch02.pdf.
36. Jan L. Jacobowitz and John G. Browning, *Legal Ethics and Social Media: A Practitioner's Handbook* (Chicago: ABA Book Publishing, 2017).
37. Southern Poverty Law Center, "Are There Limits to Prosecutorial Discretion?" http://www.splcenter.org/get-informed/intelligence-report/browse-all-issues/2007/summer/legal-brief#.UafrguDn_cs.
38. *Berger v. United States,* 295 U.S. 78 (1935).
39. *Connick v. Thompson,* 563 U.S. _____ (2011).
40. Radley Balko, "The Untouchables: America's Misbehaving Prosecutors, and the System that Protects Them," *Huffington Post,* August 1, 2013, http://www.huffingtonpost.com/2013/08/01/prosecutorial-misconduct-new-orleans-louisiana_n_3529891.html.
41. *Van De Kamp v. Goldstein,* 555 U.S. ____ (2009), quoting *Gregoire v. Biddle,* 177 F. 2d 579, 581 (2 Cir. 1949).
42. *Maine v. Moulton,* 474 U.S. 159 (1985).
43. Legal Information Institute, "Effective Assistance of Counsel," http://www.law.cornell.edu/anncon/html/amdt6frag9_user.html.
44. *Strickland v. Washington,* 104 S.Ct. 3562 (1984).
45. See California Innocence Project, "Ineffective Assistance of Counsel," http://californiainnocenceproject.org/issues-we-face/ineffective-assistance-of-counsel.
46. Ibid.; also see *Strickland v. Washington,* 466 U.S. 668 (1984).
47. William A. Mintz, "Lawyer Wouldn't Go to 'Sleazy Bar,' Client Wins Freedom From Life Term," *National Law Journal,* November 24, 1980, p. 7.
48. FindLaw, "Plea Bargaining Pros and Cons," http://criminal.findlaw.com/criminal-procedure/plea-bargain-pros-and-cons.html.
49. See, for example, Texas Fair Defense Project, "What Defense Lawyers Do," http://www.texasfairdefenseproject.org/info/right_to_counsel/defense_lawyers.
50. *Griffin v. Illinois,* 351 U.S. 12 (1956).
51. For more information about defense attorneys in general and indigent services specifically, see Cassia Spohn and Craig Hemmens, *Courts: A Text/Reader,* 2nd ed. (Los Angeles, Calif.: Sage, 2012), pp. 218–223.
52. James Eisenstein and Herbert Jacob, *Felony Justice: An Organizational Analysis of Criminal Courts* (Boston: Little, Brown, 1977).
53. U.S. Department of Labor, "Court Reporters," *Occupational Outlook Handbook, Legal,* https://www.bls.gov/ooh/legal/court-reporters.htm.
54. *Massachusetts v. Leno,* 415 Mass. 835, 616 N.E.2d 453 (1993).
55. Leo Katz, "Excuse: Duress—The Nature of the Threat, the Nature of the Crime, the Mistaken Defendant, the Semiculpable Defendant—Superior Orders: Husbands and Wives," *Law Library: American Law and Legal Information—Crime and Criminal Law,* http://law.jrank.org/pages/1128/Excuse-Duress.html.
56. *Spakes v. State,* 913 S.W.2d 597 (Tex. Crim. App. 1996).
57. *People v. Luther,* 394 Mich. 619, 232 N.W.2d 184 (1975).
58. *Sherman v. U.S.,* 356 U.S. 369 (1958).
59. *U.S. v. Russell,* 411 U.S. 423 (1973).
60. PBS, "Biography: John Hinckley, Jr.," *American Experience,* http://www.pbs.org/wgbh/americanexperience/features/biography/reagan-hinckley/.
61. David Gates, "Everybody Has Scars," *Newsweek,* October 13, 1986, p. 10; also see "Man Given 5-to-15-Year Term in Model's Slashing," *New York Times,* May 12, 1987, http://www.nytimes.com/1987/05/12/nyregion/man-given-5-to-15-year-term-in-model-s-slashing.html.
62. Carol Pogash, "Myth of the 'Twinkie Defense': The Verdict in the Dan White Case Wasn't Based on His Ingestion of Junk Food," *San Francisco Chronicle,* November 23, 2003, http://www.sfgate.com/health/article/Myth-of-the-Twinkie-defense-The-verdict-in-2511152.php.
63. Michael Perlin, *The Jurisprudence of the Insanity Defense* (Durham, N.C.: Carolina Academic Press, 1994), p. 108.

64. *M'Naghten's Case,* 8 Eng. Rep. 718 (H.L. 1843).
65. *Parsons v. State,* 2 So. 854 (Ala. 1887).
66. See, generally, James F. Hooper and Alix M. McLearen, "Does the Insanity Defense Have a Legitimate Role?" *Psychiatric Times,* April 1, 2002, http://www.psychiatrictimes.com/display/article/10168/54196.
67. Dirk Johnson, "Milwaukee Jury Says Dahmer Was Sane," *New York Times,* February 16, 1992, http://www.nytimes.com/1992/02/16/us/milwaukee-jury-says-dahmer-was-sane.html.
68. PBS, "Other Notorious Insanity Cases," *Frontline,* http://www.pbs.org/wgbh/pages/frontline/shows/crime/trial/other.html.
69. Ibid.
70. This is the Florida case of Marissa Alexander from 2010. Her defense failed because she took the time to go to the garage and get the gun, which the prosecution successfully argued meant she also had time to call the police and escape the "immediate" danger she feared. The state argued that she acted/fired out of anger and not fear, imperiling both her husband and children. She was convicted and sentenced to 20 years in prison (under a mandatory sentencing scheme triggered in part by her use of the gun). In 2013, she won a new trial based on an error at her original trial: The burden of proving the self-defense issue had been improperly shifted to Alexander and not to the prosecution. She was released from prison in November 2013 and ordered to remain on house arrest. The Florida prosecutor vowed she would retry Alexander and seek 60 years (the maximum sentence allowed on such a retrial). In November 2014, Alexander pleaded guilty in exchange for time served, to avoid retrial and the potential 60-year sentence. She was sentenced in January 2015 to two years of house arrest. See Irin Carmon, "Marissa Alexander Released From Jail," MSNBC, January 27, 2015, http://www.msnbc.com/msnbc/marissa-alexander-may-be-released.

/// CHAPTER 9

1. Quoted in Louis P. Carney, *Probation and Parole: Legal and Social Dimensions* (New York: McGraw-Hill, 1977), p. 75.
2. Lyndon Stambler, "Point Dume Man Found Guilty of Murdering His Former Girlfriend," *Los Angeles Times*, April 18, 1985, https://www.latimes.com/archives/la-xpm-1985–04–18-we-23880-story.html.
3. David Colker, "Marcella Leach Dies at 85; Advocate for Marsy's Law on Victims' Rights," *Los Angeles Times*, March 27, 2015, https://www.latimes.com/local/obituaries/la-me-marcella-leach-20150328-story.html.
4. Francis T. Cullen and Paul Gendreau, "Assessing Correctional Rehabilitation: Policy, Practice, and Prospects," *Criminal Justice 2000,* p. 111, https://www.ncjrs.gov/criminal_justice2000/vol_3/03d.pdf.
5. Brandon C. Welsh, "Monetary Costs and Benefits of Correctional Treatment Programs: Implications for Offender Reentry," *Federal Probation* (September 2004), p. 12, http://bcotn.org/subcommittees/csct/monetary_costs_and_benefits_of_correctional_treatment.pdf.
6. Ibid., pp. 9–12.
7. Jacqueline Cohen, *Incapacitating Criminals: Recent Research Findings I* (Washington, D.C.: National Institute of Justice, Research in Brief, 1983), p. 2.
8. Bureau of Prisons, "Annual Determination of Average Cost of Incarceration," *Federal Register*, April 30, 2018, https://www.federalregister.gov/documents/2018/04/30/2018–09062/annual-determination-of-average-cost-of-incarceration.
9. "California's Annual Costs to Incarcerate an Inmate in Prison," *Legislative Analyst's Office*, 2018–2019, https://lao.ca.gov/PolicyAreas/CJ/6_cj_inmatecost.
10. John J. Dilulio Jr., "Prisons Are a Bargain, by Any Measure," *New York Times,* January 16, 1996, p. A17.
11. Roush, A Desktop Guide to Good Juvenile Detention Practice, pp. 23–24.
12. The discussion of the corrections models is adapted from Todd R. Clear, George F. Cole, and Michael D. Reisig, *American Corrections,* 8th ed. (Belmont, Calif.: Thomson Wadsworth, 2009), pp. 40–64; the discussion of the accompanying prison design and operation is adapted from Steven E. Schoenherr, "Prison Reforms in American History," http://history.sandiego.edu/gen/soc/prison.html.
13. Quotes taken from the *Handbook of Correctional Institution Design and Construction* (U.S. Bureau of Prisons, 1949).
14. Amnesty International Global Report, "Death Sentences and Executions 2018," https://www.amnesty.org/download/Documents/ACT5098702019ENGLISH.PDF.
15. Ibid.
16. Saudi Arabian International Schools, "An Introduction to the Kingdom of Saudi Arabia," Spring 1992, p. 11; "U.S. Student Tells of Pain of His Caning in Singapore," *New York Times,* June 26, 1994, http://www.nytimes.com/1994/06/26/us/us-student-tells-of-pain-of-his-caning-in-singapore.html.
17. Stephanie Stullich, Ivy Morgan, and Oliver Schak, *State and Local Expenditures on Corrections and Education* (U.S. Department of Education, Policy and Program Studies Service, July 2016), https://www2.ed.gov/rschstat/eval/other/expenditures-corrections-education/brief.pdf.
18. E. Ann Carson, *Prisoners in 2016* (U.S. Department of Justice, Bureau of Justice Statistics, January 2018), https://www.bjs.gov/content/pub/pdf/p16.pdf.
19. Adapted from U.S. Department of Justice, National Institute of Corrections, and Association of Paroling Authorities International, *A Handbook for New Parole Board Members* (April 2003), p. 3, http://www.apaintl.org/documents/CEPPParoleHandbook.pdf.
20. Quoted by Sheldon G. Glueck in his "Foreword" to John V. Barry, *Alexander Maconochie of Norfolk Island* (Melbourne: Oxford University Press, 1958).
21. Mark Allenbaugh, "The Supreme Court's New Blockbuster U.S. Sentencing Guidelines Decision," FindLaw, January 14, 2005, http://writ.news.findlaw.com/allenbaugh/20050114.html.
22. At 18 U.S.C. Secs. 3551–3626 and 28 U.S.C. Secs. 991–998 (October 12, 1984).
23. Allenbaugh, "The Supreme Court's New Blockbuster U.S. Sentencing Guidelines Decision."
24. Lisa M. Seghetti and Alison M. Smith, *Federal Sentencing Guidelines: Background, Legal Analysis, and Policy Options* (Congressional Research Service Report for Congress, June 30, 2007), http://www.fas.org/sgp/crs/misc/RL32766.pdf.
25. *Blakely v. Washington,* 542 U.S. (2004).
26. *United States v. Booker,* 543 U.S. 125 (2005).
27. Allenbaugh, "The Supreme Court's New Blockbuster U.S. Sentencing Guidelines Decision," p. 2.
28. *Gall v. United States,* 552 U.S. 38 (2007).
29. *Kimbrough v. United States,* 552 U.S. 85 (2007).
30. *Beckles v. United States,* 580 U.S. ___ (2017).
31. *2018 Washington State Adult Sentencing Guidelines Manual,* http://www.cfc.wa.gov/PublicationSentencing/SentencingManual/Adult_Sentencing_Manual_2018.pdf.
32. The National Center for Victims of Crime, "Victim Impact Statements," http://www.victimsofcrime.org/help-for-crime-victims/get-help-bulletins-for-crime-victims/victim-impact-statements.
33. *Payne v. Tennessee,* 501 U.S. 808 (1991).
34. *Snyder v. Massachusetts,* 291 U.S. 97 (1934), at 122.
35. *Bosse v. Oklahoma,* 580 U.S. ___ (2016).
36. Jeanne Hruska, "'Victims' Rights' Proposals Like Marsy's Law Undermine Due Process," American Civil Liberties Union, May 3, 2018, https://www.aclu.org/blog/criminal-law-reform/victims-rights-proposals-marsys-law-undermine-due-process.
37. James Q. Wilson, *Thinking About Crime* (New York: Vintage Books, 1985), p. 260.

38. H. Naci Mocan and R. Kaj Gittings, "Getting Off Death Row: Commuted Sentences and the Deterrent Effect of Capital Punishment," *Journal of Law and Economics* 46 (October 2003), pp. 453–478.
39. Hashem Dezhbakhsh, Paul H. Rubin, and Joanna M. Shepherd, "Does Capital Punishment Have a Deterrent Effect? New Evidence From Postmoratorium Panel Data," *American Law and Economics Review* 5, no. 2 (2003), pp. 344–376.
40. Paul R. Zimmerman, "State Executions, Deterrence, and the Incidence of Murder," *Journal of Applied Economics* 7, no. 1 (May 2004), pp. 163–193.
41. Paul R. Zimmerman, "Estimates of the Deterrent Effect of Alternative Execution Methods in the United States: 1978–2000," *American Journal of Economics and Sociology* 65, no. 4 (October 2006), pp. 909–941.
42. Richard Berk, "New Claims About Executions and General Deterrence: Déjà Vu All Over Again?" *Journal of Empirical Legal Studies* 2 (2005), pp. 303–330.
43. Jeffrey Fagan, "Death and Deterrence Redux: Science, Law and Causal Reasoning on Capital Punishment," *Ohio State Journal of Criminal Law* 4 (2006), pp. 255–320.
44. Quoted in Amnesty International, "Death Penalty and Race," https://www.amnestyusa.org/issues/death-penalty/death-penalty-facts/death-penalty-and-race/.
45. Ibid.
46. Ibid.
47. *Furman v. Georgia,* 408 U.S. 238 (1972).
48. *Gregg v. Georgia,* 428 U.S. 153 (1976).
49. *Roper v. Simmons,* 543 U.S. 551 (2005).
50. *Ford v. Wainwright,* 477 U.S. 399 (1986).
51. *Kennedy v. Louisiana,* 554 U.S. 407 (2008).
52. *Strickland v. Washington,* 466 U.S. 668 (1984)
53. *Witherspoon v. Illinois,* 391 U.S. 510 (1968).
54. *Baze v. Rees,* 553 U.S. 35 (2008).
55. *Hill v. Florida,* No. SC06–2 (2006).
56. Death Penalty Information Center, "Innocence," https://deathpenaltyinfo.org/policy-issues/innocence.
57. Death Penalty Information Center: "Facts about the Death Penalty," May 2019, https://files.deathpenaltyinfo.org/legacy/documents/FactSheet.pdf.
58. National Coalition to Abolish the Death Penalty, "State by State Stats," May 2019, http://www.ncadp.org/map.
59. *Atkins v. Virginia,* 536 U.S. 304 (2002).
60. *Ring v. Arizona,* 536 U.S. 584 (2002).
61. *Roper v. Simmons,* 125 S.Ct. 1183 (2005)
62. *McKinney v. Arizona,* https://www.supremecourt.gov/docket/docketfiles/html/public/18–1109.html.
63. Devlin Barrett, "Justice Department Plans to Restart Capital Punishment After Long Hiatus," *Washington Post,* July 25, 2019, https://www.washingtonpost.com/national-security/justice-department-plans-to-restart-capital-punishment-after-long-hiatus/2019/07/25/f2cc6402-aee5-11e9-bc5c-e73b603e7f38_story.html.
64. Death Penalty Information Center, "Executions Under the Federal Death Penalty," https://deathpenaltyinfo.org/state-and-federal-info/federal-death-penalty/executions-under-the-federal-death-penalty.
65. U.S. Department of Justice, "Federal Government to Resume Capital Punishment After Nearly Two Decade Lapse," *Justice News,* July 25, 2019, https://www.justice.gov/opa/pr/federal-government-resume-capital-punishment-after-nearly-two-decade-lapse.
66. Adapted from *Nevada Revised Statutes* 200.033.
67. Adapted from *Nevada Revised Statutes* 200.035.
68. "Criminal Appeals," http://www.justia.com/criminal/criminal-appeals/.
69. *Douglas v. California,* 372 U.S. 353 (1963).
70. *Griffin v. Illinois,* 351 U.S. 12 (1956).
71. *Ross v. Moffitt,* 417 U.S. 600 (1974).

/// CHAPTER 10

1. Jack Henry Abbott, *In the Belly of the Beast: Letters From Prison* (New York: Vintage Books, 1981), p. x.
2. See, generally, UN Office on Drugs and Crime, "Why Promote Prison Reform?" https://www.unodc.org/unodc/en/justice-and-prison-reform/prison-reform-and-alternatives-to-imprisonment.html; also see Seung Min Kim and Anne Gearan, "Jared Kushner Ramps Up Push for Criminal Justice Reform," *Washington Post,* August 30, 2018, https://www.washingtonpost.com/politics/jared-kushner-ramps-up-push-for-criminal-justice-reform/2018/08/30/449e2d02-acb1–11e8–8f4b-aee063e14538_story.html.
3. Erving Goffman, *Asylums* (Garden City, N.Y.: Anchor Books, 1961), p. 5.
4. Ibid., p. 6.
5. Bureau of Justice Statistics, *Correctional Populations in the United States, 2016,* April 2018, p. 1, https://www.bjs.gov/content/pub/pdf/cpus16.pdf.
6. See, for example, Deanne Katz, "What's the Difference Between Jail and Prison?" FindLaw, December 10, 2012, http://blogs.findlaw.com/blotter/2012/12/whats-the-difference-between-jail-and-prison.html.
7. Bureau of Justice Statistics, *Correctional Populations in the United States, 2016,* p. 1.
8. Ibid., p. 16.
9. "Death Penalty Fast Facts, CNN, March 22, 2019, https://www.cnn.com/2013/07/19/us/death-penalty-fast-facts/index.html.
10. See Bureau of Justice Statistics, "Justice Expenditure and Employment Extracts, 2015—Preliminary," data table jeee15t01.csv, June 29, 2018, https://www.bjs.gov/index.cfm?ty=pbdetail&iid=6310.
11. See Bureau of Labor Statistics "Correctional Officers and Bailiffs," n.d., https://www.bls.gov/ooh/protective-service/correctional-officers.htm.
12. Timothy Williams, "U.S. Correctional Population at Lowest Level in Over a Decade," *New York Times,* December 29, 2016, https://www.nytimes.com/2016/12/29/us/us-prison-population.html?_r=0.
13. John Gramlich, "5 Facts About Crime in the U.S.," Pew Research Center, January 3, 2019, https://www.pewresearch.org/fact-tank/2019/01/03/5-facts-about-crime-in-the-u-s/.
14. Rhonda McMillion, "Boosting Bail Reform: ABA Urges Congress to Limit Use of Cash Bail," *ABA Journal,* November 2017, http://www.abajournal.com/magazine/article/ABA_urges_Congress_to_limit_use_of_cash_bail.
15. See the Marshall Project, "New York City's Bail Success Story," March 14, 2019, https://www.themarshallproject.org/2019/03/14/new-york-city-s-bail-success-story; Stephanie Wykstra, "Bail Reform, Which Could Save Millions of Unconvicted People From Jail, Explained," *Vox,* October 17, 2018, https://www.vox.com/future-perfect/2018/10/17/17955306/bail-reform-criminal-justice-inequality; Udi Ofer, "We Can't End Mass Incarceration Without Ending Money Bail," American Civil Liberties Union, December 11, 2017, https://www.aclu.org/blog/smart-justice/we-cant-end-mass-incarceration-without-ending-money-bail.
16. Ibid.
17. Ofer, "We Can't End Mass Incarceration Without Ending Money Bail."
18. Bureau of Justice Statistics, *Correctional Populations in the United States, 2016,* p. 2.
19. Quoted in D. A. Andrews, "Program Structure and Effective Correctional Practices: A Summary of the CAVIC Research," in Robert R. Ross and Paul Gendreau (eds.), *Effective Correctional Treatment* (Toronto, Canada: Butterworth, 1980), p. 42.
20. See David Garland, *Mass Imprisonment: Social Causes and Consequences* (Thousand Oaks, Calif.: SAGE), 2001; Bureau of Justice Statistics, *Correctional Populations in the United States,* 2016, p. 2.
21. Michelle Alexander, *The New Jim Crow* (New York: New Press, 2012).

22. Ibid.
23. Mark Karlin, "Michelle Alexander on the Irrational Race Bias of the Criminal Justice and Prison Systems," Truthout, http://truth-out.org/opinion/item/10629-truthout-interviews-michelle-alexander-on-the-irrational-race-bias-of-the-criminal-justice-and-prison-systems#.
24. Stephen Lurie, "The Only Man Who Can Fix Mass Incarceration Is Barack Obama," *The Atlantic,* November 12, 2014, http://www.theatlantic.com/politics/archive/2014/11/the-only-man-who-can-fix-mass-incarceration-is-barack-obama/382314/.
25. Jacob Kang-Brown, Oliver Hinds, Jasmine Heiss, and Olive Lu, *The New Dynamics of Mass Incarceration,* Vera Institute of Justice (June 2018), p. 14, https://www.documentcloud.org/documents/4516234-The-New-Dynamics-of-Mass-Incarceration-Report.html; also see Whittney Evans, "Are Prison Populations Decreasing? Depends on Where You Look," NPR, June 14, 2018, https://www.npr.org/2018/06/14/619827057/are-prison-populations-decreasing-depends-on-where-you-look.
26. Richard P. Seiter, *Correctional Administration: Integrating Theory and Practice* (Upper Saddle River, N.J.: Prentice Hall, 2002), p. 11.
27. Franklin E. Zimring and Gordon J. Hawkins, *Deterrence: The Legal Threat in Crime Control* (Chicago: University of Chicago Press, 1973).
28. Joan Petersilia, "When Probation Becomes More Dreaded Than Prison," *Federal Probation* 54 (March 1990), pp. 27.
29. Ibid.
30. Ibid.
31. Ibid., p. 25.
32. Ibid.
33. Robert B. Levinson, "Classification: The Cornerstone of Corrections," in Peter M. Carlson and Judith Simon Garrett (eds.), *Prison and Jail Administration: Practice and Theory* (Boston: Jones and Bartlett, 2006), pp. 261–267.
34. Ibid., p. 262.
35. Ibid., pp. 262–263.
36. James Austin and Patricia L. Hardyman, *Objective Prison Classification: A Guide for Correctional Agencies* (Washington, D.C.: National Institute of Corrections, July 2004); also see R. Buchanan, "National Evaluation of Objective Prison Classification Systems: The Current State of the Art," *Crime & Delinquency* 32, no. 3 (1986), pp. 272–290.
37. Ibid., p. 192.
38. James A. Inciardi, *Criminal Justice,* 7th ed. (Fort Worth, Texas: Harcourt Brace, 2001), p. 454.
39. Ibid., p. 194.
40. U.S. Bureau of Prisons, *Unit Management Manual* (Washington, D.C.: Author, 1977), p. 6.
41. Seiter, *Correctional Administration,* p. 196.
42. See National Institute of Justice, "Recidivism," https://www.nij.gov/topics/corrections/recidivism/pages/welcome.aspx.
43. Bureau of Justice Statistics, *2018 Update on Prisoner Recidivism: A 9-Year Follow-up Period (2005–2014),* May 2018, p. 1, https://www.bjs.gov/content/pub/pdf/18upr9yfup0514.pdf.
44. See "Recidivism: What It Is and How to Prevent It," Florida Tech, https://www.floridatechonline.com/blog/criminal-justice/recidivism-what-it-is-and-how-to-prevent-it/.
45. Federal Bureau of Prisons, "About Our Agency," https://www.bop.gov/about/agency/; also see ibid., "Our Locations," https://www.bop.gov/locations/list.jsp.
46. Ibid., "Statistics," https://www.bop.gov/about/statistics/population_statistics.jsp.
47. Ibid., "Our Locations," https://www.bop.gov/locations/list.jsp.
48. *Barber v. Thomas,* No. 09–5201 (9th Cir., 2010); also see U.S. Department of Justice, United States Parole Commission, *History of the Federal Parole System,* http://www.usdoj.gov/uspc/history.pdf.
49. Laura Sullivan, "Timeline: Solitary Confinement in U.S. Prisons," NPR, July 26, 2006, http://www.npr.org/templates/story/story.php?storyId=5579901.
50. Amnesty International, "Entombed: Isolation in the US Federal Prison System," 2017, http://www.amnestyusa.org/research/reports/entombed-isolation-in-the-us-federal-prison-system.
51. Ashley May, "'El Chapo' Heading to Supermax Prison? Who Else Is Inside the Hellish 'Alcatraz of the Rockies,'" *USA Today,* February 14, 2019, https://www.usatoday.com/story/news/nation/2019/02/14/el-chapo-supermax-prison-joaquin-guzman-may-face/2868219002/.
52. Terry Frieden, "Reporters Get First Look Inside Mysterious Supermax Prison," CNN, September 14, 2007, http://www.cnn.com/2007/US/09/13/supermax.btsc/index.html?_s=PM:US.
53. Craig Haney, "Mental Health Issues in Long-Term Solitary and 'Supermax' Confinement," *Crime & Delinquency* 49, no. 1 (January 2003), pp. 124–156.
54. Ibid.
55. Jesenia Pizarro and Vanja M. K. Stenius, "Supermax Prisons: Their Rise, Current Practices, and Effect on Inmates," *The Prison Journal* 84, no. 2 (June 2004), pp. 248–264.
56. Ibid., p. 260.
57. *Madrid v. Gomez,* 889 F. Supp. 1146 (1995), at p. 1229.
58. *Jones-El v. Berge,* 374 F.3d 541 (7th Cir. 2004), at p. 1118.
59. The Sentencing Project, "Capitalizing on Mass Incarceration: U.S. Growth in Private Prisons," August 2, 2018, https://www.sentencingproject.org/publications/capitalizing-on-mass-incarceration-u-s-growth-in-private-prisons/.
60. Ibid.
61. Thomas R. O'Connor, "The Debate Over Prison Privatization," North Carolina Wesleyan College, http://faculty.ncwc.edu/TOConnor/417/4171ect14.htm, p. 1.
62. Matthew D. Makarios and Jeff Maahs, "Is Private Time Quality Time?" *Prison Journal* 92, no. 3 (September 2012), p. 336.
63. The Sentencing Project, "Capitalizing on Mass Incarceration: U.S. Growth in Private Prisons."
64. Chico Harlan, "The Private Prison Industry Was Crashing—Until Donald Trump's Victory," *The Washington Post,* November 10, 2016, https://www.washingtonpost.com/news/wonk/wp/2016/11/10/the-private-prison-industry-was-crashing-until-donald-trumps-victory/?utm_term=.01cd947bef8e.
65. Bureau of Justice Statistics, *Correctional Populations in the United States, 2016, p. 2.*
66. Bureau of Justice Statistics, *Jail Inmates in 2015* (December 2016), p. 3, https://www.bjs.gov/content/pub/pdf/ji15.pdf.
67. National Institute of Corrections, *Jail Resource Issues: What Every Funding Authority Needs to Know* (February 2012), https://s3.amazonaws.com/static.nicic.gov/Library/017372.pdf.
68. National Sheriffs' Association, *The State of Our Nation's Jails, 1982* (Washington, D.C.: Author, 1982), p. 55.
69. See a description of such a jail in use and its benefits, by the Corrections Center of Northwest Ohio, "The New Generation Direct Supervision Jail," http://www.ccnoregionaljail.org/newgenerationjail.htm.
70. Linda L. Zupan, *Jails: Reform and the New Generation Philosophy* (Cincinnati, Ohio: Anderson, 1991), p. 67.
71. National Institute of Justice, *Programs and Activities: Tools for Managing Inmate Behavior* (June 2010), pp. 103–105, http://www.bscc.ca.gov/downloads/Programs_and_Activities_Tools_for_Managing_Inmate_Behavior_NIC_024368.pdf.
72. National Institute of Justice, "How We Review and Rate a Practice From Start to Finish," https://www.crimesolutions.gov/about_practicereview.aspx.
73. Ibid., "Corrections & Reentry: Inmate Programs & Treatment," https://www.crimesolutions.gov/TopicDetails.aspx?ID=31.
74. U.S. Department of Justice, National Institute of Justice Research in Brief, *Making Jails Productive* (Washington, D.C.: Author, 1987), p. 1.
75. Ibid., p. 16.
76. Courtesy Sheriff Darin Balaam, Washoe County (Nevada) Sheriff's Office, "Inmate Work Programs," https://www.washoesheriff.com/sub.php?page=inmate-work-programs.

77. Susan Miller, "Prisons Fighting Contraband Phones Get Help From Courts," GCN, July 24, 2018, https://gcn.com/articles/2018/07/24/prison-cellphone-court-disable.aspx.
78. DroneShield, "How Prisons Can Combat Drone Threats," https://www.droneshield.com/blog-content/2017/1/14/how-prisons-can-combat-drone-threats.
79. Miller, "Prisons Fighting Contraband Phones Get Help From Courts."
80. See Luke Dormehl, "Watch Lockheed Martin Burn Drones Out of the Sky With a New Laser," *Digital Trends*, September 22, 1917, https://www.digitaltrends.com/cool-tech/athena-laser-drone/; Trevor Mogg, "Drone-Catching Eagles Aren't Such a Good Idea After All," *Digital Trends*, December 11, 2017, https://www.digitaltrends.com/cool-tech/drone-catching-eagles-retired/.
81. Rob Waters, "Confronting a Broken Juvenile Justice System," *Psychotherapy Networker*, n.d., https://www.psychotherapynetworker.org/blog/details/1293/confronting-a-broken-juvenile-justice-system.
82. Pew Charitable Trusts, *Commission on Children in Foster Care, 2009,* http://www.pewtrusts.org/en/archived-projects/commission-on-children-in-foster-care.
83. Dean John Champion, Alida V. Merlo, and Peter J. Benekos, *The Juvenile Justice System: Delinquency, Processing, and the Law,* 6th ed. (New York: Prentice Hall, 2009), p. 435.
84. Ibid., p. 436.
85. Ibid., p. 437.
86. Ibid., pp. 438–439.
87. Ibid., pp. 440–441.
88. American Bar Association, "Abuse and Inefficiency in Juvenile Offender Boot Camps: Is Regulation the Answer?" http://www.americanbar.org/content/dam/aba/publishing/criminal_justice_section_newsletter/crimjust_juvjust_newsletterjune09_june09_pdfs_bootcamps.authcheckdam.pdf.
89. Cox et al., *Juvenile Justice,* p. 446.
90. Molly Kneffel, "Trying to Fix America's Broken Juvenile Justice System," *Rolling Stone,* January 22, 2015, https://www.rollingstone.com/politics/politics-news/trying-to-fix-americas-broken-juvenile-justice-system-241177/.
91. Michael Rocque and Raymond Paternoster, "Understanding the Antecedents of the 'School-to Jail' Link: The Relationship Between Race and School Discipline," *Journal of Criminal Law & Criminology* 101, no. 2 (2011), pp. 633–643; see also Carla Amurao, "Fact Sheet: How Bad Is the School-to-Prison Pipeline?" *Tavis Smiley Reports,* http://www.pbs.org/wnet/tavissmiley/tsr/education-under-arrest/school-to-prison-pipeline-fact-sheet/.
92. Ibid.
93. Mary Ellen Flannery, "The School-to-Prison Pipeline: Time to Shut It Down," *NEA Today,* January 5, 2015, http://neatoday.org/2015/01/05/school-prison-pipeline-time-shut/.
94. Yoona Ha, "Case Study: How a Maine Correctional Facility Reduced Contraband Instances to Zero," October 26, 2018, https://www.correctionsone.com/products/facility-products/body-scanners/articles/481700187-Case-study-How-a-Maine-correctional-facility-reduced-contraband-instances-to-zero/.
95. Shane Peterson, "The Internet Moves Behind Bars," *Government Technology* (Supplement: *Crime and the Tech Effect*) (April 2001), p. 18.
96. Jim McKay, "Virtual Visits," *Government Technology* (October 2001), p. 46.

/// CHAPTER 11

1. Bureau of Justice Statistics, *Probation and Parole in the United States, 2016,* April 2018, https://www.bjs.gov/content/pub/pdf/ppus16_sum.pdf.
2. Peter J. Benekos, "Beyond Reintegration: Community Corrections in a Retributive Era," *Federal Probation* 54 (March 1990), p. 53.
3. See the President's Commission on Law Enforcement and Administration of Justice, *Task Force Report: Corrections* (Washington, D.C.: U.S. Government Printing Office, 1967), p. 7.
4. Benekos, "Beyond Reintegration," p. 53.
5. Belinda R. McCarthy, *Intermediate Punishments: Intensive Supervision, Home Confinement, and Electronic Surveillance* (Monsey, N.Y.: Criminal Justice Press, 1987), p. 3.
6. Barry J. Nidorf, "Community Corrections: Turning the Crowding Crisis Into Opportunities," *Corrections Today* (October 1989), p. 85.
7. Benekos, "Beyond Reintegration," p. 54.
8. Although neither the word *probation* nor today's definition of it appear in the Bible, it was historically assumed that all mankind was under the sentence of eternal death—an endless life in misery during this life. However, humans could escape this sentence through repentance and faith in Christ; this was man's "probation," an opportunity to escape hell and secure heaven. If he failed to improve this opportunity, and died impenitent, the sentence was irrevocably executed and the man was eternally lost. See Tentmaker, "Probation," http://www.tentmaker.org/books/SpiritOfTheWord/014Probation.htm.
9. Howard Abadinsky, *Probation and Parole: Theory and Practice,* 3rd ed. (Englewood Cliffs, N.J.: Prentice Hall, 1987), p. 18.
10. Lawrence M. Friedman, *A History of American Law* (New York: Simon and Schuster, 1973), p. 518.
11. Paul F. Cromwell Jr., George C. Killinger, Hazel B. Kerper, and Charles Walker, *Probation and Parole in the Criminal Justice System,* 2nd ed. (St. Paul, Minn.: West, 1985).
12. John Augustus, *John Augustus, First Probation Officer* (Montclair, N.J.: Patterson Smith, 1972), pp. 4–5.
13. John Augustus, *A Report of the Labors of John Augustus* (Boston: Wright & Hasty, 1852), republished in 1984 by the American Probation and Parole Association, Lexington, Ky., pp. 96–97.
14. Abadinsky, *Probation and Parole,* p. 143.
15. Torsten Eriksson, *The Reformers: An Historical Survey of Pioneer Experiments in the Treatment of Criminals* (New York: Elsevier, 1976), p. 81.
16. Abadinsky, *Probation and Parole,* pp. 146–147.
17. The issues concerned whether parole (1) infringed on the power of the judiciary to sentence, the governor to pardon, or the legislature to determine penalty levels; (2) denied prisoners due process; or (3) constituted cruel and unusual punishment.
18. U.S. Attorney General's Survey of Release Procedures, *Parole,* Vol. 4 (New York: Arno Press, 1974), p. 20.
19. Sheldon L. Messinger, "Introduction," in Andrew von Hirsch and Kathleen J. Hanrahan (eds.), *The Question of Parole: Retention, Reform, or Abolition?* (Cambridge, Mass.: Ballinger, 1970), pp. xviii–xix.
20. U.S. Attorney General's Survey of Release Procedures, *Parole,* p. 20.
21. Bureau of Justice Statistics, *Probation and Parole in the United States, 2016.*
22. Robert M. Regoli and John D. Hewitt, *Exploring Criminal Justice: The Essentials* (Sudbury, Mass.: Jones and Bartlett, 2010), p. 316.
23. Bureau of Justice Statistics, *Probation and Parole in the United States, 2016,* p. 17.
24. *Mempa v. Rhay,* 389 U.S. 128 (1967).
25. *Gagnon v. Scarpelli,* 411 U.S. 778 (1973).
26. Bureau of Justice Statistics, *Probation and Parole in the United States,* 2016, April 2018, https://www.bjs.gov/content/pub/pdf/ppus16.pdf .
27. Park Dietz, "Hypothetical Criteria for the Prediction of Individual Criminality," in Christopher Webster, Mark Ben-Aron, and Stephen Hucker (eds.), *Dangerousness: Probability and Prediction, Psychiatry, and Public Policy* (Cambridge, U.K.: Cambridge University Press, 1985), p. 32.
28. Bureau of Justice Statistics, *Probation and Parole in the United States, 2016.*
29. *Morrissey v. Brewer,* 92 S.Ct. 2593 (1972).

30. Ibid.
31. Council of State Governments Justice Center, *The Impact of Probation and Parole Populations on Arrests in Four California Cities* (January 2013), p. 4, http://www.cdcr.ca.gov/Reports/docs/External-Reports/CAL-CHIEFS-REPORT.pdf.
32. Ibid., p. 1.
33. Ibid., p. 17.
34. Ibid., p. 6.
35. U.S. Bureau of Labor Statistics, "Occupational Outlook Handbook: Probation Officers and Correctional Treatment Specialists," http://www.bls.gov/ooh/Community-and-Social-Service/Probation-officers-and-correctional-treatment-specialists.htm#tab-2.
36. Ibid.
37. H. S. Ntuli, V. I. Khoza, J. M. Ras, and P. J. Potgieter, "Role Conflict in Correctional Supervision," *Acta Criminologica* 20, no. 4 (2007), pp. 85–95.
38. Ibid.
39. John Conrad, "The Pessimistic Reflections of a Chronic Optimist," *Federal Probation* 55 (June 1991), pp. 4–9.
40. Matthew T. DeMichele, *Probation and Parole's Growing Caseloads and Workload Allocation: Strategies for Managerial Decision Making* (Lexington, Ky.: American Probation and Parole Association), http://www.appa-net.org/eweb/docs/appa/pubs/SMDM.pdf, pp. 14–15.
41. M. Claxton, N. Sinclair, and R. Hanson, "Felons on Probation Often Go Unwatched," *Detroit News,* December 10, 2002, p. 2.
42. DeMichele, *Probation and Parole's Growing Caseloads and Workload Allocation*, pp. 13, 33.
43. Shawn E. Small and Sam Torres, "Arming Probation Officers: Enhancing Public Confidence and Officer Safety," *Federal Probation* 65, no. 3 (2001), pp. 24–28.
44. Richard P. Seiter, *Correctional Administration: Integrating Theory and Practice* (Upper Saddle River, N.J.: Prentice Hall, 2002), p. 387.
45. U.S. Department of Justice, Bureau of Justice Statistics, *Jail Inmates in 2017*, April 2019, https://www.bjs.gov/content/pub/pdf/ji17.pdf.
46. Howard Abadinsky, *Probation and Parole: Theory and Practice,* 7th ed. (Upper Saddle River, N.J.: Prentice Hall, 2000), p. 410.
47. Joan Petersilia and Susan Turner, *Evaluating Intensive Supervision Probation/Parole: Results of a Nationwide Experiment* (Washington, D.C.: National Institute of Justice, 1993).
48. Lawrence A. Bennett, "Practice in Search of a Theory: The Case of Intensive Supervision—An Extension of an Old Practice," *American Journal of Criminal Justice* 12 (1988), pp. 293–310.
49. Ibid., p. 293.
50. This information was compiled from ISP brochures and information from the Oregon Department of Corrections by Joan Petersilia.
51. Todd R. Clear and Patricia R. Hardyman, "The New Intensive Supervision Movement," *Crime & Delinquency* 36 (January 1990), pp. 42–60.
52. Ibid., p. 44.
53. Brittany Anas, "Boulder Sex Offender Balks at 'Intensive,' Asks to Be Sent to Prison," *Daily Camera,* August 19, 2012, http://www.dailycamera.com/ci_20891466/boulder-sex-offender-balks-at-intensive-probation-asks.
54. Ibid.
55. Pew Research Center, "Public Opinion on Sentencing and Corrections Policy in America," March 2012, https://www.prisonpolicy.org/scans/PEW_NationalSurveyResearchPaper_FINAL.pdf.
56. Barbara A. Sims, "Questions of Corrections: Public Attitudes Toward Prison and Community-Based Programs," *Corrections Management Quarterly* 1, no. 1 (1997), p. 54.
57. Jeffery T. Ulmer, "Intermediate Sanctions: A Comparative Analysis of the Probability and Severity of Recidivism," *Sociological Inquiry* 71, no. 2 (Spring 2001), pp. 164–193.
58. Ibid., p. 184.
59. Ibid., p. 185.
60. Bureau of Justice Assistance, *Offender Supervision With Electronic Technology: Community Corrections Resource,* 2nd ed. (Washington, D.C.: U.S. Department of Justice, 2009), p. 16, http://www.appa-net.org/eweb/docs/APPA/pubs/OSET_2.pdf.
61. John K. Roman, Akiva M. Lieberman, Samuel Taxy, and P. Mitchell Downey, "The Costs and Benefits of Electronic Monitoring for Washington, D.C.," Urban Institute, September 2012, https://www.urban.org/sites/default/files/alfresco/publication-pdfs/412678-The-Costs-and-Benefits-of-Electronic-Monitoring-for-Washington-D-C-.PDF.
62. Ibid.
63. Annesley K. Schmidt, "Electronic Monitors: Realistically, What Can Be Expected?" *Federal Probation* 59 (June 1991), pp. 47–53.
64. Abadinsky, *Probation and Parole,* 7th ed., p. 428.
65. David Brauer, "Satellite 'Big Brother' Tracks Ex-Inmates," *Chicago Tribune,* December 18, 1998, p. 31.
66. National Institute of Justice, "Standard Offender Tracking Systems," U.S. Department of Justice, July 2016, https://www.ncjrs.gov/pdffiles1/nij/249810.pdf.
67. Jeanne B. Stinchcomb and Vernon B. *Fox, Introduction to Corrections,* 5th ed. (Upper Saddle River, N.J.: Prentice Hall, 1999), p. 165.
68. Gaylene Styve Armstrong, Angela R. Gover, and Doris Layton MacKenzie, "The Development and Diversity of Correctional Boot Camps," *Turnstile Justice: Issues in American Corrections,* ed. Rosemary L. Gido and Ted Alleman (Upper Saddle River, N.J.: Prentice Hall, 2002), pp. 115–130.
69. Doris Layton MacKenzie, "Boot Camp Prisons and Recidivism in Eight States," *Criminology* 33, no. 3 (1995), pp. 327–358.
70. National Institute of Justice, crimesolutions.gov.
71. Doris Layton MacKenzie and Alex Piquero, "The Impact of Shock Incarceration Programs on Prison Crowding," *Crime & Delinquency* 40, no. 2 (April 1994), pp. 222–249.
72. John Ashcroft, Deborah J. Daniels, and Sarah V. Hart, *Correctional Boot Camps: Lessons from a Decade of Research* (Washington, D.C.: U.S. Department of Justice, Office of Justice Programs, June 2003), p. 2. Note: A comprehensive list of correctional boot camps in operation is not currently available.
73. Ibid.
74. Joshua A. Jones, "A Multi-State Analysis of Correctional Boot Camp Outcomes: Identifying Vocational Rehabilitation as a Complement to Shock Incarceration," *Inquiries* 4, no. 9 (2012), pp. 1–3.
75. Dale G. Parent, "Day Reporting Centers: An Evolving Intermediate Sanction," *Federal Probation* 60 (December 1996), pp. 51–54.
76. Liz Marie Marciniak, "The Addition of Day Reporting to Intensive Supervision Probation: A Comparison of Recidivism Rates," *Federal Probation* 64 (June 2000), pp. 34–39.
77. Nathan James, *Offender Reentry: Correctional Statistics, Reintegration into the Community, and Recidivism*, Congressional Research Service, January 2015, https://fas.org/sgp/crs/misc/RL34287.pdf.
78. Janet Portman, "When Is a Prisoner Released to a Halfway House?" Lawyers.com, 2017, http://criminal.lawyers.com/criminal-law-basics/when-is-a-prisoner-released-to-a-halfway-house.html.
79. Federal Bureau of Prisons, "Completing the Transition: Reentry Assistance Reduces Recidivism," https://www.bop.gov/about/facilities/residential_reentry_management_centers.jsp.
80. See, for example, Federal Bureau of Prisons, "Inmate Furloughs," January 11, 2011, http://www.prisonology.com/files/1914/1514/8069/BOP_Inmate_Furloughs.pdf.
81. See Washington State Institute for Public Policy, *Predicting Criminal Recidivism: A Systematic Review of Offender Risk Assessments in Washington State,* February 2014, http://www.wsipp.wa.gov/ReportFile/1554/Wsipp_Predicting-Criminal-Recidivism-A-Systematic-Review-of-Offender-Risk-Assessments-in-Washington-State_Final-Report.pdf.
82. Bureau of Justice Assistance and the Council of State Governments, *Integrated Reentry and Employment Strategies: Reducing Recidivism and Promoting Job Readiness*, September 2013, https://www.bja.gov/Publications/CSG-Reentry-and-Employment.pdf, p. 10.

83. Ibid., p. 12.
84. Jan Looman and Jeffrey Abracen, "The Risk Need Responsivity Model of Offender Rehabilitation: Is There Really a Need for a Paradigm Shift?" *International Journal of Behavioral Consultation and Therapy* 8 (34) (2013), p. 32.
85. See Guy Bourgon and James Bonta, "Reconsidering the Responsivity Principle: A Way to Move Forward," *Federal Probation* 78(2), September 2014, https://www.uscourts.gov/sites/default/files/78_2_1_0.pdf
86. Stephanie Stullich, Ivy Morgan, and Oliver Schak, *State and Local Expenditures on Corrections and Education* (U.S. Department of Education, Policy and Program Studies Service, July 2016), https://www2.ed.gov/rschstat/eval/other/expenditures-corrections-education/brief.pdf. .

/// CHAPTER 12

1. U.S. Department of Homeland Security, "Fusion Center Locations and Contact Information," https://www.dhs.gov/fusion-center-locations-and-contact-information.
2. Elaine Pittman, Jim McKay, "Las Vegas Fusion Center Is a Model for Public-Private Collaboration," *Government Technology*, May 24, 2011, https://www.govtech.com/em/safety/Las-Vegas-Fusion-Center-Public-Private-Collaboration-052411.html.
3. See, for example, a discussion of the "lone-wolf model" of terrorism now in use here and abroad—and discussed in al-Qaeda publications—in Lori Hinnant, "Intel Dilemma in Boston, London, Paris Attacks," *Associated Press*, May 31, 2013, http://abcnews.go.com/International/wireStory/intel-dilemma-boston-london-paris-attacks-19294987#.UazbZuDn_cs.
4. See Marcella Corona, "'Pill-Mill' Case: Where Are the Defendants Now?" *Reno Gazette-Journal*, April 27, 2017, https://www.rgj.com/story/news/crime/2017/04/28/pill-mill-case-where-defendants-now/307384001/.
5. National Institute on Drug Abuse, "What Is the Scope of Heroin Use in the United States?" June 2018, https://www.drugabuse.gov/publications/research-reports/heroin/scope-heroin-use-in-united-states.
6. Francis Diep, "Some Officials Worry a Cocaine Epidemic Is About to Hit the U.S.," *Pacific Standard*, May 9, 2018, https://psmag.com/news/some-officials-worry-a-cocaine-epidemic-is-about-to-hit-the-u-s.
7. See National Institute on Drug Abuse, "What Is Fentanyl," February 2019, https://www.drugabuse.gov/publications/drugfacts/fentanyl; also see Meilan Solly, "Fentanyl Has Outpaced Heroin as Drug Implicated Most Often in Fatal Overdoses," *Smithsonian Magazine*, December 17, 2018, https://www.smithsonianmag.com/smart-news/fentanyl-has-outpaced-heroin-drug-implicated-most-often-fatal-overdoses-180971050/.
8. Abby Goodnough, Josh Katz, and Margot Sanger-Katz, "Drug Overdose Deaths Drop in U.S. for First Time Since 1990," *New York Times*, July 17, 2019, https://www.nytimes.com/interactive/2019/07/17/upshot/drug-overdose-deaths-fall.html.
9. U.S. Department of Health and Human Services, "What Is the U.S. Opioid Epidemic?" https://www.hhs.gov/opioids/about-the-epidemic/index.html.
10. Brett Molina, "Odds of Opioids Death Higher Than Car Crash," *USA Today*, January 15, 2019, https://epaper.daytondailynews.com/popovers/dynamic_article_popover.aspx?guid=1eaed9bc-711e-4ed2-bac0–70e9822e0c42&pbid=66ab59ea-5cfc-438d-83e4-dc9c4a34f79d&utm_source=app.pagesuite&utm_medium=app-interaction&utm_campaign=pagesuite-epaper-htm15_share-article.
11. Ioan Grillo, "Where the Legend of El Chapo Was Born, *Newsweek*, May 21, 2018, pp. 31–36.
12. Don Winslow, quoted in George Will, "Worse Living Through Chemistry," *Reno Gazette Journal*, March 17, 2019, p. 5D.
13. Daniela Silva, "El Chapo's Conviction Is 'Great Moral Victory,' but Has 'Done Nothing' Against Sinaloa Cartel, Experts Say," NBC News, February 13, 2019, https://www.nbcnews.com/news/us-news/el-chapo-s-conviction-great-moral-victory-has-done-nothing-n971201.
14. Jerry Mitchell, "Is There a Solution to the Opioid Epidemic?" *USA Today*, January 30, 2018, https://www.usatoday.com/story/news/nation-now/2018/01/29/175-americans-dying-day-what-solutions-opioid-epidemic/1074336001/.
15. Police Executive Research Forum, *The Unprecedented Opioid Epidemic: As Overdoses Become a Leading Cause of Death, Police, Sheriffs, and Health Agencies Must Step Up Their Response*, September 2017, http://www.policeforum.org/assets/opioids2017.pdf, p. 14; Mitchell, "Is There a Solution to the Opioid Epidemic?"; also see U.S. Department of Health and Human Services Secretary Thomas E. Price's recommendations, https://www.hhs.gov/about/leadership/secretary/speeches/2017-speeches/secretary-price-announces-hhs-strategy-for-fighting-opioid-crisis/index.html.
16. Ibid.
17. Substance Abuse and Mental Health Services Administration, *Adult Drug Courts and Medication-Assisted Treatment for Opioid Dependence*, Summer 2014, https://store.samhsa.gov/system/files/sma14–4852.pdf.
18. "State Marijuana Laws in 2018 Map," *Governing*, http://www.governing.com/gov-data/safety-justice/state-marijuana-laws-map-medical-recreational.html.
19. Tom Angell, "These States Are Most Likely to Legalize Marijuana in 2019," *Forbes*, December 26, 2018, https://www.forbes.com/sites/tomangell/2018/12/26/these-states-are-most-likely-to-legalize-marijuana-in-2019/#20af12555add
20. Sara Brittany Somerset, "Attorney General Barr Favors a More Lenient Approach to Cannabis Prohibition," *Forbes*, April 15, 2019, https://www.forbes.com/sites/sarabrittanysomerset/2019/04/15/attorney-general-barr-favors-a-more-lenient-approach-to-cannabis-legalization/#46e92297c4c8.
21. "Estimating Enforcement Costs in Legal States and Correlation to Marijuana Policy: Whack-a-Mole Cannabis Enforcement," WM Policy, August 14, 2017, http://www.wmpolicy.com/wp-content/uploads/sites/17/2017/11/11Estimating-Enforcement-Costs.pdf.
22. U.S. Department of Justice, "Sex Trafficking Defined," https://www.justice.gov/sextrafficking.
23. Aryn Barker, "The Survivor," *Time*, January 28, 2019, p. 38.
24. Ibid.
25. See U.S. Department of Health and Human Services, "Trafficking Victims Protection Act of 2000: Fact Sheet," http://www.markwynn.com/trafficking/trafficking-victims-protection-act-of-2000-fact-sheet.pdf.
26. U.S. Citizenship and Immigration Services, "Victims of Sex Trafficking: T Nonimmigrant Status," https://www.uscis.gov/humanitarian/victims-human-trafficking-other-crimes/victims-human-trafficking-t-nonimmigrant-status.
27. U.S. Department of Health & Human Services, Office of Refugee Resettlement, "Fact Sheet: Certification for Adult Victims of Trafficking," https://www.acf.hhs.gov/orr/resource/fact-sheet-certification-for-adult-victims-of-trafficking.
28. See, for example, International Association of Chiefs of Police, "Toolkit: Child Sex Trafficking: A Training Series for Frontline Officers," https://www.theiacp.org/resources/document/toolkit-child-sex-trafficking-a-training-series-for-frontline-officers.
29. Jovana Lara, "Sex-Trafficking Crackdown: 510 Arrested, 56 Rescued in California," ABC News, January 30, 2018, http://abc7news.com/hundreds-arrested-in-sex-trafficking-crackdown-in-california/3008362/.
30. Nicole Chavez, "More Than 1,000 Arrests in Sex Trafficking Operation," CNN, August 4, 2017, https://www.cnn.com/2017/08/04/us/sex-trafficking-sting/index.html.

31. Erin Calabrese, Todd Miyazawa, and Andrew Kozak, "84 Children Rescued, 120 Sex Traffickers Arrested Across U.S., FBI Says," *U.S. News*, October 18, 2017, https://www.nbcnews.com/news/us-news/84-children-rescued-120-sex-traffickers-arrested-across-u-s-n812156.
32. Federal Bureau of Investigations, "Terrorism," https://www.fbi.gov/investigate/terrorism.
33. 18 U.S.C. § 2331, states that international terrorism involves: violent acts or acts dangerous to human life that violate federal or state law; that appear to be intended (i) to intimidate or coerce a civilian population; (ii) to influence the policy of a government by intimidation or coercion; or (iii) to affect the conduct of a government by mass destruction, assassination, or kidnapping; and occur primarily outside the territorial jurisdiction of the U.S., or transcend national boundaries in terms of the means by which they are accomplished, the persons they appear intended to intimidate or coerce, or the locale in which their perpetrators operate or seek asylum.
34. Christal Hayes, Kevin Johnson, and Candy Woodall, "Who is Robert Bowers? Accused Pittsburgh Synagogue Shooter Left Anti-Semitic Trail," *USA TODAY*, https://www.cnn.com/2018/10/27/us/synagogue-attack-suspect-robert-bowers-profile/index.htmlhttps://www.cnn.com/2018/10/27/us/synagogue-attack-suspect-robert-bowers-profile/index.html.
35. Nicole Chavez, "Cesar Sayoc was a DJ, bodybuilder and pizza delivery man before he became a bomb suspect," CNN, October 27, 2018, https://www.cnn.com/2018/10/27/politics/cesar-sayoc-mail-bomb-suspect/index.html.
36. Nathan Bomey, "Suspect Devices Hard for Delivery Services to Detect," *USA Today*, October 27, 2018, https://www.usatoday.com/story/money/2018/10/25/suspicious-package-handling-ups-fedex-usps/1763186002/.
37. International Association of Chiefs of Police, "Homegrown Violent Extremism Awareness Brief," January 1, 2014, https://www.theiacp.org/resources/document/homegrown-violent-extremism-awareness-brief.
38. Peter Bergen, "Can We Stop Homegrown Terrorists?" *Wall Street Journal*, January 23–24, 2016, pp. C1–C2.
39. See, generally, Senate Committee on Homeland Security and Governmental Affairs, *Zachary Chesser: A Case Study in Online Islamist Radicalization and Its Meaning for the Threat of Homegrown Terrorism*, February 2012, https://www.hsgac.senate.gov/imo/media/doc/CHESSER%20FINAL%20REPORT(1)2.pdf.
40. Sam O'Kane, "Chinese hackers charged with stealing data from NASA, IBM, and others," The Verge, December 18, 2018, https://www.theverge.com/2018/12/20/18150275/chinese-hackers-stealing-data-nasa-ibm-charged; also see Sam Bocette, "Chinese State Sponsored Hacking," The Bridge, November 23, 2017, https://www.realcleardefense.com/articles/2017/11/23/chinese_state_sponsored_hacking_112675.html.
41. "China overtakes Russia as world's biggest state hacker," The Week, October 10, 2018, https://www.theweek.co.uk/96999/china-overtakes-russia-as-world-s-biggest-state-hacker.
42. Donie O'Sullivan and Joshua Berlinger, "North Korea Hackers Targeting US 'Critical Infrastructure,' Cybersecurity Firm," CNN, March 5, 2019, https://www.cnn.com/2019/03/04/politics/north-korea-hackers-intl/index.htmlsayshttps://www.cnn.com/2019/03/04/politics/north-korea-hackers-intl/index.html.
43. Interpol, "Cybercrime," https://www.interpol.int/en/Crimes/Cybercrime.
44. Privacy Rights Clearinghouse, "Data Breaches," https://www.privacyrights.org/data-breaches.
45. Rob Marvin, "How Much Does a Data Breach Cost?" *PC Magazine*, July 19, 2018, https://www.pcmag.com/news/362543/how-much-does-a-data-breach-cost.
46. Dana A. Shea and Frank Gottron, *Small-Scale Terrorist Attacks Using Chemical and Biological Agents: An Assessment Framework and Preliminary Comparisons*, Congressional Research Service, Report for Congress, May 20, 2004, http://www.fas.org/irp/crs/RL32391.pdf.
47. See Department of Homeland Security, "About DHS," https://www.dhs.gov/about-dhs.
48. See FEMA, "National Incident Management System," https://www.fema.gov/national-incident-management-system.
49. Billy Max, "How to Prevent Hackers in a Company," *AZ Central*, April 13, 2018, https://yourbusiness.azcentral.com/prevent-hackers-company-23336.html.
50. Steve Ranger, "What Is Cyberwar? Everything You Need to Know About the Frightening Future of Digital Conflict," ZDNet, December 4, 2018, https://www.zdnet.com/article/cyberwar-a-guide-to-the-frightening-future-of-online-conflict/.
51. Tod Newcombe, "3 Ways Governments Are Fighting Hackers," *Governing*, September 2016, https://www.governing.com/columns/tech-talk/gov-cybersecurity-hackers-government.html.
52. Alex Bennett, "Teaching Our Police Force How to Hack Like the Pros," *Wired*, June 12, 2017, https://www.wired.co.uk/article/teaching-police-hacking.
53. D. G. Bolgiano, "Military Support of Domestic Law Enforcement Operations: Working Within Posse Comitatus," *FBI Law Enforcement Bulletin* (December 2001), pp. 16–24.
54. Erin Kelly, "Senate Approves USA Freedom Act," *USA Today*, June 2, 2015, http://www.usatoday.com/story/news/politics/2015/06/02/patriot-act-usa-freedom-act-senate-vote/28345747/.
55. Jurist: Legal News and Research, "Bush Signs Military Commissions Act," http://jurist.law.pitt.edu/paperchase/2006/10/bush-signs-military-commissions-act.php.
56. Ibid.
57. Daniel Victor, "School Shootings Have Already Killed Dozens in 2018," *New York Times*, May 18, 2018, https://www.nytimes.com/2018/05/18/us/school-shootings-2018.html.
58. For more information, see, generally, Center for Homeland Defense and Security, "K-12 School Shooting Database," https://www.chds.us/ssdb/incidents-by-month/.
59. Sue Riseling, quoted in *The Police Response to Active Shooter Incidents*, Police Executive Research Forum, March 2014, https://www.policeforum.org/assets/docs/Critical_Issues_Series/the%20police%20response%20to%20active%20shooter%20incidents%202014.pdf, p. 30.
60. Abigail Abrams and Melissa Chan, "Special Report: Guns in America," *Time*, November 5, 2018, p. 28.
61. Ibid., p. 29.
62. Gun Violence Archive, "Past Summary Ledgers," https://www.gunviolencearchive.org/past-tolls.
63. Sean Gregory, Chris Wilson, Alice Park, and Aric Jenkins, "Guns in America," *Time*, November 5, 2018, p. 32.
64. Ibid.
65. Lisa Marie Pane, "Parkland Attack Fueled Big Shift in America's Gun Politics," *Associated Press*, February 7, 2019, http://www.startribune.com/parkland-attack-fueled-big-shift-in-america-s-gun-politics/505504142/.
66. Katherine A. Vittes, Jon S. Vernick, and Daniel W. Webster, "Legal Status and Source of Offenders' Firearms in States With the Least Stringent Criteria for Gun Ownership," Center for Gun Policy and Research, Johns Hopkins Bloomberg School of Public Health, Baltimore, Maryland.
67. Ibid.
68. Abrams and Chan, "Special Report: Guns in America," p. 30.
69. Adapted from Police Executive Research Forum, *Key Findings and Actions That Will Have the Largest Impact in Reducing Gun Violence*, https://www.policeforum.org/assets/GunViolenceActionPlan.pdf.
70. Kayla Dwyer, "Guns in School? Here's a List of States That Allow Armed Teachers," *Morning Call*, February 14, 2019, https://www.mcall.com/news/education/mc-nws-guns-in-schools-list-20181108-story.html.

71. "U.S. unauthorized immigrant population estimates by state, 2016," Pew Research Center, February 5, 2019, https://www.pewresearch.org/hispanic/interactives/u-s-unauthorized-immigrants-by-state/.
72. Miriam Jordan, "8 Million People Are Working Illegally in the U.S. Here's Why That's Unlikely to Change," *New York Times,* December 11, 2018, https://www.nytimes.com/2018/12/11/us/undocumented-immigrant-workers.html.
73. National Immigration Forum, "Chiefs and Sheriffs Oppose Immigration Enforcement Policies Undermining Community Policing," July 20, 2015, https://immigrationforum.org/blog/chiefs-and-sheriffs-oppose-immigration-enforcement-policies-undermining-community-policing/.
74. Edwards, "Bye Dad, I Love You," p. 39.
75. Mike James, "Trump Seeks $18 Billion to Extend Border Wall Over 10 Years," *USA Today,* January 6, 2018, https://www.usatoday.com/story/news/2018/01/05/trump-border-wall-proposal/1009584001/.
76. Christal Hayes, "Not a 2,000-Mile Concrete Structure From Sea to Sea': Is Trump Scaling Back Border Wall Plan?" *USA Today,* January 19, 2019, https://www.usatoday.com/story/news/politics/2019/01/19/trump-wall-wont-2–000-mile-concrete-structure-sea-sea/2627378002/.
77. See Nina Totenberg, "Trump's National Emergency Sets Up Legal Fight Over Spending Authority," NPR, February 16, 2019, https://www.npr.org/2019/02/16/695321387/trumps-national-emergency-sets-up-legal-fight-over-spending-authority.
78. See Jackson Rollings, "Eight Border Wall Prototypes Are Unveiled Along the U.S.-Mexico Border," *The Architect's Newspaper,* October 25, 2017, https://archpaper.com/2017/10/eight-border-wall-prototypes-unveiled/#gallery-0-slide-0.
79. See Migration Policy Institute, "President Signs DHS Appropriations and Secure Fence Act, New Detainee Bill Has Repercussions for Noncitizens," https://www.migrationpolicy.org/article/president-signs-dhs-appropriations-and-secure-fence-act-new-detainee-bill-has-repercussions.
80. "Suspect in Four Northern Nevada Murders Arrested," KOLO TV, January 24, 2019, https://www.kolotv.com/content/news/Two-people-found-dead-in-south-Reno-home-504465091.html.
81. Alan Gomez, "Mollie Tibbetts Murder Case: Here Are the Facts on Immigrants Committing Crimes in US," *USA Today,* August 18, 2018, https://www.usatoday.com/story/news/nation/2018/08/22/mollie-tibbitts-murder-reignites-debate-over-immigrant-crime/1060792002/.
82. Kaitlyn Schallhorn, "Sanctuary Cities: What Are They?" Fox News, March 22, 2018, http://www.foxnews.com/politics/2018/03/22/sanctuary-cities-what-are.html.
83. Ibid.
84. Ibid.
85. "U.S. System for Refugee, Asylum Seekers Explained," NBC News, http://www.nbcnews.com/id/43442030/ns/us_news-life/t/us-system-refugee-asylum-seekers-explained/#.XGsJz_ZFzcs.
86. "Which Countries Do Most People Granted Asylum in the U.S. Come From?" NOLO, 2018, https://www.nolo.com/legal-encyclopedia/which-countries-do-most-people-granted-asylum-the-us-come-from.html
87. See Sofia Martinez, "Today's Migrant Flow Is Different," *The Atlantic,* June 26, 2018, https://www.theatlantic.com/international/archive/2018/06/central-america-border-immigration/563744/.
88. W. J. Hennigan, "The Long Wait," *Time,* November 26/December 3, 2018, pp. 44–45.
89. Ibid.
90. Ibid.
91. For a fuller discussion of this topic, see Federation for American Immigration Reform, "The Role of State & Local Law Enforcement in Immigration Matters and Reasons to Resist Sanctuary Policies," January 2016, https://www.fairus.org/issue/illegal-immigration/role-state-local-law-enforcement-immigration-matters-and-reasons-resist .
92. See Division C of Public Law 104–208, 110 Stat. 3009–546, enacted September 30, 1996.
93. Alan Gomez, "Trump May Empower Local Police to Round Up Immigrants," *USA Today,* November 25, 2016.
94. See U.S. Immigration and Customs Enforcement, "Delegation of Immigration Authority Section 287(g) Immigration and Nationality Act," https://www.ice.gov/287g.

/// INDEX

POLICING

COURTS

BOOKING

Released

INVESTIGATION

ARREST

UNSOLVED/ NO ARREST

CHARGES FILED

Prosecutor rejects case

INITIAL APPEARANCE

Charges Dismissed

Preliminary Hearing

Grand Jury

Charges Dismissed

Bail/Detention Hearing

INFORMATION OR INDICTMENT

ARRAIGNMENT

Charges Dismissed

Diversion Program

Pleads Not Guilty

Pleads Guilty

PLEA BARGAIN

TRIAL

Jury Selection

Victim/ Witness Impac Statement

Convicted

Acquitted

Police Agencies

FEDERAL AGENCIES

- Department of Homeland Security
- Department of Justice
- Other Federal Agencies With Law Enforcement Authority

STATE AGENCIES

- State Bureaus of Investigation
- State Police and Highway Patrol Agencies
- Special-Purpose (e.g., Campus-, Wildlife-, and Alcohol-Related) Agencies

LOCAL AGENCIES

- County Police Departments
- Municipal Police Departments

PRIVATE POLICE

NOTE: This chart gives a simplified view of caseflow through the criminal justice system. Procedures vary among jurisdictions.